T0136964

Studies in Computational Intelligence

Volume 645

Series editor

Janusz Kacprzyk, Polish Academy of Sciences, Warsaw, Poland
e-mail: kacprzyk@ibspan.waw.pl

About this Series

The series "Studies in Computational Intelligence" (SCI) publishes new developments and advances in the various areas of computational intelligence—quickly and with a high quality. The intent is to cover the theory, applications, and design methods of computational intelligence, as embedded in the fields of engineering, computer science, physics and life sciences, as well as the methodologies behind them. The series contains monographs, lecture notes and edited volumes in computational intelligence spanning the areas of neural networks, connectionist systems, genetic algorithms, evolutionary computation, artificial intelligence, cellular automata, self-organizing systems, soft computing, fuzzy systems, and hybrid intelligent systems. Of particular value to both the contributors and the readership are the short publication timeframe and the worldwide distribution, which enable both wide and rapid dissemination of research output.

More information about this series at http://www.springer.com/series/7092

Jörg Lässig · Kristian Kersting
Katharina Morik
Editors

Computational Sustainability

Editors
Jörg Lässig
Department of Computer Science
University of Applied Sciences
Zittau/Görlitz
Görlitz
Germany

Katharina Morik
Department of Computer Science
TU Dortmund University
Dortmund
Germany

Kristian Kersting
Department of Computer Science
TU Dortmund University
Dortmund
Germany

ISSN 1860-949X ISSN 1860-9503 (electronic)
Studies in Computational Intelligence
ISBN 978-3-319-81138-3 ISBN 978-3-319-31858-5 (eBook)
DOI 10.1007/978-3-319-31858-5

Printed on acid-free paper

This Springer imprint is published by Springer Nature
The registered company is Springer International Publishing AG Switzerland

Contents

Sustainable Development and Computing—An Introduction

Jörg Lässig

Abstract Computational Sustainability is the computer scientific branch of the interdisciplinary field of sustainability research, an applied science about the research in sustainable solutions and their implementation. This introductory chapter describes the origins and the development of common and current sustainability goals and the development of sustainability science as separate field of research. It points out the relevance of Computer Science in many fields and gives an overview of the state of the art research in Computational Sustainability as well as about the content of this edited volume and about the case studies that are addressed in subsequent chapters.

1 Introduction

The world wide situation concerning resource consumption and environmental impact of human actions becomes most apparent when looking at the development of the world population and the world energy consumption. Both entered a steep increase at around 1900, growing steeper and steeper, faster and faster. While the world population has tripled in about 100 years, we consumed around the year 2000 around 15 times the amount of energy compared to 1900 [8]. In particular the growth-based development of the recent decades was accomplished by various environmental threats such as pollution, acid rain, deforestation, the destruction of the ozone layer, climate change, etc. The refinement of natural resources such as coal, oil, natural gas or uranium produce pollutants, toxins and other residues into the earth, air and water causes serious environmental effects, risks and problems.

In the past, technical developments have mainly been evaluated by technical aspects as functionality, quality, safety and economic aspects, such as profitability. Looking at the apparent developments as mentioned above, sustainable development in terms of sustainable technologies means to think in terms of environment, economy

J. Lässig (✉)
Department of Computer Science, University of Applied Sciences Zittau/Görlitz, Görlitz, Germany
e-mail: joerg.laessig@hszg.de

J. Lässig et al. (eds.), *Computational Sustainability*,
Studies in Computational Intelligence 645,
DOI 10.1007/978-3-319-31858-5_1

and society together, including besides the mentioned evaluations, also an evaluation of environmental, human and social aspects as well as inter- and intragenerational justice.

The term "Sustainable Development" goes back to the 1987 report of the Brundtland Commission [1], which has been established by the United Nations (UN) in 1983, having its name from the chairman of the commission, former Prime Minister of Norway, Gro Harlem Brundtland. The decision to work on this topic was caused by the insight of the UN in the 1980s, that the natural resources would not be available for future generations if they are not managed and applied in a more sustainable way to protect the environment and the natural resources themselves.

Economic development has to be either strongly limited or decoupled from resource consumption and the connected environmental damage. Hence, given this insight, major directions of research and development in the subsequent years and decades were the analysis of reasons for the observed negative developments, the discussion and development of shared sustainability goals, and the suggestion of solutions for sustainability problems to resolve this situation and to gain control.

The outcome of this development are the following central aspects of sustainable development, which are also up-to-date today[1]:

1. long-term environmental strategies for achieving sustainable development
2. co-operation among developing countries and between countries at different stages of economic and social development
3. common and mutually supportive objectives which take account of the interrelationships between people, resources, environment and development
4. ways and means by which the international community can deal more effectively with environmental concerns
5. efforts to deal successfully with the problems of protecting and enhancing the environment
6. a long-term agenda for action during the coming decades.

Of central importance are the areas of population, food security, the loss of species and genetic resources, energy, industry, and human settlements [1]. Poverty reduction, gender equity, and wealth redistribution have been identified as important prerequisite for environmental conservation.

The central and most common definition is to consider *sustainable development* as *development that meets the needs of the present without compromising the ability of future generations to meet their own needs*. More systematic definitions of sustainability are e.g. given by Jischa, Fig. 1.

[1] http://www.un.org/documents/ga/res/38/a38r161.htm.

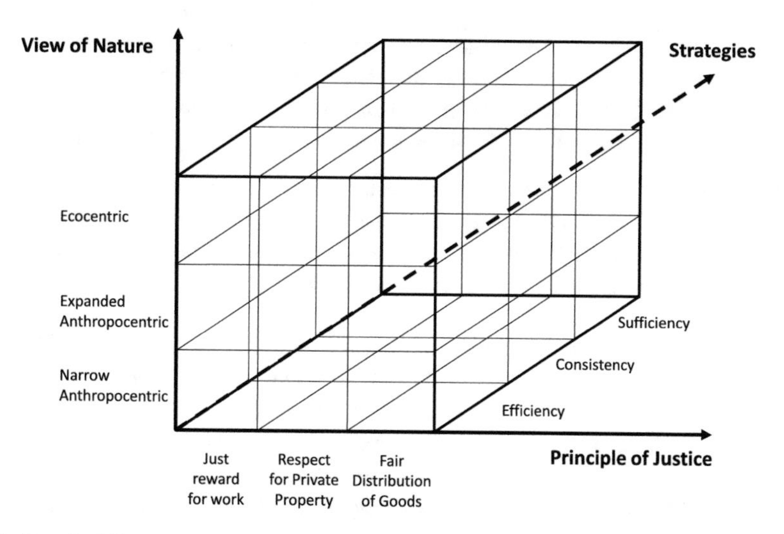

Fig. 1 Sustainability matrix according to Jischa [8]

The developments of COP 21 in December 2015 in Paris (United Nations Framework Convention on Climate Change, 21st Conference of the Parties, see[2]) further push current initiatives all over the world and establish the 2° goal—if possible 1.5°—in an international climate contract involving 194 countries.

The edited volume at hand shows best practice examples of contemporary sustainability science in particular in the field of Computer Science and data driven methods. In Sect. 2 we describe the initial need for sustainability science as a separate field of research and how it has been established. Section 3 in particular describes the role of engineering as a subject that gained increasing importance in this context and why Computer Science plays a key role today. Also we show key research work in the field of Computational Sustainability in Sect. 4. We continue with an overview of the chapters in this volume in Sect. 5. After a short conclusion in Sect. 6 the technical part of the book starts.

2 Sustainability Science as Research Field

Based on the ideas of the Brundtland commission, sustainability research has been build up, an interdisciplinary and applied science about the research in sustainable solutions and their implementation. In June 1992, at the United Nations Conference on Environment and Development in Rio de Janeiro, the role of research in the endeavor for sustainable development has been emphasized the first time [6]. Important in the context of sustainability science is the need for new paradigms and disciplines in order to master the challenges of sustainable development. Going back

[2]https://unfccc.int/resource/docs/2015/cop21/eng/l09r01.pdf.

to Snow (1959) [18], sustainability research also has to connect the two cultures, human and social sciences, and on the other hand, the natural and engineering sciences, thinking about solutions creating a sustainable future.

Sustainability Science has the following corner stones and basic concepts [6]:

1. Sustainability Science develops a scientific foundation for sustainable practice and actions. Concerning the goals, values and norms and to balance them, it is normative, while Sustainability Science is a descriptive research concerning the instruments of implementation or when it comes to causes for missing sustainability.
2. Sustainability Science is multi- or transdisciplinary. Sustainable development surmounts by far the potential of one single scientific discipline.
3. Sustainability Science is primarily a practical science. It is meant to solve existential problems of the world society and the system earth. Practically the management of sustainability and sustainable development are central issues.
4. Sustainability Science is based on the responsibility of science and of the single scientist regarding future generations in the life system earth.

The necessity for the existence of Sustainability Science is given by the fact, that a researcher of one discipline usually hits the borders of the discipline. The need for Sustainability Science comes from this multidisciplinarity and the challenge to secure assertions profoundly by the involved research fields.

According to the Forum on Science and Technology for Sustainability (Harvard University), "The world's present development path is not sustainable. Efforts to meet the needs of a growing population in an interconnected but unequal and human-dominated world are undermining the earth's essential life-support systems. Meeting fundamental human needs while preserving the life-support systems of planet earth will require a world-wide acceleration of today's halting progress in a transition towards sustainability."

Looking for the most urgent problems to be solved, the German Advisory Council on Global Change [22] a mentions the following Sustainable Development Goals:

- climate change
- ocean acidification
- loss of biological diversity and ecosystem services
- land and soil degradation
- risks posed by long-lived and harmful anthropogenic substances
- loss of phosphorus as an essential resource for agriculture and therefor also for food security.

3 The Role of Engineering and Computer Science

According to Jischa [7], engineering takes place in three dimensions: process—how is something processed, tool—how is the processing taken into action, and medium—

is material, energy or information processed. In human history, various developments have changed our way of action, from being hunters, farmers, industrialized societies and finally the globalized service society. Accordingly, the source of value and manipulation was changing as well—from nature, to farmland, to capital, and finally, to information. The development takes place on a logarithmic scale concerning capital involved and concerning the dimension time, considered as acceleration. The effect is according to Lübbe [12] also known as shrinking of the present. Defining the present as *length of time of constant life and working conditions*, our stay in the present is constantly decreasing. The unknown future moves constantly closer and closer to the present. This brings decision makers in the modern world in a challenging situation: While we never before in history had a time in which societies knew as little as today about their near future, the number of innovations, which change our life situation structurally and irreversibly, is constantly growing.

This goes in engineering conform with the Collingridge dilemma, according to which, the (social) consequences of technology cannot be predicted early in the life of a technology. Instead, the consequences are discovered over time, but, unfortunately, by that time they are often already so much part of the economy and social fabric, that its control is difficult: When change is easy, the need for control cannot be foreseen, but when the need for change becomes apparent, it has become expensive, difficult and time consuming, Jischa [7].

Leggewie and Welzer [11] declare in this sense the end of the world we have known: "Our perception limps behind the speed of transformation in a globalized world. This is obvious at all levels of our existence, regarding to critical developments concerning energy, environment and climate as well as economic and financial crises".

How this situation can be resolved is unclear, but in general the role of Computer Science in the multi-disciplinary field of sustainability research—called Computational Sustainability—comes from its increasing importance and from being involved in many fields. Kenneth Birman, Cornell University, describes it as follows: "The importance of Computer Science has never been greater. We are discovering ways to build just about everything out of small, simple mechanisms glued together with software, so no matter what you do, Computer Science tends to be inside. And the scope of this new Computer Science is amazing: We are at the center of the action in biology, nanotechnology, particle physics. If society is ever going to slash medical costs, Computer Science will play the key role. I see Computer Science as a sort of universal science. We are beginning to pervade everything." Looking at the recent developments in the Silicone Valley and the large IT companies that constantly invent the innovations of tomorrow, this is very true. So the conclusion, that Computer Science can play an important role in guiding the progress and research towards a sustainable development and the development goals as described, is coherent.

In this sense, the vision of Carla P. Gomes about the role of Computer Science in the sustainable development process, which she formulated in 2008, is straightforward: "Computer scientists can—and should—play a key role in increasing the efficiency and effectiveness of the way we manage and allocate our natural resources, while enriching and transforming Computer Science." In this sense, Gomes addresses only a subset of the sustainable development efforts, but probably the subset where Computer Science is most helpful. Interesting and important at the same time is the fact, that she also mentions developments within Computer Science. This includes

- energy efficient hardware combined with centralization, consolidation and virtualization
- better algorithms and processes for more efficiency, e.g. fleet management and logistics
- in place work and communication, new work models
- monitoring and sensors in buildings, awareness as key
- autonomic cars and new approaches in logistics
- dematerialization: e.g. storage of information in files, cloud and service computing.

Worth to mention in the struggle to reduce energy consumption is the progress in optimization science and algorithms; here an example: In the 2010 Report to the President and Congress [*of the USA*], Designing a Digital Future: Federally Funded Research and Development in Networking and Information Technology, on page 71, algorithmic progress is mentioned. Martin Grötschel (Zuse Institute Berlin) reports about a benchmark for production planning with LP algorithms, that it took current hardware and LP algorithms of 1988 82 years to find the solution. In 2003, 15 years later, the problem is solved in roughly one minute. This is equivalent to a factor of 1,000, which is due to better hardware and a factor of 43,000, due to algorithm improvement.

4 Development in the Field of Computational Sustainability

Since the term Computational Sustainability has been coined, there is vivid research going on in the field. This edited volume gives an overview.

Watson et al. [20] describe energy informatics and new directions for the community, proposing ways for the Information Systems (IS) community to engage in the development of environmentally sustainable business practices. IS researchers, educators, journal editors, and association leaders are dedicated to demonstrate how the transformative power of IS can be leveraged to create an ecologically sustainable society. Substantial research has been done in this application context of energy informatics, covering various directions towards better energy efficiency, energy management, and control approaches as in the smart grid context [16]. The authors argue that the value of smart grids will become fully visible once the collected data in the grid is ingested, processed, analyzed and translated into meaningful decisions. As applications of the information, forecast electricity demand, respond to peak load events, and improved sustainable use of energy by consumers are described. As technologies pattern mining and machine learning over complex events as well as integrated semantic information processing, distributed stream processing, cloud platforms and privacy policies to mitigate information leaking are addressed. See also [24] for more results in the smart grid context.

Also in the context of single buildings IT applications are considered, e.g. by Lawrence et al. [10]. Leveraging energy informatics the authors propose a new concept for how buildings and their systems are designed and operated. The authors break down the approach to the simple formula 'Energy + Information = Less Energy'. Different scales of application are considered, ranging from single equipment items up to whole regions or countries. It is in particular also considered how cross-realty concepts can be applied in the training of students and professionals in the operation of complex building systems. A good overview on energy informatics as a field is given by [4].

Modern IT technologies with different focus have been utilized in the context of energy informatics and computational sustainability. While many works have a particular focus on machine learning and data mining, also the application of other techniques is reported. E.g. [17] develop a cloud based software platform for data driven smart grid management. Their article describes the application of cloud technologies used in a scalable software platform for the Smart Grid. Dynamic Demand Response (D2R) is reported as challenging problem with the goals to perform intelligent demand-side management and to relieve peak load. Considered are hybrid cloud setups, including IaaS and PaaS. Typical requirements are on-demand provisioning, massive scaling, and manageability.

Pernici et al. [15] describe, what IS can do for environmental sustainability in a report from the CAiSE'11 panel on green and sustainable information systems. The panel report describes the panelists' views on using information systems for improving sustainability and on improving the energy efficiency of the data centres on which information systems are based. The current topics of research, possible contributions of the IS community, and future directions are discussed. In a broader view, the study searches an information strategy for environmental sustainability, seeking solutions for a problem of change [21].

Of increasing importance today is also the establishment of e-mobility solutions, e.g. described by Wagner et al. [19]. In the smart grid context, renewable energies such as wind and solar power increase fluctuations in the grid due to changing weather conditions. This raises the need for additional supply and demand, reserved to compensate for fluctuations. The authors describe an approach to use electric vehicles as distributed storage devices that draw or supply power to the grid during frequency fluctuations. Based on simulations and real data the article shows that this approach is able to support power grid stability while generating substantial revenues for the operating intermediary.

Also more theoretically oriented work has been done in the field, describing, e.g. dynamic sustainability games for renewable resources [3]. The authors consider a dynamic Nash game among firms harvesting a renewable resource and propose a differential variational inequality (DVI) framework for modeling and solving such a game. The results consider myopic planning versus long-term perspectives. Computationally efficient algorithms are applied for the solution of the game.

The field of Computational Sustainabiliy is very broad and there is an ample number of publications in the field, e.g. on methodologies and systems for green house gas inventories and the related software systems [5, 23] or energy efficiency benchmarking [9]. For further topics and work we refer to surveys in the field. Milano et al. [13] conclude that Computational Sustainability specifically considers the major problem domains that impact global sustainability. They describe key sustainability issues and how they translate into decision and optimization problems that fall within the realm of computing and information science. Chande [2] argues that Computational Sustainability is an emerging research and academic discipline. The author makes the attempt to view Computational Sustainability as an academic and research subject and suggests how it could be incorporated as a course in academic programmes.

An introduction particularly to data mining for sustainability is given by Morik et al. [14]. The article describes challenges in the field of data mining for sustainability such as scalability, integration of data, distributed data mining, real-time prediction or the processing of spatio-temporal data as well as the understandability of data and the analysis itself. Also typical application fields as disaster management, climate applications and the conservation of natural resources are mentioned. The work continues describing (research) tasks and techniques in more detail. Furthermore, organizations, projects, repositories and tools in the field of data mining for sustainability are described. In particular for the field of computational intelligence also a book covering various research directions and approaches is available [25].

5 Organization of the Book

The volume at hand is now organized as follows:

In Chap. 2, wind power prediction with machine learning is described. The chapter focuses on prediction models for a reliable forecast of wind power production to successfully integrate wind power plants into the power grid. The prediction task is formulated as regression problem. Different regression techniques such as linear regression, k-nearest neighbors and support vector regression are tested and evaluated. By analyzing predictions for individual turbines as well as entire wind parks, the machine learning approach yields feasible results for short-term wind power prediction.

Chapter 3 focuses on the application of statistical learning for short-term photovoltaic power predictions. Due to changing weather conditions, e.g. clouds and fog, a precise forecast in a few hour range can be a difficult task. On the basis of data sets of PV measurements, methods from statistical learning based on data with hourly resolution are applied. Nearest neighbor regression and support vector regression are applied, based on measurements and numerical weather predictions. Also an hybrid predictor that uses forecasts of both employed models is presented.

Again renewable energy prediction is the topic of Chap. 4, but in this case for improved utilization and efficiency in datacenters and backbone networks. Datacenters are one of the important global energy consumers and carbon producers. However, their tight service level requirements prevent easy integration with highly variable renewable energy sources. Short-term green energy prediction can mitigate this variability. Predictions are leveraged to allocate and migrate workloads across geographically distributed datacenters to reduce brown energy consumption costs. The results show that prediction enables up to 90 % green energy utilization.

A completely different field is addressed in Chap. 5. A hybrid machine learning and knowledge based approach is utilized to limit the combinatorial explosion in biodegradation prediction. The work focuses on the prediction of the environmental fate of products, i.e., their degradation products and pathways of organic environmental pollutants. In the chapter, a hybrid knowledge-based and machine learning-based approach to overcome limitations of current approaches in the context of the University of Minnesota Pathway Prediction System (UM-PPS) is proposed.

Feeding the world with big data by uncovering spectral characteristics and dynamics of stressed plants is the topic of Chap. 6. The work addresses the problem of mining hyperspectral images to uncover spectral characteristic and dynamics of drought stressed plants to showcase that current challenges in the field can be met and that big data mining can—and should—play a key role for feeding the world, while enriching and transforming data mining.

Global monitoring of inland water dynamics is the application field of Chap. 7. Inland water is an important natural resource that is critical for sustaining marine and terrestrial ecosystems as well as supporting a variety of human needs. Remote sensing datasets provide opportunities for global-scale monitoring of the extent or surface area of inland water bodies over time. The authors present a survey of existing remote sensing based approaches for monitoring the extent of inland water bodies and discuss their strengths and limitations. Furthermore, an outline of the major challenges that need to be addressed for monitoring the extent and dynamics of water bodies at a global scale are addressed, discussing how to overcome them and motivating future research in global monitoring of water dynamics.

The installation of electric vehicle charging stations—how many and where—is the topic of Chap. 8. Electric Vehicles (EVs) are touted as the sustainable alternative to reduce our overreliance on fossil fuels and stem our excessive carbon emissions. The chapter describes a KDD framework to plan the design and deployment of EV charging stations over a city. The framework integrates user route trajectories, owner characteristics, electricity load patterns, and economic imperatives in a coordinated clustering framework to optimize the locations of stations and assignment of user trajectories to (nearby) stations.

Chapter 9 has its focus on the computationally efficient design optimization of compact microwave and antenna structures. While conventional optimization algorithms (both gradient-based and derivative-free ones such as genetic algorithms) normally require large numbers of simulations of the structure under design, which may be prohibitive, considerable design speedup can be achieved by means of surrogate-based optimization (SBO) where a direct handling of the expensive high-fidelity model is replaced by iterative construction and re-optimization of its faster representation, a surrogate model. In this chapter, some of the recent advances and applications of SBO techniques for the design of compact microwave and antenna structures are reviewed.

Next, in Chap. 10, we focus on sustainable industrial processes that are achieved by embedded real-time quality prediction. Sustainability of industrial production focuses on minimizing greenhouse gas emissions and the consumption of materials and energy. The chapter describes how embedding data analysis (data mining, machine learning) enhances steel production such that resources are saved, where a framework for processing data streams is used for real-time processing. New algorithms that learn from aggregated data and from vertically distributed data as well as two real-world case studies are described.

From steel production we change to relational learning for sustainable health in Chap. 11. Sustainable healthcare is a global need and predictive models have the opportunity to greatly increase value without increasing cost. Concrete examples include reducing heart attacks and reducing adverse drug events by accurately predicting them before they occur. The chapter examines how accurately such events can be predicted presently and discusses a machine learning approach that produces accurately such predictive models.

The final chapter of the book does focus on Compute Science itself as application domain. In Chap. 12, ARM clusters for performant and energy-efficient storage are considered. Low power hardware—such as ARM CPUs—combined with novel storage concepts—such as Ceph—promise scalable storage solutions at lower energy consumptions than today's standard solutions like network attached storage (NAS) systems. Its performance as well as its energy consumption is compared to typical NAS storages. The goal of the study is to outline paths for energy efficient storage systems.

6 Conclusion

This first chapter gave a brief introduction to Computational Sustainability and its development as a separate research field. Sustainability in general is by construction an interdisciplinary science—so it is a bit surprising that something like *Computational Sustainability* exists. The term is pragmatically born from the desire of the protagonists to positively influence the process of sustainable development in the world. Most problems addressed in this book of course cannot be solely solved by the application of computer scientific methods, but it is indeed illustrated how comprehensive and diverse computer scientific methods can be in order to push sustainable development forward. Computational Sustainability is understood as the share that Computer Science can contribute to intelligent, often data driven approaches to reduce energy and resource consumption in certain processes. This volume *Computational Sustainability* understands itself as compilation of case studies and best practices that show, how Computer Science and Data Mining are key concepts on the way to a more sustainable world. This is our all goal and gave rise to this book project. The covered application fields, methods and approaches do not claim completeness—there may be many more application fields and scientific methodologies which would be worth to mention here.

Acknowledgments The author particularly thanks his editor colleagues Katharina Morik and Kristian Kersting as well as all other contributors and chapter authors to the book very much for their great work, the support, and their patience during the process of finishing the volume. We hope to contribute a humble share to the solutions that address the challenge of world wide sustainable development—while enriching and transforming Computer Science and its methods.

References

1. Brundtland, G., Khalid, M., Agnelli, S., Al-Athel, S., Chidzero, B., Fadika, L., Hauff, V., Lang, I., Shijun, M., de Botero, M.M., et al.: Our Common Future (\'Brundtland report\') (1987)
2. Chande, S.V.: Computational sustainability: an emerging research and academic discipline. Int. J. **3**(8) (2014)
3. Chung, S.H., Friesz, T., Weaver, R.: Dynamic sustainability games for renewable resources—a computational approach. IEEE Trans. Comput. **63**(1), 155–166 (2014)
4. Goebel, C., Jacobsen, H.-A., Del Razo, V., Doblander, C., Rivera, J., Ilg, J., Flath, C., Schmeck, H., Weinhardt, C., Pathmaperuma, D., Appelrath, H.-J., Sonnenschein, M., Lehnhoff, S., Kramer, O., Staake, T., Fleisch, E., Neumann, D., Stricker, J., Erek, K., Zarnekow, R., Zieklow, H., Lässig, J.: Energy informatics. Bus. Inf. Syst. Eng. **6**(1), 25–31 (2014)
5. Heider, J., Tasche, D., Lässig, J., Will, M.: A greenhouse gas accounting tool for regional and municipal climate change management. Sustain. Internet ICT Sustain. (SustainIT) **2013**, 1–3 (2013)
6. Heinrichs, H., Martens, P., Michelsen, G., Wiek, A.: Sustainability Science: An Introduction. Springer (2015)
7. Jischa, M.: Die Mechanik in der Geschichte. In: Hartmann, S., Brenner, G. (Hrsg): Jahresbericht, 2008/2009. pp. 7–17. TU Clausthal (2010)
8. Jischa, M.F.: Herausforderung Zukunft–Technischer Fortschritt und Globalisierung. Chemie Ingenieur Technik **78**(9), 1161–1161 (2006)
9. Lässig, J., Will, M., Heider, J., Tasche, D., Riesner, W.: Energy efficiency benchmarking system for industrial enterprises. In: 2014 IEEE International Energy Conference (ENERGYCON), pp. 1069–1075, May 2014
10. Lawrence, T.M., Watson, R.T., Boudreau, M.-C., Johnsen, K., Perry, J., Ding, L.: A new paradigm for the design and management of building systems. Energy Build. **51**, 56–63 (2012)
11. Leggewie, C., Welzer, H.: Das Ende der Welt, wie wir sie kannten: Klima. Fischer Verlag, Zukunft und die Chancen der Demokratie. S (2010)
12. Lübbe, H.: Im Zug der Zeit: Verkürzter Aufenthalt in der Gegenwart. Springer (2003)
13. Milano, M., O'Sullivan, B., Sachenbacher, M.: Guest editors' introduction: special section on computational sustainability: where computer science meets sustainable development. IEEE Trans. Comput. **1**, 88–89 (2014)
14. Morik, K., Bhaduri, K., Kargupta, H.: Introduction to data mining for sustainability. Data Min. Knowl. Disc. **24**(2), 311–324 (2012)
15. Pernici, B., Aiello, M., vom Brocke, J., Donnellan, B., Gelenbe, E., Kretsis, M.: What IS can do for environmental sustainability: a report from CAiSE'11 panel on Green and sustainable IS. Commun. Assoc. Inf. Syst. **30**(1), 18 (2012)
16. Simmhan, Y., Aman, S., Cao, B., Giakkoupis, M., Kumbhare, A., Zhou, Q., Paul, D., Fern, C., Sharma, A., Prasanna, V.: An informatics approach to demand response optimization in smart grids. Natl. Gas **31**, 60 (2011)
17. Simmhan, Y., Aman, S., Kumbhare, A., Liu, R., Stevens, S., Zhou, Q., Prasanna, V.: Cloud-based software platform for data-driven smart grid management. IEEE/AIP Comput. Sci. Eng. (2013)
18. Snow, C.P.: Two cultures. Science **130**(3373), 419–419 (1959)

19. Wagner, S., Brandt, T., Neumann, D.: Beyond mobility-an energy informatics business model for vehicles in the electric age. In: ECIS, p. 154 (2013)
20. Watson, R.T., Boudreau, M.-C., Chen, A.J.: Information systems and environmentally sustainable development: energy informatics and new directions for the is community. Manag. Inf. Syst. Q. **34**(1), 4 (2010)
21. Watson, R.T., Corbett, J., Boudreau, M.-C., Webster, J.: An information strategy for environmental sustainability. Commun. ACM **55**(7), 28–30 (2012)
22. WBGU.: Sondergutachten—Klimaschutz als Weltbürgerbewegung. WBGU Wissenschaftlicher Beirat der Bundesregierung Globale Umweltveränderungen Berlin (2014)
23. Will, M., Lässig, J., Tasche, D., Heider, J.: Regional carbon footprinting for municipalities and cities. In: U. V. A. W. B. R. N. G. Marx Gomez, J., Sonnenschein, M. (eds.) BIS-Verlag, pp. 653–660. BIS-Verlag (2014). ISBN: 978-3-8142-2317-9
24. Woon, W.L., Aung, Z., Madnick, S.: Data Analytics for Renewable Energy Integration: Third ECML PKDD Workshop, DARE 2015, Porto, Portugal, September 11, 2015. Revised Selected Papers, vol. 9518. Springer (2016)
25. Yu, T., Chawla, N., Simoff, S.: Computational Intelligent Data Analysis for Sustainable Development. CRC Press (2013)

Wind Power Prediction with Machine Learning

Nils André Treiber, Justin Heinermann and Oliver Kramer

Abstract Better prediction models for the upcoming supply of renewable energy are important to decrease the need of controlling energy provided by conventional power plants. Especially for successful power grid integration of the highly volatile wind power production, a reliable forecast is crucial. In this chapter, we focus on short-term wind power prediction and employ data from the National Renewable Energy Laboratory (NREL), which are designed for a wind integration study in the western part of the United States. In contrast to physical approaches based on very complex differential equations, our model derives functional dependencies directly from the observations. Hereby, we formulate the prediction task as regression problem and test different regression techniques such as linear regression, k-nearest neighbors and support vector regression. In our experiments, we analyze predictions for individual turbines as well as entire wind parks and show that a machine learning approach yields feasible results for short-term wind power prediction.

1 Introduction

The strong increase in renewable energy causes some problems induced by high fluctuations in production. A precise prediction is the key technology for successful integration of the wind power into the grid because it allows planning reserve plants, battery loading strategies and scheduling of consumers.

Generally, there are two model classes for prediction tasks. Most prediction approaches are based on physical models employing numerical weather simulations, see e.g., [3]. These models are used for short- and long-term forecasts in the

N.A. Treiber (✉) · J. Heinermann · O. Kramer
University of Oldenburg, 26111 Oldenburg, Germany
e-mail: nils.andre.treiber@uni-oldenburg.de

J. Heinermann
e-mail: justin.philipp.heinermann@uni-oldenburg.de

O. Kramer
e-mail: oliver.kramer@uni-oldenburg.de

J. Lässig et al. (eds.), *Computational Sustainability*,
Studies in Computational Intelligence 645,
DOI 10.1007/978-3-319-31858-5_2

range of hours. The other class of prediction methods is formed by machine learning algorithms that are implemented more frequently in recent years. Since these statistical models derive functional dependencies directly from the observations, they are also known as *data-driven* models. They can be used for predictions that target a time horizon from seconds to hours and therefore are important for balancing the electrical grid with its different authorities.

In this chapter, we build models that are exclusively based on wind power time series measurements. We formulate the prediction task as a regression problem and compare the accuracy of different regression techniques by employing the two simple regression methods *linear regression* and *k-nearest neighbors* (*k*NN), and the state-of-the-art technique *support vector regression* (SVR). In our studies we make predictions for individual turbines and then for entire wind parks. The latter predictions can be made with various feature aggregation combinations of the corresponding time series that we also compare. This chapter extends our preliminary workshop paper [21] on the workshop *Data Analytics for Renewable Energy Integration* on the *European Conference on Machine Learning (ECML 2013).*

2 Related Work

In their review, Costa et al. [2] present a broad overview of various methods and mathematical, statistical and physical models employed in the last 30 years for short-term prediction. Soman et al. [19] give an extensive survey of the possible techniques for different forecast horizons. Past results have shown that methods from statistical learning are powerful approaches for short-term energy prediction. For example, Juban et al. [8] presented a kernel density estimation approach for a probabilistic forecasting for different wind parks. Foresti et al. [5] employed multiple kernel learning regression as an extended support vector model that autonomously detects the relevant features for wind speed predictions. Also neural networks have been applied to wind power prediction in the past, e.g., by Mohandes et al. [13], who compared an autoregressive model with a classical backpropagation network. In this line of research, Catalao et al. [1] trained a three-layered feedforward network with the Levenberg-Marquardt algorithm for short-term wind power forecasting, which outperformed the persistence model and ARIMA approaches. Further, Han et al. [7] focused on an ensemble method of neural networks for wind power prediction. Regarding the aggregation of wind turbines, Focken et al. [4] studied the decrease of the prediction error of an aggregated power output caused by spatial smoothing effects. From the perspective of electrical engineers, Pöller and Achilles [16] explored, how different wind turbines can be aggregated to a single generator.

The spatio-temporal wind power prediction approach that is basis of our line of research has been introduced in [10] with a more extensive depiction in [11]. In [20], we presented an approach for preselection of turbines for *k*NN-based prediction. As the optimization problem is difficult to solve, we proposed an evolutionary blackbox method for an efficient feature selection, which corresponds to a selection of appropriate turbines. In [6], we proposed an ensemble approach for SVR, where small subsets of training data are randomly sampled and the predictions of multiple SVRs are combined to a strong classifier. As wind power ramps are difficult events for the integration into the grid, we considered this problem in a separate work [12]. We treat ramp prediction as classification problem, which we solve with SVMs. Recursive feature selection illustrates how the number of neighbored turbines affects this approach. The problem of imbalanced training and test sets is analyzed with regard to the number of no-ramp events. In practice, sensors might fail for various reasons and the prediction models cannot be applied. In [17], we compared various missing data methods for the imputation problem. A new contribution of this work is a kNN-based regression method, which is used as geo-imputation preprocessing step by taking into account the time series of the neighbored turbines. Last, in [22] we extended the repertoire of prediction methods with a cross-correlation weighted k-nearest neighbor regression (x-kNN) variant. The kNN-based similarity measure employs weights that are based on the cross-correlation of the time series of the neighboring turbines and the target. If the cross-correlation coefficient is high, the turbine gets a major influence for the prediction by expanding the corresponding dimension in the regression model.

3 Wind Data Set

The models are evaluated based on the *National Renewable Energy Laboratory* (NREL) western wind resources data set [14], which is part of the *Western Wind and Solar Integration Study*, a large regional wind and solar integration study initiated by the United States. The data set has been designed to perform temporal and spatial comparisons like load correlation or estimation of production from hypothetical (i.e., simulated) wind turbines for demand analysis and planning of storage based on wind variability. The data set consists of three years of wind energy data from numerical simulations that are mainly based on real-world wind measurements. It consists of 32,043 turbines in the western area of the US, and can be downloaded from the NREL website.[1] The whole model employs a total capacity of 960 GW of wind energy. A GUI allows to select turbines, and to download their corresponding time series data. Based on a time-resolution of ten minutes, 52,560 entries per year and per turbine are available for 2004, 2005, and 2006, respectively.

In Fig. 1, four different wind situations in a park near Tehachapi are illustrated. One can observe spatio-temporal correlations between the wind speed of the turbines. But it can also be noticed, that occasionally high wind speeds can occur locally.

[1] http://www.nrel.gov/.

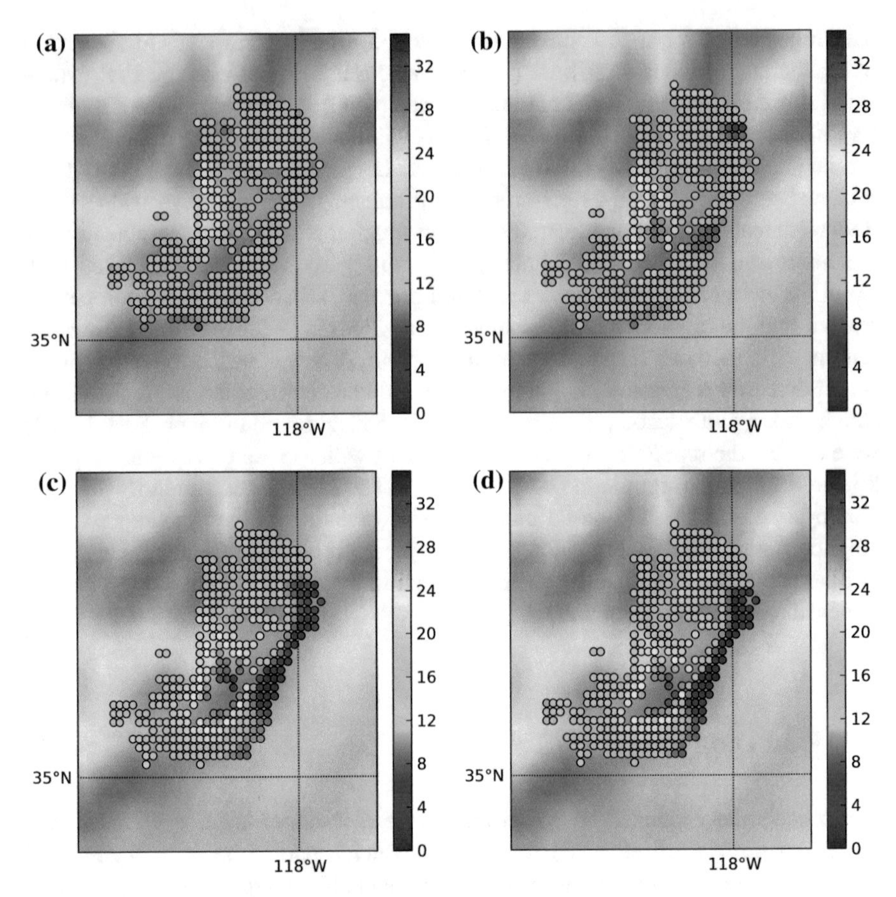

Fig. 1 Visualization of the wind speeds (m/s) in a park near Tehachapi (California) at four time steps with a temporal difference of 20 min. The turbines are colorized with regard to the wind strength, from low in *blue* to strong in *red*. **a** 2004-01-25 00:40, **b** 2004-01-25 01:00, **c** 2004-01-25 01:20, **d** 2004-01-25 01:40

4 WINDML

For addressing the challenge to couple machine learning and data mining methods to wind power time series data, we developed the Python-based framework WINDML [9], which aims at minimizing the obstacles for data-driven research in the wind power domain. It allows the simplification of numerous steps like loading and preprocessing large scale wind data sets or the effective parameterization of machine learning and data mining approaches. With a framework that bounds specialized mining algorithms to data sets of a particular domain, frequent steps of the data mining process chain can be re-used and simplified. The WINDML framework is released under the open source *BSD 3-Clause License*.

The WINDML framework provides a data server, which automatically downloads requested data sets to a local cache when used for the first time. The system only downloads the data for the requested wind turbines and associated time range. The local copies on the user's hard disk are stored in the NUMPY [24] binary file format, allowing an efficient storage and a fast (recurrent) loading of the data. The data set interface allows the encapsulation of different data sources, resulting in a flexible and extendible framework. A complete documentation describing the modules and the overall framework in more detail is openly available. Further, documentations and examples are available, e.g., for the tasks of wind power prediction or visualization, whose outputs are rendered with text and graphics on the WINDML project website.

The application of machine learning and data mining tools to raw time series data often requires various preprocessing steps. For example, wind power prediction for a target wind turbine using turbines in the neighborhood affords the composition of the given power values as feature vector matrix. The WINDML framework simplifies such tasks and offers various methods for assembling corresponding patterns. It also effectively supports the imputation of missing data as well as other sophisticated processes including the computation of high-level features. In addition, the framework also provides various small helper functions, which, for instance, address the tasks of calculating the Haversine distance between given coordinates or selecting the wind turbines in a specified radius. The WINDML framework contains various supervised and unsupervised learning models. Most of the employed machine learning and data mining implementations are based on SCIKIT-LEARN [15], which offers a wide range of algorithms. The methods are continuously extended with own developments, see Sect. 2.

5 General Times Series Model

Our model makes predictions with past wind power measurements. For this task, we formulate the prediction as regression problem. Let us first assume, we intend to predict the power production of a single turbine with its time series. The wind power measurement $\mathbf{x} = p(t)$ (pattern) is mapped to the power production at a target time $y = p_T(t + \lambda)$ (label) with $\lambda \in \mathbb{N}^+$ being the forecast horizon. For our regression model we assume to have N of such pattern-label pairs that are the basis of the training set $T = \{(\mathbf{x}_1, y_1), \ldots, (\mathbf{x}_N, y_N)\}$ and allow, via a regression, to predict the label for a unknown pattern $\mathbf{x}'$.

One can assume that this model yields better predictions if more information of the time series will be used. For this reason, we extend the pattern $\mathbf{x}$ by considering past measurements $p_T(t-1), \ldots, p_T(t-\mu)$ with $\mu \in \mathbb{N}^+$, see Fig. 2. In addition, it might be helpful to take differences of measurements $p_T(t) - p_T(t-1), \ldots p_T(t - (\mu - 1)) - p_T(t-\mu)$ into account. Since we aim to catch spatio-temporal correlations, we add further information to our patterns from m neighboring turbines, see Fig. 3. Their attributes can be composed in the same way like for the target turbine

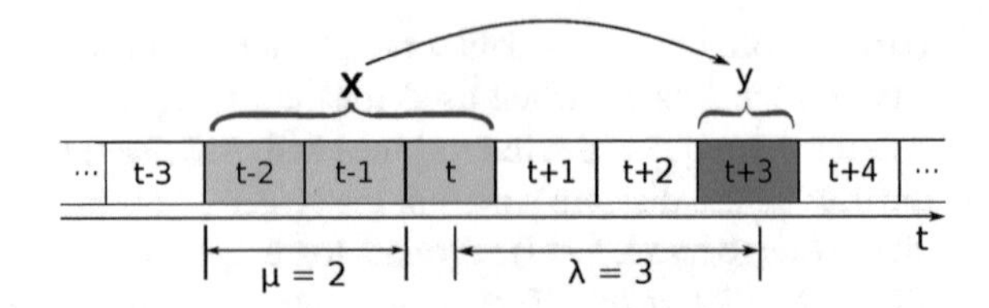

Fig. 2 Section of a time series. The pattern **x** is mapped to label y. The time horizon of the prediction is λ, and the number of additional past measurements is μ

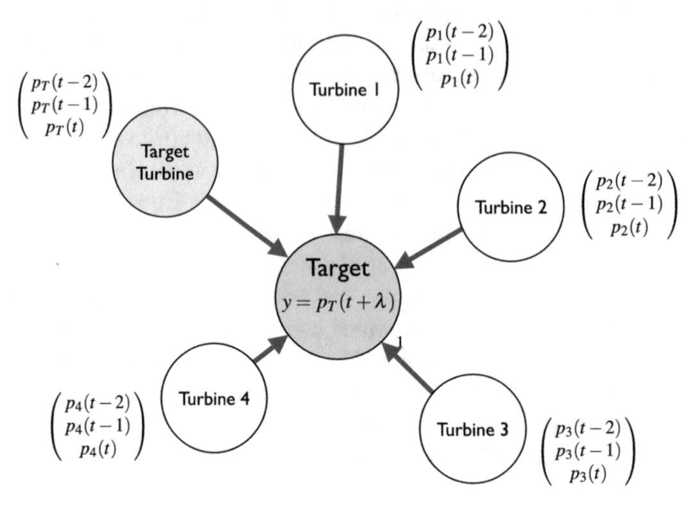

Fig. 3 Setup with four neighboring turbines and the target turbine itself. For each turbine, the current and two past measurements are considered, resulting in a $(4+1) \cdot 3 = 15$-dimensional input vector **x**

(e.g., power values and differences). Finally, we determine the accuracy of the prediction model by computing the error E of the prediction, which is measured as mean squared error (MSE):

$$E = \frac{1}{N} \sum_{i=1}^{N} (f(\mathbf{x}_i) - y_i)^2 \tag{1}$$

6 Regression Techniques

In this section, we introduce the implemented regression techniques. Generally, the goal is to find a function f that provides appropriate predictions to unseen patterns $\mathbf{x}'$. In the following, we explain linear regression, the basic idea of SVR and k-nearest neighbor regression.

6.1 Linear Regression

First, we focus on linear regression. In this model, the prediction value $f(\mathbf{x})$ is expected to be a linear combination of the input variables:

$$f(\mathbf{w}, \mathbf{x}) = w_0 + w_1 x_1 + \cdots + w_N x_N \tag{2}$$

The goal is to find coefficients $\mathbf{w} = (w_1, \ldots, w_N)^T$ that minimize the residual sum of the squares between the observed labels y and the responses predicted by the linear approximation. The problem has the form:

$$\min_{\mathbf{w}} \|\mathbf{X}\mathbf{w} - y\|^2 \quad \text{with} \quad \mathbf{X} = \begin{pmatrix} x_{1,1} & \cdots\cdots & x_{1,N} \\ \cdots & \cdots\cdots & \cdots \\ x_{d,1} & \cdots\cdots & x_{d,N} \end{pmatrix} \tag{3}$$

6.2 Support Vector Regression

Support vector regression is one of the state-of-the-art techniques for prediction tasks. It is based on support vector machines (SVMs) that were proposed by Vapnik [23] in 1995. For the training of the regressor, we aim at finding weights $\mathbf{w}$ by minimizing the following problem, formulated by Vapnik with an ε-sensitive loss function:

$$\text{minimize } \frac{1}{2}\|\mathbf{w}\|^2 + C \sum_{i=1}^{N} (\xi_i + \xi_i^*) \tag{4}$$

$$\text{subject to } \begin{cases} y_i - \langle \mathbf{w}, \mathbf{x}_i \rangle - b & \leq \varepsilon + \xi_i \\ \langle \mathbf{w}, \mathbf{x}_i \rangle + b - y_i & \leq \varepsilon + \xi_i^* \\ \xi_i, \xi_i^* & \geq 0 \end{cases}$$

In this formulation, $C > 0$ is a constant chosen by the user that is used as a parameter that penalizes only those errors which are greater than ε. The so-called slack variables ξ_i^* are introduced to provide a soft margin instead of a hard decision border.

To give good results on non-linear separable data as well, kernel functions are used. A kernel function can be seen as a similarity measure between patterns and is especially useful for non-linear regression tasks. In our experiments, we employ an RBF-kernel:

$$k(\mathbf{x}, \mathbf{x}') = \exp\left(-\frac{||\mathbf{x} - \mathbf{x}'||^2}{2\sigma^2}\right) \tag{5}$$

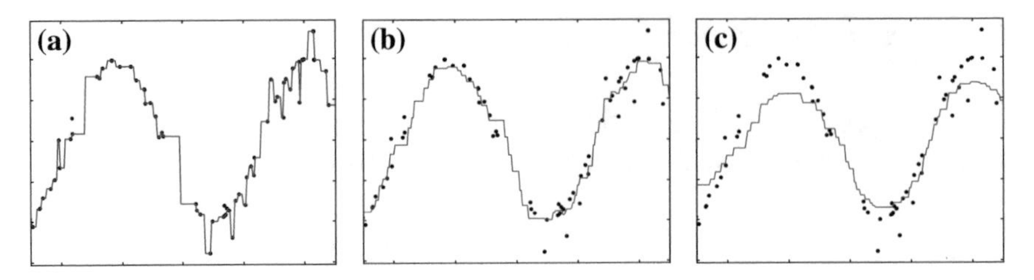

Fig. 4 Visualization of kNN regression. The figures show how the choice of the neighborhood size k determines the character of the regression. **a** $k = 1$, **b** $k = 6$, **c** $k = 18$

6.3 k-Nearest Neighbor Regression

In opposition to the SVR, the kNN model is one of the simplest of all machine learning algorithms. Nevertheless, this technique is very effective for many applications and often offers a better performance than using SVR. The outcome of kNN given a new pattern $\mathbf{x}'$ depends on the k-nearest neighbored patterns in a training set. These patterns are found by calculating the distance between the pattern and all existing patterns $dist(\mathbf{x}, \mathbf{x}')$ using the Euclidean metric:

$$dist(\mathbf{x}, \mathbf{x}') = \left(\sum_{i=1}^{d} (x_i - x_i')^2 \right)^{1/2} \tag{6}$$

With set $\mathcal{N}_k(\mathbf{x}')$ that contains the indices of the k-nearest neighbors of $\mathbf{x}'$, the target value is given by:

$$f_{k\mathrm{NN}}(\mathbf{x}') = \frac{1}{k} \sum_{i \in \mathcal{N}_k(\mathbf{x}')} y_i, \tag{7}$$

if we intend to calculate the arithmetic average of the other k target values. Here, the number of considered neighbors determines the form of the resulting regression function. Given a small value, the model fits the data but is also strongly influenced by outliers, whereas large k might build up models that are too simple, see Fig. 4.

7 Wind Power Prediction for a Single Turbine

In this first experimental section, we want to predict the power output of a single wind turbine for a time horizon of 30 min.

7.1 Setup and Evaluation

For this prediction, neighboring turbines must be selected whose time series features are used in our multivariate time series model. For this sake, we arbitrarily pick 15 turbines that surround the target turbine within a radius of 10 km. For the determination of the distance between turbines, we use the Haversine formula [18]. Finally, every model is trained by using the data from the year of 2005. To accelerate the training process, only every fourth time-step is taken into account. Despite a smaller training set, it is guaranteed that wind conditions at different seasons are included. For the evaluation we test our models on the year 2006 by determining the MSE of the forecasts $f(\mathbf{x})$ with the measured power outputs y_i for N forecasts, see Eq. 1. In our experiments we use five turbines at different locations that are distributed over the western part of the United States.

7.2 Persistence Model

In our studies, forecasts are compared to the persistence model (PST). This model is based on the assumption that the wind power in the time horizon λ does not change and is as strong as at the present time, i.e. $f(p_\alpha(t)) = p_\alpha(t+\lambda)$. Although this naive approach seems to be trivial, its predictions are quite successful for short time horizons and temporally relative constant wind situations, shown for example by Wegley et al. [25]. In our studies, this model achieves mean squared errors, shown in Table 1.

Since it is an important challenge for balancing the grid to predict ramps of changing wind, our models should outperform the persistence model, which understandably fails in such situations.

7.3 Accuracy of the Linear Regression Prediction Model

The experimental results of the linear regression are shown in Table 2. Here the mean squared errors are given for predictions that, on the one hand, are achieved

Table 1 Mean squared error E of the PST model in $[MW^2]$ for the five target turbines

Turbine	$E\,[MW^2]$
Tehachapi	9.499
Lancaster	11.783
Palmsprings	7.085
Cheyenne	9.628
Casper	12.555

It is noticeable that the errors of the individual turbines are quite different

Table 2 Mean squared errors E of the linear regression model in $[MW]^2$

Turbine	$E\ [MW^2]$					
	Absolute			Absolute + changes		
	$\mu = 2$	$\mu = 3$	$\mu = 4$	$\mu = 2$	$\mu = 3$	$\mu = 4$
Tehachapi	7.441	7.334	**7.321**	7.470	7.348	7.365
Lancaster	8.610	**8.479**	8.488	8.631	8.536	8.536
Palmsprings	5.533	5.372	5.347	5.533	5.374	**5.346**
Cheyenne	**7.666**	7.691	7.701	7.702	7.704	7.716
Casper	9.984	10.026	10.067	**9.982**	10.039	10.108

The results shown on the left side are achieved with features only presenting past absolute measurements of the wind power. On the right side, the corresponding differences of the measurements are also taken into account

only with the absolute measurements of the generated power and, on the other hand, are produced with the absolute values and their corresponding changes. From the table one cannot decide which of the two feature vectors is more suitable for the prediction. It is also not clear how many past steps should be considered. The main result is that our linear regression predictions can clearly outperform the persistence model.

7.4 *Results of the SVR Prediction*

The SVR model requires an appropriate choice of its parameters. We implement a grid search with a three-fold cross validation and test values of the trade-off parameter $C = 0.1,\ 10,\ 100,\ 1000,\ 2000$ and the kernel bandwidth $\sigma = 10^{-i}$ with $i = 2, 3, \ldots, 7$. The Fig. 5 shows an arbitrarily chosen section of the time series of a turbine near Tehachapi and allows to compare the persistence and the SVR prediction. The quantitative results are concluded in Table 3.

The results show that the SVR technique even achieves better results than the linear regression if one takes only the absolute measurements as features and selects the number of past steps carefully.

7.5 *Results of the kNN Model*

The kNN model is tested with three different values for k. The mean squared errors of the predictions are shown in Table 4. The accuracy of kNN model is not as high as in the other two methods. In contrast to the SVR technique, this model requires inputs for its best predictions that also contain the corresponding differences of the measurements.

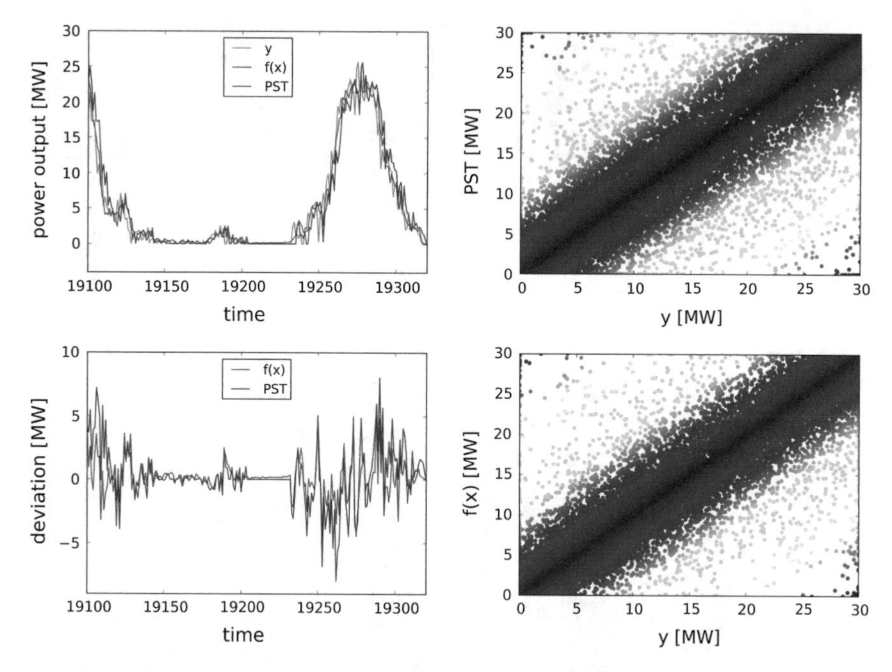

Fig. 5 Visualization of the SVR prediction for a turbines in Tehachapi. The *top left plot* compares the SVR and the persistence predictions to the actual wind power measurements. The *bottom left plot* shows the differences of the corresponding models. The plots on the *right hand side* show the actual and predicted measurement pairs. The absolute prediction errors is the deviation to the main diagonal

Table 3 Results of the prediction using the SVR model

Turbine	$E\,[MW^2]$					
	Absolute			Absolute + changes		
	$\mu = 2$	$\mu = 3$	$\mu = 4$	$\mu = 2$	$\mu = 3$	$\mu = 4$
Tehachapi	7.093	7.074	**6.984**	7.131	7.096	7.012
Lancaster	8.364	**8.108**	8.157	8.397	8.339	8.342
Palmsprings	5.303	5.231	**5.209**	5.387	5.251	5.256
Cheyenne	7.670	7.713	**7.646**	7.679	7.702	7.694
Casper	9.644	**9.493**	9.507	9.798	9.628	9.539

In comparison to the previous results, a higher accuracy is achieved

8 Prediction for a Wind Park

Since the power output of a single turbine is generally not high enough to cause relevant effects on the electric grid, we want to analyze the power output of an entire wind park. The objective of this section is to compare different features and various feature aggregation combinations of the turbines belonging to the park.

Table 4 Prediction errors of the kNN model

Turbine	k	$E\ [MW^2]$					
		Absolute			Absolute + changes		
		$\mu = 2$	$\mu = 3$	$\mu = 4$	$\mu = 2$	$\mu = 3$	$\mu = 4$
Tehachapi	5	9.957	10.101	10.115	9.380	9.785	9.932
	25	9.296	9.720	9.906	**8.887**	9.399	9.725
	50	9.643	10.117	10.339	9.241	9.799	10.133
Lancaster	5	11.340	11.266	11.541	10.569	10.665	11.099
	25	10.837	10.848	11.016	**10.058**	10.296	10.698
	50	10.929	11.216	11.488	10.229	10.628	11.115
Palmsprings	5	7.292	7.403	7.628	6.903	7.176	7.396
	25	6.812	7.136	7.404	**6.521**	6.864	7.250
	50	6.900	7.340	7.666	6.649	7.102	7.523
Cheyenne	5	9.749	10.014	10.171	9.526	9.897	10.181
	25	9.041	9.388	9.633	**8.864**	9.308	9.636
	50	9.135	9.539	9.860	9.010	9.480	9.858
Casper	5	13.033	13.371	13.698	12.409	12.951	13.440
	25	12.440	12.985	13.414	**11.943**	12.619	13.189
	50	12.835	13.440	13.925	12.348	13.127	13.723

It is expected that a tuning of parameter k should partially lead to slightly better predictions

8.1 *Definition of a Wind Park*

In our experiments, we intend to predict the power production of an entire wind park. Since parks are not explicitly given in the data set, we define a park by a central turbine, identified with a certain ID, and some neighboring turbines. To get these further turbines, we determine all turbines in a particular radius r_{in} around the central turbine. Since at least one of the considered aggregations should benefit from spatio-temporal relationships in a park, we choose a radius in an interval of $r_{in} = 7 - 12$ km to ensure that the park covers substantial region. Finally, we select $z = 25$ from all turbines within the selected radius.

8.2 *Feature Aggregations*

We test the three feature aggregation types.

8.2.1 Park Prediction as Sum of All Single Predictions

One possibility to make predictions for a park is to sum up the predictions of all z wind turbines. While here many different approaches for the individual predictions are possible, we only concentrate on two. First, we use the implementation for each turbine in the park by considering only some features of its own time series, resulting in a pattern with a dimension d_{st}. This setup is called AG-1.

Second, we also respect additional information of m arbitrarily selected neighboring turbines in the park for the individual predictions to get spatio-temporal information. Hereby, the extra features x_j are generated in the same way as in our studies of the individual prediction in the sections before. Eventually, one has a pattern with a dimension $d = (1 + m)\, d_{st}$, which is still mapped only to the power output $p_T(t + \lambda)$ of the target turbine T. This implementation is defined as setup AG-2.

8.2.2 Park Prediction with All Features of All Turbines

In this implementation, all time series features of all z turbines in the park are considered at once. The high-dimensional pattern includes the features of each turbine and thus has a dimension of $d = z \cdot d_{st}$. The label y is no longer the output power of an individual turbine, but rather the accumulated power of the park $y^{park} = \sum_{j=1}^{z} p_j(t + \lambda)$. This setup defines AG-3.

8.2.3 Park Prediction with One Aggregated Time Series

Finally, we use the aggregated power not only as label, but also for the patterns by computing one overall time series whose features are extracted as before. We label this model with AG-4:

$$\mathbf{x}^{park} = \begin{pmatrix} \sum_{j=1}^{z} p_j(t) \\ \cdots \\ \sum_{j=1}^{z} p_j(t-\mu) \end{pmatrix} \longrightarrow \quad y^{park} = \sum_{j=1}^{z} p_j(t+\lambda)$$

A question is which feature aggregations yield the best predictions with regard to the MSE (see Eq. 1). We will experimentally compare the introduced aggregations after the explanation of the evaluation details.

8.3 *Prediction with Support Vector Regression*

For the SVR training process we employ a three-fold cross-validation and a grid search for the parameters in the interval $C = 10^i$ with $i = 2, 3, 4$ and $\sigma = 10^{-i}$ with $i = 4, 5$. These intervals have been manually restricted to accelerate the training process. In our first studies, we employ the SVR model with different aggregations. We compare the precision of the forecast with patterns that consist only of absolute values of wind measurements by using the actual measurement ($\mu = 0$), the actual and two past measurements ($\mu = 2$), and the actual and six past measurements ($\mu = 6$).

Table 5 Employing a 30 min forecast for a park with 25 turbines with the SVR model

	E^* $[MW^2]$					
	Absolute			Absolute + changes		
	$\mu = 0$	$\mu = 2$	$\mu = 6$	$\mu = 1$	$\mu = 2$	$\mu = 3$
AG-1	280.7	282.3	275.8	289.1	280.6	275.7
AG-2	281.1	237.9	237.4	243.1	238.0	238.0
AG-3	278.6	223.8	223.4	224.0	220.9	218.6
AG-4	282.1	213.7	213.3	219.3	212.4	**211.2**

The values show the validation errors with regard to the square loss $E^* = E/10^5$ [MW^2]. The persistence model achieves an error of $E^* = 280.6$

Furthermore, we test patterns that again contain the differences of the corresponding measurements for $\mu = 1, 2, 4$. In this way, it is possible to compare the results for $\mu = 2$ with and without wind power changes. In addition, we can compare the error when the patterns have the same dimensionality, which is the case for $\mu = 6$ and $\mu = 3$ when using only absolute values and changes respectively.

We train every model using the first nine months of 2006. For the quantitative evaluation on the last quarter of 2006, we determine the MSE E of the forecasts $f(\mathbf{x}^{park})$ with the measured power outputs y^{park}. Again, we compare our results with the persistence model PST that achieves an error of $E = 280.6 \cdot 10^5 [MW^2]$.

The two parts of the Table 5 show that the prediction is more precise for the aggregations AG-1, AG-3 and AG-4 and for ($\mu \geq 2$) if the patterns include the corresponding changes. It appears that the sum of the individual simple predictions (AG-1) leads to the worst results, comparable with those of the persistence model. If the individual predictions are made with neighboring turbines (AG-2), the persistence model is clearly outperformed. The aggregation considering all features at once (AG-3) leads to an even better accuracy than AG-2. The best results for prediction is achieved by aggregation AG-4, which builds up one time series for the entire park by summing the power values of the single turbines. As we focus on prediction models for a park, this aggregation step pays off. Another advantage is the fact that the data set is reduced and the used regression technique needs a much shorter runtime than using all individual features.

8.4 Prediction with k-Nearest Neighbors

The best parameter for the kNN model is determined with a four-fold cross-validation by testing neighborhood sizes of $k = \{3, 10, 20, 40, 80\}$. Table 6 serves as a direct comparison to Table 5. For the aggregations AG-1, AG-3 and AG-4, including the changes of measurements into the patterns is helpful when using kNN regression. While the results for AG-2 are slightly worse than for AG-1 with $\mu \geq 2$, no good predictions are achieved by the setup AG-3.

Table 6 Employing a 30 min forecast for a park with 25 turbines with kNN

	E^* $[MW^2]$					
	Absolute			Absolute + changes		
	$\mu = 0$	$\mu = 2$	$\mu = 6$	$\mu = 1$	$\mu = 2$	$\mu = 3$
AG-1	292.1	281.4	290.6	283.0	278.8	280.4
AG-2	275.0	295.1	339.0	269.3	281.5	296.0
AG-3	288.8	332.5	394.1	292.9	315.2	335.9
AG-4	282.7	215.7	241.5	220.0	**213.5**	216.6

The table shows the validation error $E^* = E/10^5$ [MW^2]. For the error of the persistence model, see caption of Table 5

In conclusion, one should note that here the AG-4 achieves the highest accuracy, too. The errors achieved are almost as low as for the SVR technique. This is remarkable, since kNN performed significantly worse in the prediction for a single turbine, see Sect. 7.

9 Conclusions

A precise short-term wind power prediction is important for a safe and sustainable balancing of the electricity grid. This work focuses on the statistical wind power forecast for an individual turbine and an entire park with a horizon of 30 min. The most important result is that predictions with the highest accuracy are achieved for both setups with the SVR technique. For the park, it is with respect to the accuracy and the performance of the prediction advantageous to build one aggregated times series, formed by the summation of the power outputs of all individual turbines.

A direct comparison of our results to other models is a quite difficult undertaking, in particular, because other models would have to use exactly the same data. Further, there is no standard for measuring the performance of wind energy prediction models, which has often been criticized in literature [2]. However, the objective of prediction models is to generally outperform the persistence model in terms of MSE. We do this in a convincing way: Our final approach for the park achieves an accuracy that is 24 % better than the persistence model. We expect that tuning parameters of the SVR models further improves the results. This will be subject to our future research activities.

Acknowledgments We thank the presidential chair of the University of Oldenburg, the EWE research institute NextEnergy, and the Ministry of Science and Culture of Lower Saxony for partly supporting this work. Further, we thank the US National Renewable Energy Laboratory (NREL) for providing the wind data set.

References

1. Catalao, J.P.S., Pousinho, H.M.I., Mendes, V.M.F.: An artificial neural network approach for short-term wind power forecasting in Portugal. In: 15th International Conference on Intelligent System Applications to Power Systems (2009)
2. Costa, A., Crespo, A., Navarro, J., Lizcano, G., Feitosa, H.M.E.: A review on the young history of the wind power short-term prediction. Renew. Sustain. Energy Rev. **12**(6), 1725–1744 (2008)
3. Ernst, B., Oakleaf, B., Ahlstrom, M., Lange, M., Moehrlen, C., Lange, B., Focken, U., Rohrig, K.: Predicting the wind. Power Energy Mag. **5**(6), 78–89 (2007)
4. Focken, U., Lange, M., Mönnich, K., Waldl, H., Beyer, H., Luig, A.: Short-term prediction of the aggregated power output of wind farms—a statistical analysis of the reduction of the prediction error by spatial smoothing effects. J. Wind Eng. Ind. Aerodyn. **90**(3), 231–246 (2002)
5. Foresti, L., Tuia, D., Kanevski, M., Pozdnoukhov, A.: Learning wind fields with multiple kernels. Stoch. Env. Res. Risk Assess. **25**(1), 51–66 (2011)
6. Heinermann, J., Kramer, O.: Precise wind power prediction with SVM ensemble regression. In: Artificial Neural Networks and Machine Learning—ICANN 2014, pp. 797–804. Springer, Switzerland (2014)
7. Han, S., Liu, Y., Yan, J.: Neural network ensemble method study for wind power prediction. In: Asia Pacific Power and Energy Engineering Conference (APPEEC) (2011)
8. Juban, J., Fugon, L., Kariniotakis, G.: Probabilistic short-term wind power forecasting based on kernel. In: Density Estimators. European Wind Energy Conference, pp. 683–688. IEEE (2007)
9. Kramer, O., Gieseke, F., Heinermann, J., Poloczek, J., Treiber, N.A.: A framework for data mining in wind power time series. In: Proceedings of ECML Workshop DARE (2014)
10. Kramer, O., Gieseke, F.: Short-term wind energy forecasting using support vector regression. In: 6th International Conference on Soft Computing Models in Industrial and Environmental Applications (2011)
11. Kramer, O., Gieseke, F.: Analysis of wind energy time series with kernel methods and neural networks. In: 7th International Conference on Natural Computation (2011)
12. Kramer, O., Treiber, N.A., Sonnenschein, M.: Wind power ramp event prediction with support vector machines. In: 9th International Conference on Hybrid Artificial Intelligence Systems (2014)
13. Mohandes, M.A., Rehmann, S., Halawani, T.O.: A neural networks approach for wind speed prediction. Renew. Energy **13**(3), 345–354 (1998)
14. Lew, D., Milligan, M., Jordan, G., Freeman, L., Miller, N., Clark, K., Piwko, R.: How do wind and solar power affect grid operations: the western wind and solar integration study. In: 8th International Workshop on Large Scale Integration of Wind Power and on Transmission Networks for Offshore Wind Farms (2009)
15. Pedregosa, F., Varoquaux, G., Gramfort, A., Michel, V., Thirion, B., Grisel, O., Blondel, M., Prettenhofer, P., Weiss, R., Dubourg, V., Vanderplas, J., Passos, A., Cournapeau, D., Brucher, M., Perrot, M., Duchesnay, E.: Scikit-learn: machine learning in Python. J. Mach. Learn. Res. **12**, 2825–2830 (2011)
16. Pöller, M., Achilles, S.: Aggregated wind park models for analyzing power system dynamics. In: 4th International Workshop on Large-scale Integration of Wind Power and Transmission Networks for Offshore Wind Farms, Billund (2003)
17. Poloczek, J., Treiber, N.A., Kramer, O.: KNN regression as geo-imputation method for spatio-temporal wind data. In: 9th International Conference on Soft Computing Models in Industrial and Environmental Applications (2014)
18. Robusto, C.C.: The Cosine-Haversine formula. Am. Math. Mon. **64**(1), 38–40 (1957)
19. Soman, S.S., Zareipour, H., Malik, O., Mandal, P.: A review of wind power and wind speed forecasting methods with different time horizons. In: North American Power Symposium (NAPS), pp. 1–8 (2010)
20. Treiber, N.A., Kramer, O.: Evolutionary turbine selection for wind power predictions. In: 37th Annual German Conference on AI, pp. 267–272 (2014)

21. Treiber, N.A., Heinermann, J., Kramer, O.: Aggregation of features for wind energy prediction with support vector regression and nearest neighbors. In: European Conference on Machine Learning, DARE Workshop (2013)
22. Treiber, N.A., Kramer, O.: Wind power prediction with cross-correlation weighted nearest neighbors. In: 28th International Conference on Informatics for Environmental Protection (2014)
23. Vapnik, V.: The Nature of Statistical Learning Theory. Springer, New York (1995)
24. Van der Walt, S., Colbert, S.C., Varoquaux, G.: The numpy array: a structure for efficient numerical computation. Comput. Sci. Eng. **13**(2), 22–30 (2011)
25. Wegley, H., Kosorok, M., Formica, W.: Subhourly wind forecasting techniques for wind turbine operations. Technical report, Pacific Northwest Lab., Richland, WA (USA) (1984)

Statistical Learning for Short-Term Photovoltaic Power Predictions

Björn Wolff, Elke Lorenz and Oliver Kramer

Abstract A reliable prediction of photovoltaic (PV) power plays an important part as basis for operation and management strategies for a efficient and economical integration into the power grid. Due to changing weather conditions, e.g., clouds and fog, a precise forecast in a few hour range can be a difficult task. The growing IT infrastructure allows a fine screening of PV power. On the basis of big data sets of PV measurements, we apply methods from statistical learning for one- to six-hour ahead predictions based on data with hourly resolution. In this work, we employ nearest neighbor regression and support vector regression for PV power predictions based on measurements and numerical weather predictions. We put an emphasis on the analysis of feature combinations based on these two data sources. After optimizing the settings and comparing the employed statistical learning models, we build a hybrid predictor that uses forecasts of both employed models.

1 Introduction

In Germany, with an installed capacity of more than 38.5 GW at the end of 2014, PV power prediction services are already an essential part of the grid control. On the local scale, storage management and smart grid applications define a sector with increasing need for PV power forecasting to mitigate the impact of the highly fluctuating PV power production.

B. Wolff (✉) · E. Lorenz
EHF Laboratory, Energy Meteorology, Department of Physics, Carl von Ossietzky University of Oldenburg, Carl-von-Ossietzky-Straße 9-11, Oldenburg, Germany
e-mail: bjoern.wolff@uni-oldenburg.de

E. Lorenz
e-mail: elke.lorenz@uni-oldenburg.de

O. Kramer
Department of Computing Science, Computational Intelligence, Carl von Ossietzky University of Oldenburg, Uhlhornsweg 84, Oldenburg, Germany
e-mail: oliver.kramer@uni-oldenburg.de

J. Lässig et al. (eds.), *Computational Sustainability*,
Studies in Computational Intelligence 645,
DOI 10.1007/978-3-319-31858-5_3

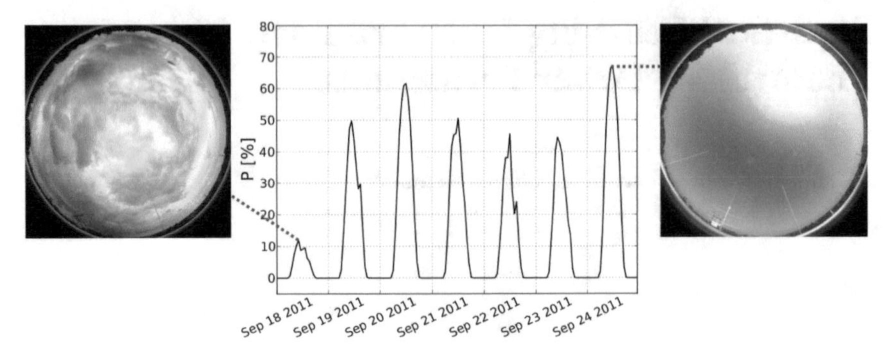

Fig. 1 PV measurement time series example of one week with different weather conditions. There are example pictures from a sky imager camera showing corresponding cloud situations on both sides of the time series

An example PV measurement time series of one week with changing PV power outputs is given in Fig. 1. These fluctuations are caused by varying weather conditions, mainly due to the formation and movement of clouds and by the deterministic course of the sun that causes the typical diurnal and seasonal variation. Two typical weather scenarios that can bee seen in Fig. 1 are overcast (on Sep 18, 2011) and clear sky conditions (on Sep 24, 2011).

As the benefit of using a forecast is directly related to the forecast accuracy, continuous research is performed to enhance PV power predictions.

The aim of this work is the analysis of training sets, features and model parameters for accurate short-term (up to six hours) predictions with hourly resolution based on nearest neighbors regression and support vector machines on 87 PV systems with real-world data.

An early approach for PV power prediction is the autoregressive integrated moving average (ARIMA) model [5]. A related comparison of models is given by Reikard [15]. Popular prediction models are based on neural networks, see Mellit [13]. Similar to our approach, Cao and Cao [3] and Bacher et al. [1] employ forecasts from numerical weather predictions as input of their models. Chakraborty et al. [4] use nearest neighbor regression in combination with a Bayes model to predict PV power output; Fonseca et al. [6] employ support vector regression for power predictions of a 1-MW photovoltaic power plant in Kitakyushu, Japan. In preceding works by Kramer and Gieseke [10], they use support vector regression on wind energy outputs with promising prediction results. Many approaches are based on cloud motion vectors [9] and numerical weather predictions [11]. For a comprehensive survey, we refer the interested reader to the overview by Lorenz and Heinemann [12].

In contrast to the mentioned approaches, we analyze the performance of support vector machines and nearest neighbors on a larger and spatially more distributed set of PV systems. While nearest neighbor regression is one of the simplest machine learning algorithms, it is able to create relatively accurate models in a short amount of time by adjusting only one parameter. Support vector regression is a prominent choice for non-linear modeling. One of its advantages is the ability to handle high dimensional input, without the need of dimensionality reduction efforts in preprocessing steps (unlike, e.g., neural networks), but in return needs proper adjustment to the given task. The flexibility of nearest neighbors and the robustness of support vector regression is helpful as we are concentrating on the combination of meteorological and statistical knowledge by selecting features from different data sources.

This chapter is structured as follows: In Sect. 2, we present the data sets we employ in our experimental analysis. In Sect. 3, we shortly introduce the employed methods, i.e., uniform and weighted K-nearest neighbors (KNN) and support vector regression (SVR). An analysis of training sets, model parameters and features for both models are presented in Sect. 4. A comparison of both methods is introduced in Sect. 5. In Sect. 6, we present a hybrid model of KNN and SVR, and the most important results are summarized in Sect. 7. This chapter is an extension of a contribution to the ECML DARE workshop [18].

2 Data Sets

Our forecasts are based on the combination of past PV power measurements with numerical weather predictions (NWP). Bacher et al. [1] have shown that for forecasts under a two-hour horizon, the most important features are measurements while, for longer horizons, NWPs are appropriate. We combine both types of features for our analysis to produce forecasts up to six hours ahead. The data sets we employ consist of hourly PV power measurements (from PV monitoring provider Meteocontrol GmbH [14]) and predictions from the Integrated Forecast System (IFS), a global NWP model of the European Centre for Medium-Range Weather Forecasts (ECMWF) [7]. Here, we use the IFS output with a temporal resolution of three hours and a spatial resolution of approximately 25 km $\times$ 25 km. Predictions of the NWP model are converted to hourly data with a simple linear interpolation.

For PV prediction, we use the following features organized by their origin:

- PV measurements
  - **P**: relative PV power in % of nominal power
- NWP data
  - **Temp**: temperature forecast
  - **I**: irradiance forecast
  - **CC**: cloud cover forecast
- Additional features
  - **T**: time of day
  - **CS**: modeled clear sky power

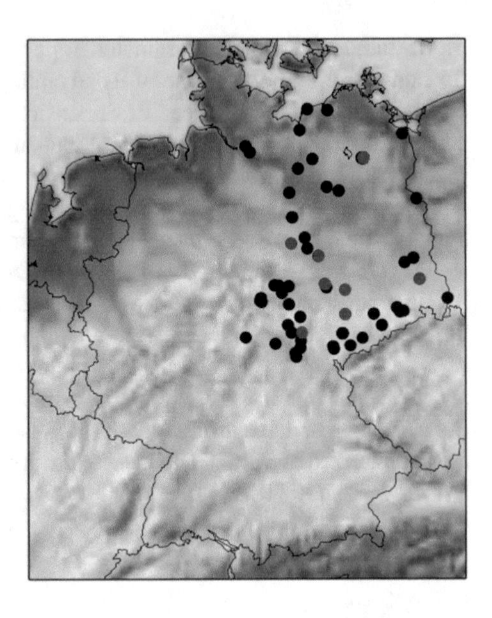

Fig. 2 Locations of PV systems in the eastern part of Germany. *Red points* mark the systems that are basis of the 10-systems model parameter experiments

The 87 PV systems that are basis of our analysis are plotted in Fig. 2. The red dots mark the positions that are employed in a 10-systems experiments for detailed model setup. All PV power measurements of these PV systems have passed an automated quality control so that, for each time step, reasonable measurement data can be assumed. For each of these 87 PV systems, a forecast for a predefined time period is generated and the average of resulting errors is analyzed.

The clear sky PV power (CS) describes the expected power for cloudless weather conditions, which is calculated for every location with a clear sky model. An overview on clear sky modeling methods is given in Lorenz and Heinemann [12].

The test set consists of three distinct weeks of hourly PV time series, i.e., from 2011-07-13 to 2011-07-19, from 2011-08-20 to 2011-08-26, and from 2011-09-18 to 2011-09-24 (see Fig. 1). These weeks, covering different possible weather conditions from cloudy to clear, have been selected to reduce possible overfitting effects on the predictors. In total, the test time series has a length of 504 hours/data points. The different training sets for our experiments are defined in Sect. 4.

For training of the regression models that aim at predicting the PV power at time t for time step $t + n$, where $1 \leq n \leq 6$ denotes the prediction horizon in hours, we can only employ the information we have at time t. This is, the time information and the power at time t, as well as clear sky power and NWP predictions (temperature, irradiance and cloud coverage) for time $t + n$. For the construction of a training pattern, the corresponding label is the PV power at time $t + n$. In the test setting, this is the target value we want to predict. In Sect. 4, we concentrate on the selection of feature subsets that result in an optimal forecast.

3 Statistical Learning Methods

In this section, we introduce the machine learning methods KNN and SVR that are basis of our PV prediction system. In general, the basis for training appropriate models is a test set $T = \{(\mathbf{x}_1, y_1), \ldots, (\mathbf{x}_N, y_N)\} \subset \mathbb{R}^d \times \mathscr{Y}$ consisting of patterns $\mathbf{x}_i = (x_1, \ldots, x_d)$ and their associated labels y_i. For classification settings, the space $\mathscr{Y}$ of labels is discrete (e.g., $\mathscr{Y} = \{-1, +1\}$ for the binary case). For regression scenarios, the space $\mathscr{Y}$ is given by $\mathbb{R}$; here, the goal of the learning process is to find a prediction function $f : \mathscr{X} \rightarrow \mathbb{R}$ that maps new patterns $\mathbf{x} \in \mathbb{R}^d$ to reasonable real-valued labels, in our case PV power outputs.

3.1 *Nearest Neighbors*

Nearest neighbor predictions are based on the labels of the K-nearest patterns in data space. The prediction of the uniform KNN regression model is defined as $\mathbf{f}_{KNN}(\mathbf{x}') = \frac{1}{K} \sum_{i \in \mathscr{N}_K(\mathbf{x}')} \mathbf{y}_i$ with the set $\mathscr{N}_K(\mathbf{x}')$ containing the indices of the K-nearest neighbors of $\mathbf{x}'$. A distance-weighted variant has been introduced by Bailey and Jain [2] to smooth the prediction function by weighting the prediction with the similarity $\Delta(\mathbf{x}', \mathbf{x}_i)$ of the nearest patterns $\mathbf{x}_i$, with $i \in \mathscr{N}_K(\mathbf{x}')$, to the target $\mathbf{x}'$

$$\mathbf{f}_{wKNN}(\mathbf{x}') = \sum_{i \in \mathscr{N}_K(\mathbf{x}')} \frac{\Delta(\mathbf{x}', \mathbf{x}_i)}{\sum_{j \in \mathscr{N}_K(\mathbf{x}')} \Delta(\mathbf{x}', \mathbf{x}_j)} \mathbf{y}_i. \tag{1}$$

Patterns close to the target should contribute more to the prediction than patterns that are further away. Here, we define the similarity with the distance between patterns by

$$\Delta(\mathbf{x}', \mathbf{x}_i) = 1/\|\mathbf{x}' - \mathbf{x}_i\|^2, \tag{2}$$

where the model $\mathbf{f}_{wKNN}$ produces a continuous output.

3.2 *Support Vector Regression*

For PV power prediction, we also employ one of the most popular tools in the field of machine learning, i.e., support vector machines, which can be used for classification, regression, and a variety of other learning settings [8, 16, 17]. The resulting prediction model f can be seen as a special instance of problems having the form

$$\inf_{f \in \mathscr{H}, b \in \mathbb{R}} \frac{1}{N} \sum_{i=1}^{N} L_\varepsilon\big(y_i, f(\mathbf{x}_i + b)\big) + \lambda ||f||^2_{\mathscr{H}}, \tag{3}$$

with $\lambda = 1/2NC$, where $C > 0$ is a user-defined real-valued parameter, $L : \mathbb{R} \times \mathbb{R} \to [0, \infty)$, a loss function (e.g., the ε-insensitive loss: $L_\varepsilon(y, t) = \max(0, |y - t| - \varepsilon)$, $\varepsilon \in \mathbb{R}$, $\varepsilon > 0$), and $||f||^2_{\mathcal{H}}$ the squared norm in a so-called *reproducing kernel Hilbert space* $\mathcal{H} \subseteq \mathbb{R}^{\mathcal{X}} = \{f : \mathcal{X} \to \mathbb{R}\}$ induced by an associated *kernel function* $k : \mathcal{X} \times \mathcal{X} \to \mathbb{R}$. In this case, the RBF-kernel with another user-defined parameter $\gamma \in \mathbb{R}$, is applied. The space $\mathcal{H}$ contains all considered models, and the term $||f||^2_{\mathcal{H}}$ is a measure for the *complexity* of a particular model f [8]. Forecasts based on machine learning models require large data sets and advanced knowledge about the underlying data mining process chain. Important aspects are model selection, data preprocessing, definition of learning sets, and parameter tuning.

4 Experimental Study

In the following section, we present the results of our experimental study. We want to answer the questions, (1) how many hours/days of data are appropriate for the training sets, (2) which model parameters to choose, and (3) which features are important for accurate short-term PV power predictions.

To measure the quality of a forecast, we use the root mean square error (RMSE):

$$RMSE(\mathbf{z}, \mathbf{z}') = \frac{1}{\sqrt{N}} \sqrt{\sum_{i=1}^{N} (z_i - z'_i)^2}, \tag{4}$$

with measured PV power output time series $\mathbf{z}$ and predicted PV power output time series $\mathbf{z}'$. The RMSE is commonly used in the evaluation of solar and wind power predictions. It is more sensitive to large forecast errors and hence suitable for applications where small errors are more tolerable and larger errors cause disproportionately high costs, which is the case for many applications in the energy market and for grid management issues.

The bias, corresponding to systematic prediction errors, is expected to be neglectable when applying statistical learning algorithms for time series prediction.

By using the relative power in percentage of the nominal power, the power output of different PV systems becomes comparable. Concerning the analysis, it is important to mention that the training and test sets include patterns at night-time with zero PV power.

The changes in performance of the two machine learning models in comparison to previous optimization steps is evaluated with improvement scores. They are defined as the difference of the RMSE for the model before ($RMSE_{ref}$) and after ($RMSE_{opt}$) the optimization divided by $RMSE_{ref}$, so that the RMSE improvement score IS is given as:

$$IS = \frac{RMSE_{ref} - RMSE_{opt}}{RMSE_{ref}} \tag{5}$$

Furthermore, we use the IS in Sect. 5 to compare the different prediction models. The experiments are performed with a prediction horizon of one hour for all 87 single PV systems. For each one hour time step a new model is trained and a prediction is calculated.

4.1 Training Set

This section concentrates on the choice of a proper training set size for KNN and SVR. For this task, we choose default parameter settings for both models.

Considering KNN, we employ the uniform model with $K = 5$; the default settings for the SVR model are choosen as follows: $C = 100$, $\gamma = 0.0$ and $\varepsilon = 0.1$. In both cases, all available features are used, i.e., time of day (T), PV power measurement of the previous hour (P), clear sky power (CS), temperature (Temp), irradiance (I) and cloud cover (CC). For this analysis the predictions of all 87 PV systems are considered and the mean RMSE for each training set size is calculated. The results are presented in Fig. 3, showing the runtime and the corresponding RMSE for increasing training set sizes. The runtime describes the average time needed to build the regression models and perform the prediction of one week.

In both cases, the RMSE can be reduced while the runtime is increasing for a growing training set size. Especially the overproportional runtime increase of the SVR can be problematic for large PV data set sizes. The training set should at least cover 500 hours of data to enable both models to calculate reasonable predictions. A good trade-off between prediction quality and runtime is attained by selecting the past 1680 hours (70 days) as training data for both models.

The yielding results w.r.t the RMSE and runtime are shown in Table 1. At this stage—without model parameter tuning—the KNN model is able to generate better PV power predictions than the SVR.

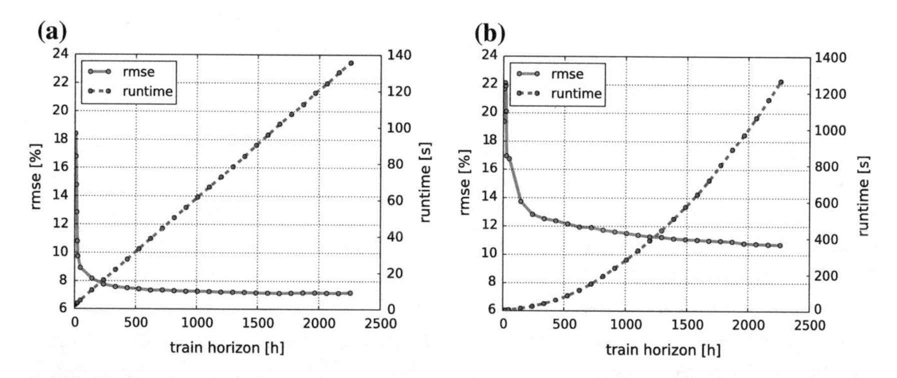

Fig. 3 The RMSE and runtime w.r.t. an increasing training set size using (**a**) KNN and (**b**) SVR. The RMSE is rapidly decreasing in the beginning and the improvement in prediction quality saturates with bigger training set sizes, the runtime steadily increases in both cases

Table 1 Resulting RMSE and runtime with a training set size of 1680 hours (70 days) for both models

Training set, 87-systems	RMSE (%)	Runtime (s)
KNN	7.16	102
SVR	10.99	718

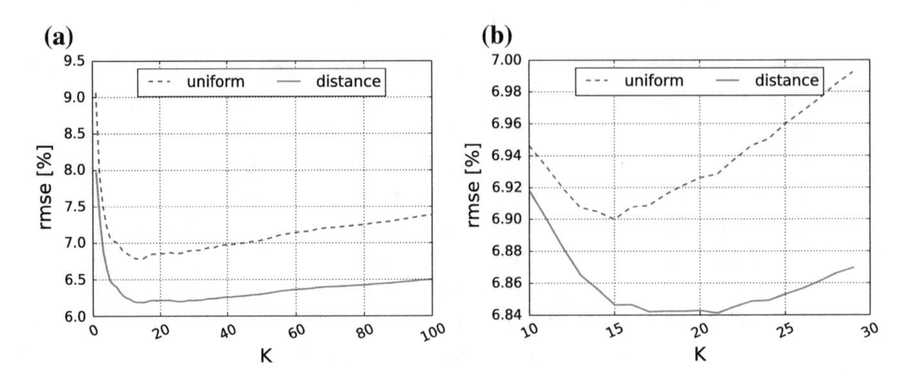

Fig. 4 Analysis of neighborhood size K in terms of RMSE for uniform and distance-weighted KNN on (**a**) the 10-systems setting and (**b**) on all 87 PV systems

4.2 Model Parameters

After choosing an appropriate training set size, the model parameters of KNN and SVR are adjusted for PV power prediction. Again, we are using all available features in our analysis and applying the training set size of 70 days as mentioned above. For both models, we are performing a wide parameter search on the 10-systems setting to reduce the search space of the experiments on all 87 PV systems. In the model parameter analysis of KNN, we compare uniform and distance-weighted KNN and various settings for K. In Fig. 4, increasing neighborhood sizes K are tested regarding uniform and distance-weighted KNN for (a) 10-systems with $K \in [1; 100]$ and (b) 87-systems on a more restricted search space where the optimum is expected ($K \in [10; 30]$). The distance-weighted model is outperforming the uniform model in both experiments, although the difference between the two models is less significant for the 87-systems setting. This shows the importance to use as many PV systems as possible for parameter tuning, as the individual PV systems perform quite differently. The distance-weighted KNN model shows almost equal results w.r.t. the RMSE in a range from $K = 17$ to 22, with an optimum for $K = 21$.

For the SVR, we tune three different model parameters: C, γ and ε. To minimize the complexity of this task, ε is tuned after optimal settings for C and γ are found. In Fig. 5, the RMSE is plotted as a function of C and γ on logarithmic scales. As before,

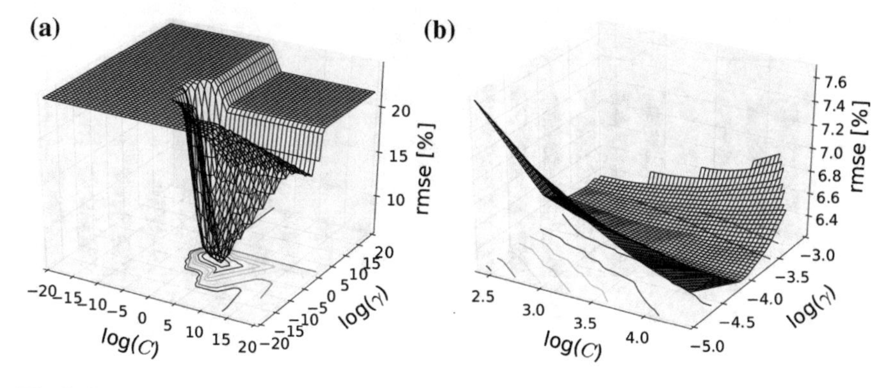

Fig. 5 SVR model parameter study of C and γ for the (**a**) 10-systems setting and (**b**) on a reduced search space with 87-systems

Table 2 Resulting RMSE and runtime after tuning the corresponding model parameters of both models

Model parameters, 87-systems		
	RMSE (%)	Runtime (s)
KNN	6.84 (0.04)	113 (−0.11)
SVR	6.26 (0.43)	515 (0.28)

The IS values, in comparison to the results after determining training set sizes, are denoted in parentheses

we initially analyze the model parameters on (a) the 10-systems setting (Fig. 5 (a)) to reduce the search space for an analysis on (b) all 87 PV systems (Fig. 5 (b)).

The RMSE space employs large plateaus of high errors for improper parameter combinations, but good results for C in the range of 10^2 to 10^5 and $\gamma = 10^{-3}$ to 10^{-5} for the 10-systems setting. This parameter range is tested on all 87 PV systems, resulting in an array of parameter combinations that produce applicable predictions. From these optimized settings, the combination $C = 10^4$ and $\gamma = 10^{-4}$ is selected due to runtime considerations. The analysis on the parameter ε leads to an optimal setting of 0.15.

Detailed information about the resulting errors and runtimes of both models are presented in Table 2. The large improvement (43 %) of the SVR w.r.t. the RMSE indicates the parameter sensitivity of the SVR predictor. With tuned model parameters, the SVR model is now able to produce better predictions than KNN.

4.3 Features

The next step is to analyze which features are useful for PV power predictions. For this sake, we employ the training set size and model parameters from the last experiments and test all $2^6 - 1 = 63$ possible feature combinations. The ten best

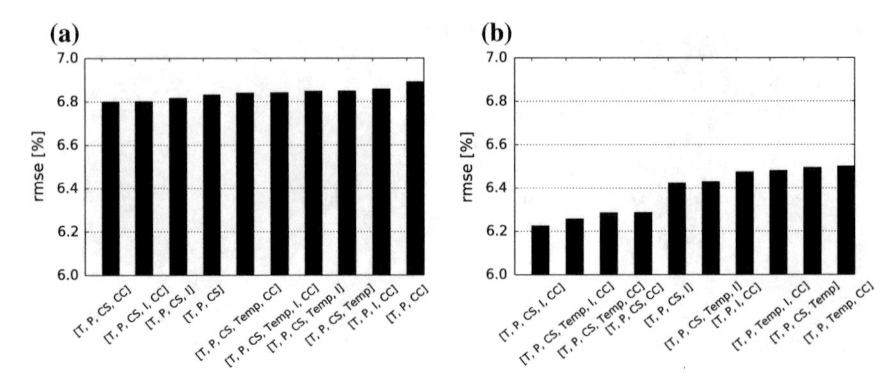

Fig. 6 Ranking of the ten best (out of 63 possible) feature combinations for (**a**) KNN and (**b**) SVR utilizing the model parameters of the prior analysis

Table 3 Resulting RMSE and runtime of KNN and SVR with optimized patterns

Features, 87-systems		
	RMSE (%)	Runtime (s)
KNN	6.80 (0.01)	63 (0.44)
SVR	6.22 (0.01)	365 (0.29)

The IS values, in comparison to the results after tuning model parameters, are denoted in parentheses

feature combinations for PV predictions sorted w.r.t. the resulting RMSE for (a) KNN and (b) SVR are listed in Fig. 6.

The best results are achieved by using the feature set T, P, CS, CC for the KNN model and T, P, CS, I, CC for SVR. In both cases, the pattern containing all available features (i.e., T, P, CS, $Temp$, I, CC) is not the best choice. Hence, by utilizing less features, we are able to improve the runtime of the models considerably and even score a slightly better RMSE (see Table 3). Furthermore, we can observe that either time of day (T) or clear sky power (CS) or both—in combination with the PV power of the previous hour (P)—are present in all of the best feature combinations. It seems that these features are directly related to the diurnal course of PV energy.

Aside the overall best feature combinations, we selected the best patterns with either (1) only measurement data or (2) only NWP data features, i.e., (1) T, P, CS and (2) T, I, CC for both models. All these settings are employed in the model comparison of the following section.

5 Comparison

After the analysis of training sets, parameterization of the models and features, we compare KNN and SVR on a larger evaluation time series that consists of 4.5 months in 2011, i.e., from 2011-06-15 to 2011-10-29, with optimized settings. We further compare our SVR models a comparatively simple model called persistence. This persistence model uses a constant ratio between power measurement P and clear

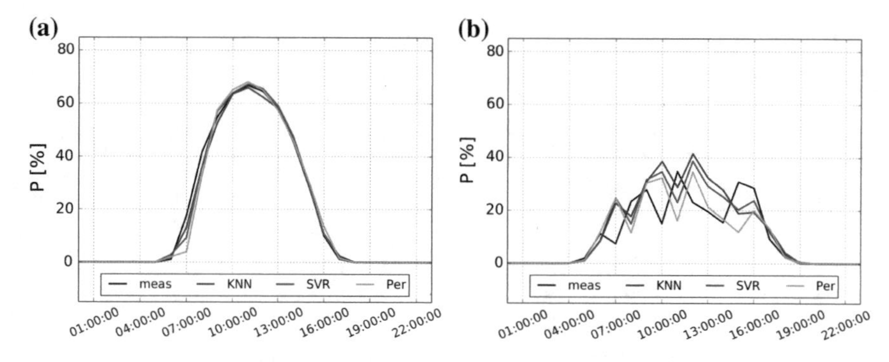

Fig. 7 Comparison between PV measurements (meas), the KNN and SVR prediction with optimized settings and the persistence model (Per) on (**a**) clear sky and (**b**) cloudy weather conditions for a single PV system

sky power CS, thus the expected PV power output at the time step to predict can be described as:

$$P_{per}(t+n) = \frac{P(t)}{CS(t)} \cdot CS(t+n),\ n \in \mathbb{N} \tag{6}$$

For predictions before sunrise with zero P and CS, we calculate the mean PV measurement at the same time of day with data of the previous 15 days and use this as $P_{per}(t+n)$.

Figure 7 demonstrates example PV measurement time series (black) with the corresponding KNN (blue) and SVR (red) predictions, as well as the persistence model (grey), for a prediction horizon of one hour. We can observe that all predictions are able to capture the main curvature of the PV energy on clear sky scenarios (Fig. 7a) and, with some delay that is adjusted to the diurnal course, on cloudy weather conditions (Fig. 7b).

First, we compare the different data sources. In Fig. 8, predictions using the three sources (1) PV power measurements, (2) NWP model, and (3) combinations of both are compared on all six prediction horizons. Similar to the results of Bacher et al. [1], for forecasts under a two- to three-hour horizon, the most important features are measurements while, for horizons over two or three hours, the use of NWP models is appropriate. The combined features produce good forecasts on all applied horizons, however, are not able to generate better results than the NWP models on longer prediction horizons. This most likely derives from optimizing the models for a prediction horizon of one hour. In the following comparisons we focus on the combined features.

The RMSE of all three models on all 87 PV systems, with prediction horizons up to six hours, is presented in Table 4. As expected, the persistence model is only competitive for a prediction horizon of one hour due to the decreasing autocorrelation of cloud situations. On all prediction horizons, the SVR can compute the best predictions w.r.t. the RMSE.

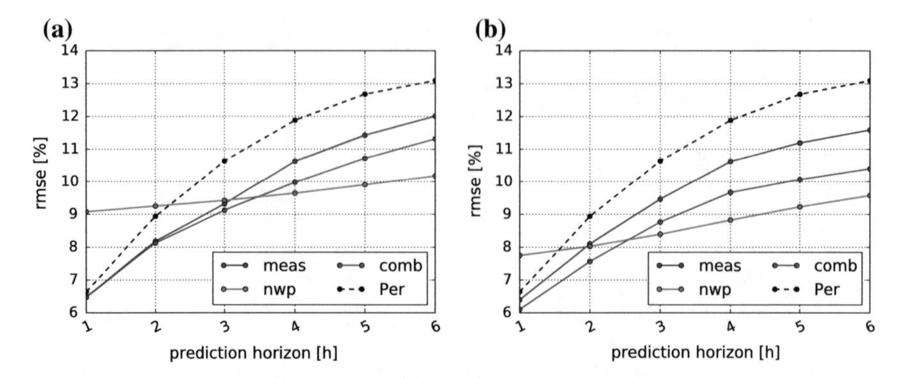

Fig. 8 Comparison of patterns with different data origins, e.g., measurements (meas), NWP model (nwp) and combinations of both (comb), for (**a**) KNN and (**b**) SVR on different prediction horizons. The persistence (Per) is used as a reference model

Table 4 Comparison of RMSE values of the persistence model (Per), KNN and SVR prediction with optimal settings on all 87 PV systems available

RMSE (%)

	Prediction horizons					
	1	2	3	4	5	6
Per	6.64	8.94	10.63	11.88	12.68	13.10
KNN	6.47(0.03)	8.12(0.09)	9.12(0.14)	9.98(0.16)	10.71(0.16)	11.32(0.14)
SVR	**6.10(0.08)**	**7.56(0.15)**	**8.76(0.18)**	**9.67(0.19)**	**10.07(0.21)**	**10.40(0.21)**

The comparison is done for a prediction horizon up to six hours, the improvement (IS) over the persistence model is inside the parentheses

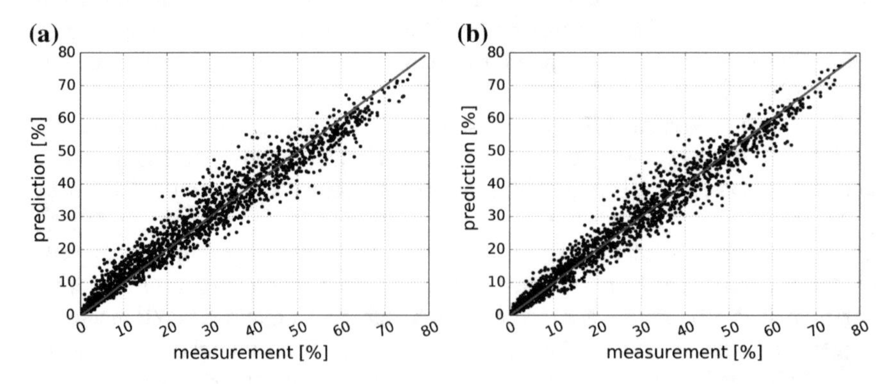

Fig. 9 Scatter plots of (**a**) KNN and (**b**) SVR predictions versus corresponding PV measurements with a prediction horizon of one hour

Scatter plots of KNN and SVR predictions with a prediction horizon of one hour (see Fig. 9) indicate a tendency of the KNN predictor to overestimate low PV power outputs, while underestimating higher outputs. This causes more generalized predictions in comparison to the SVR predictor and results in higher RMSE error values (see Table 4). The highest deviations between PV power measurements and predictions are observed in a range from 20 % to 60 % of the measurement for both predictors (Fig. 9), which is most likely the result of changing weather situations due to cloud movements that are difficult to capture (compare Fig. 7b).

6 Hybrid Model

To utilize advantages of the two models and to produce a more robust predictor, we combine KNN and SVR predictions creating a hybrid model of both. The predictions of a hybrid model f_{hybrid} at time step t, consisting of the models f_{KNN} and f_{SVR}, can be defined as

$$f_{hybrid}(t) = \eta f_{KNN}(t) + (1 - \eta) f_{SVR}(t), \tag{7}$$

with $0 \leq \eta \leq 1$. The parameter η is optimized w.r.t. the RMSE on the test time series (see Sect. 2). After the optimization, the obtained settings are used on the evaluation time series.

Table 5 Comparison of the RMSE values of the hybrid model with optimized settings for all 87 PV systems available

	Prediction horizons					
	1	2	3	4	5	6
η	0.21	0.20	0.37	0.44	0.36	0.32
RMSE (%)	6.08(0.08)	7.53(0.16)	8.61(0.19)	9.31(0.22)	9.80(0.23)	10.16(0.22)

The comparison is done on prediction horizons up to six hours. The IS values, in comparison to the persistence model, are denoted in parentheses

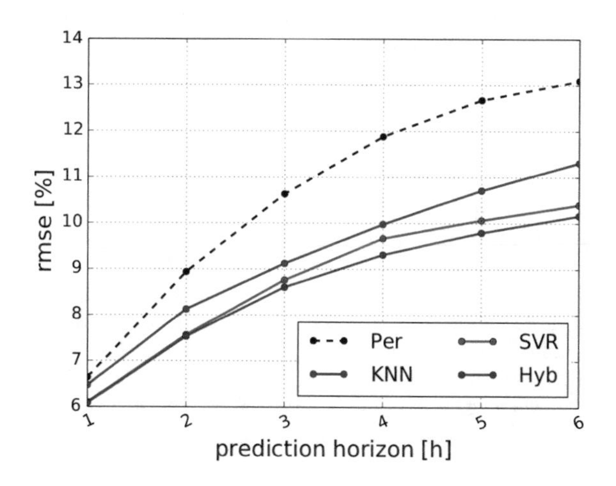

Fig. 10 Comparison of KNN, SVR and hybrid model predictions (Hyb) as well as the results of the persistence model (Per) on all six prediction horizons w.r.t. the RMSE with predictions of all 87 PV systems

Table 5 shows the results of the hybrid model with an optimized η for all six prediction horizons. At longer prediction horizons, the hybrid model is increasingly profiting from the KNN predictions, hence the increase of η.

Figure 10 demonstrates the improvement and stabilizing effect of the hybrid model on the RMSE on all six prediction horizons.

7 Conclusion and Outlook

The prediction of PV power plays an important part in smart grids. Our experiments have demonstrated that the often employed persistence model based on clear sky models already achieves good results for very short-term horizons, but we could improve short-term PV power forecasts with statistical learning methods. The quality of the predictions significantly depend on the proper choice of training sets, features, and model parameters. We analyzed the predictions with this regard and identified recommendations, in particular for model parameters K for KNN and C, γ and ε in case of SVR. Furthermore, we compared different data sources, i.e., PV power measurements and NWP features, and selected suitable patterns for PV power predictions. These recommendations are tailored to the employed data sets and may vary for other settings and scenarios. Still, these parameters and features were comparatively robust in our experimental line of research. The comparison of both models indicates that the SVR outperforms KNN and the persistence model w.r.t. the RMSE, but for fast results KNN is still a reasonable choice. A hybrid model takes advantage of the capabilities of both implemented models and combines them to build a robust model that produces more accurate PV power predictions.

As future work, we plan to perform experimental studies on the different prediction horizons using larger data sets with higher temporal resolution. Furthermore, we plan to increase the set of NWP features and test different data sources like satellite images and cloud-motion vectors. To create even better predictions, we will build hybrid models by directly integrating physical models, e.g., PV simulation models. This also enables the comparison between more complex physical and machine learning models on longer time series with a higher temporal resolution.

Another step is to use these models at the operational level, where the prediction system has to deal with problems like missing or false data. This requires the development of a preprocessing unit that is able to identify such occurrences and handle them accordingly. All these further tasks can be processed by statistical learning methods similar to the ones already implemented.

Acknowledgments We thank Meteocontrol—Energy and Weather Services (http://www.meteo\discretionary-control.com) for providing the PV system measurements, the ECMWF for the numerical weather predictions that are basis of our experimental analysis and the Lower Saxony Ministry for Science and Culture (MWK) for promoting the Ph.D. program "System Integration of Renewable Energies" (SEE) of the University of Oldenburg.

References

1. Bacher, P., Madsen, H., Nielsen, H.A.: Online short-term solar power forecasting. Sol. Energy **83**(10), 1772–1783 (2009)
2. Bailey, T., Jain, A.: A note on distance-weighted k-nearest neighbor rules. IEEE Trans. Syst. Man Cybern. **8**(4), 311–313 (1978)
3. Cao, J., Cao, S.: Study of forecasting solar irradiance using neural networks with preprocessing sample data by wavelet analysis. Energy **31**(15), 3435–3445 (2006)
4. Chakraborty, P., Marwah, M., Arlitt, M., Ramakrishnan, N.: Fine-grained photovoltaic output prediction using a Bayesian ensemble. In: Twenty-Sixth AAAI Conference on Artificial Intelligence, pp. 274–280 (2012)
5. Chowdhury, B.: Short-term prediction of solar irradiance using time-series analysis. Energy Sources **12**(2), 199–219 (1990)
6. da Silva Fonseca, J., Oozeki, T., Takashima, T., Koshimizu, G., Uchida, Y., Ogimoto, K.: Photovoltaic power production forecasts with support vector regression: a study on the forecast horizon. In: 2011 37th IEEE Photovoltaic Specialists Conference (PVSC), pp. 2579–2583 (2011)
7. European Centre for Medium-Range Weather Forecasts (ECMWF). http://www.ecmwf.int
8. Hastie, T., Tibshirani, R., Friedman, J.: The Elements of Statistical Learning. Springer, Berlin (2009)
9. Heinemann, D., Lorenz, E., Girodo, M.: Forecasting of solar radiation. In: Proceedings of the International Workshop on Solar Resource from the Local Level to Global Scale in Support of the Resource Management of Renewable Electricity Generation. Institute for Environment and Sustainability, Joint Research Center, Ispra, Italy (2004)
10. Kramer, O., Gieseke, F.: Short-term wind energy forecasting using support vector regression. In: 6th International Conference SOCO 2011 Soft Computing Models in Industrial and Environmental Applications, pp. 271–280 (2011)
11. Lorenz, E., Hurka, J., Heinemann, D., Beyer, H.G.: Irradiance forecasting for the power prediction of grid-connected photovoltaic systems. IEEE J. Sel. Top. Appl. Earth Obs. Remote Sens. **2**, 2–10 (2009)
12. Lorenz, E., Heinemann, D. (eds.): Prediction of solar irradiance and photovoltaic power. In: Comprehensive Renewable Energy, vol. 1, pp. 239–292. Springer (2012)
13. Mellit, A.: Artificial intelligence technique for modelling and forecasting of solar radiation data—a review. Int. J. Artif. Intell. Soft Comput. **1**(1), 52–76 (2008)
14. Meteocontrol—Energy and Weather Services GmbH. http://www.meteocontrol.com
15. Reikard, G.: Predicting solar radiation at high resolutions: a comparison of time series forecasts. Solar Energy **83**(3), 342–349 (2009)
16. Schölkopf, B., Smola, A.J.: Learning with Kernels: Support Vector Machines, Regularization, Optimization, and Beyond. MIT Press, Cambridge, MA, USA (2001)
17. Smola, A.J., Schölkopf, B.: A tutorial on support vector regression. In: Statistics and Computing, vol. 14, pp. 199–222. Kluwer Academic Publishers, Hingham, MA, USA (2004)
18. Wolff, B., Lorenz, E., Kramer, O.: Statistical learning for short-term photovoltaic power predictions. In: European Conference on Machine Learning (ECML), Workshop Data Analytics for Renewable Energy Integration (DARE) (2013)

Renewable Energy Prediction for Improved Utilization and Efficiency in Datacenters and Backbone Networks

Baris Aksanli, Jagannathan Venkatesh, Inder Monga and Tajana Simunic Rosing

Abstract Datacenters are one of the important global energy consumers and carbon producers. However, their tight service level requirements prevent easy integration with highly variable renewable energy sources. Short-term green energy prediction can mitigate this variability. In this work, we first explore the existing short-term solar and wind energy prediction methods, and then leverage prediction to allocate and migrate workloads across geographically distributed datacenters to reduce brown energy consumption costs. Unlike previous works, we also study the impact of wide area networks (WAN) on datacenters, and investigate the use of green energy prediction to power WANs. Finally, we present two different studies connecting datacenters and WANs: the case where datacenter operators own and manage their WAN and the case where datacenters lease networks from WAN providers. The results show that prediction enables up to 90 % green energy utilization, a 3× improvement over the existing methods. The cost minimization algorithm reduces expenses by up to 16 % and increases performance by 27 % when migrating workloads across datacenters. Furthermore, the savings increase up to 30 % compared with no migration when servers are made energy-proportional. Finally, in the case of leasing the WAN, energy proportionality in routers can increase the profit of network providers by 1.6×.

B. Aksanli (✉) · J. Venkatesh · T.S. Rosing
University of California, San Diego, La Jolla, CA, USA
e-mail: baksanli@ucsd.edu

J. Venkatesh
e-mail: jvenkate@ucsd.edu

T.S. Rosing
e-mail: tajana@ucsd.edu

I. Monga
Lawrence Berkeley National Laboratory, Energy Sciences Network, Berkeley, CA, USA
e-mail: imonga@es.net

J. Lässig et al. (eds.), *Computational Sustainability*,
Studies in Computational Intelligence 645,
DOI 10.1007/978-3-319-31858-5_4

1 Introduction

As the demand for computing increases globally, the number of datacenters has increased to meet the need. Recent studies indicate that the total power consumption of all datacenters in the world has increased by 56 % from 2005 to 2010 [1], with associated global carbon emissions and an estimated annual growth rate of 11 %. Their energy needs are supplied mainly by non-renewable, or brown energy sources, which are increasingly expensive as a result of availability and the introduction of carbon emissions taxes [2]. Consequently, several datacenter operators have turned to renewable energy to offset the energy cost.

The integration of renewable energy is complicated by the inherent variability of its output. Output inconsistency typically leads to inefficiency due to lack of availability or sub-optimal proportioning, which carries an associated financial cost. These costs are mitigated in various ways: several datacenter owners, such as Emerson Networks, AISO.net, and Sonoma Mountain Data Center supplement their solar arrays with utility power, and other datacenter owners, such as Baronyx Corporation and Other World Corporation, have been forced to augment their input power with other forms of energy or through over-provisioning, respectively [3]. Previous investigation into the existing methods in datacenter green energy demonstrates that variability results in low utilization, on average 54 %, of the available renewable energy [4].

A number of publications investigated the best strategy to manage renewable energy as a part of datacenter operation. The work in [3] reduces the peak datacenter power with local renewable sources and power management algorithms. They investigate power capping, both of individual servers using dynamic frequency scaling, and of server pools by reducing the number of machines utilized in each pool. However, they have significant quality-of-service (QoS) violations when limiting peak power. The study in [4] explores brown energy capping in datacenters, motivated by carbon limits in cities such as Kyoto. The authors leverage distributed Internet services to schedule workloads based on electricity prices or green energy availability. By defining workload distribution as a local optimization problem, the authors demonstrated 35 % lower brown energy consumption with a nominal (10 %) hit on service level agreement (SLA) violations. Similarly, [5] optimizes for energy prices, to reduce overall energy consumption by distributing workloads to datacenters with the lowest current energy prices. The insight is that renewable sources such as solar energy are actually cheapest during the day, when workloads are at the highest and utility sources are most expensive. Job migration is then modeled as an optimization problem, and the authors identify a local minimum energy cost among the available datacenters that still meets deadlines. The results demonstrate that their algorithm performs within 5.7 % of the optimum distribution, a significant improvement over established greedy algorithms. The authors of [6] analyze the opportunities and problems of using supply-following loads to match green energy availability. When green energy is insufficient, workloads are terminated or suspended, restarting or resuming when availability returns. However, the results show very low green energy efficiency and a failure to meet required service-level guarantees.

The above datacenter examples demonstrate the high cost and necessary precautions needed to successfully use highly variable green energy, at the cost of efficient utilization. However, an important means of reducing such variability remains overlooked in datacenters: green energy prediction. In [7], we investigated the existing methods in solar and wind energy prediction, developing prediction algorithms suitable for the datacenter domain. We implemented and evaluated our algorithms in a case study, leveraging prediction to improve green energy utilization in datacenters by 90 %, a $3\times$ improvement over the existing methods.

Previous publications concerned with energy costs primarily propose a "follow the sun" cost-management strategy [3, 6, 8–11] and generally neglect the cost of wide area net-working (WAN) incurred by job migration between datacenters. This assumption is reasonable for small datacenter networks that own the WAN and incur low network costs. Consequently, related work has WANs used to increase system performance via load balancing [12–14] or improve energy efficiency by migrating jobs [8, 9, 11]. However, these arguments are not applicable for large WAN costs and datacenters that lease the network.

Datacenters lease the WAN by agreeing to pay a certain price for a fixed bandwidth usage. However, as WAN usage increases, network owners [15, 16] offer Bandwidth-on-Demand services, especially for datacenter applications [17]. Additionally, the WAN may take up to 40 % of the total IT energy cost, and is expected to continue growing as demand for distributed data processing continues to rise [18] and as the server hardware becomes more energy efficient [19]. With the increasing importance of managing energy consumption in the network, WAN providers can charge users not just on the amount of bandwidth they use, but also the time of day when they use it. For example, using the network in a peak hour may be more expensive than when it is idle, reflecting electricity market prices [20]. Additionally, with the introduction of carbon taxes, WAN providers can also vary energy prices depending on the energy source. Consequently, datacenters might be open to longer, less expensive paths on the network. For example, a datacenter may request a path that uses green energy to avoid paying extra carbon emission taxes, or a less-utilized path to avoid extra utilization costs. Our work uniquely considers both the costs of geographically distributed datacenters and the profits of the network provider. We analyze different network cost functions, along with the analysis of new technologies that would allow using more energy-proportional routers in the future.

In this chapter, we first evaluate the advantages of short-term green energy prediction on the datacenter scale. We explore the existing in short-term solar and wind energy prediction methods, applying each to real power traces to analyze the accuracy. Using green energy prediction in local renewable energy sites and varying brown energy prices, we propose an online job migration algorithm among datacenters to reduce the overall cost of energy. While such job migration has been studied extensively before, we uniquely consider network constraints such as availability, link capacity and transfer delay at the same time. By including these constraints in our framework, we model the impact of the network and create a more holistic multi-datacenter model. Additionally, we investigate the impact of two aspects of datacenter operation typically overlooked in previous work: tiered power pricing, which penalize the datacenter for exceeding certain level of power restrictions with as much as 5× higher energy costs [21], and WAN leasing costs/cost models, which leverage energy-aware routing. Both play a significant impact in datacenter job scheduling, reflected in our results.

We also analyze the impact of new technologies in datacenter WAN, such as energy-proportional routing, green energy aware routing, and analyze leasing versus owning the WAN. Our work is the first analyzing different WAN properties in a job migration algorithm involving both mixed energy sources and prediction. We observe that green energy prediction helps significantly increase the efficiency of energy usage and enables network provisioning in a more cost effective way. Similarly, we show that using a WAN to transfer workloads between datacenters increases the performance of batch jobs up to 27 % with our performance maximization algorithm, and decreases the cost of energy by 30 % compared to no data migration with our cost minimization algorithm. Unlike previous works, we show the potential for green energy to go beyond simply cost reduction to improving performance as well. Our analysis of leasing WAN shows that network providers can increase profits by charging datacenter owners by bandwidth, but datacenters can still benefit by using dynamic routing policies to decrease their energy costs. Furthermore, as servers and routers become more energy proportional, we demonstrate increases in both datacenter cost savings and network provider profits.

2 Green Energy Prediction and Datacenters

2.1 Solar Energy Prediction

Solar energy algorithms exploit the daily pattern of solar irradiance, a primary factor in determining power output. The simplest algorithms are based on exponential weighted moving average (EWMA) [22]. Several extensions to the EWMA algorithm have been proposed, incorporating multiple days' predictions to derive a more representative average value in variable weather [23, 24]. Extended EWMA, eEWMA, [23] uses previous days' measurements to account for the error of each slot. The weather-conditioned moving average (WCMA) algorithm [24] takes into account the actual values from previous D days and the current day's previous N measurement slots. It averages the values for the predicted slot from previous days and scales it with a GAP factor, which represents the correlation of the current day against the previous days:

$$E(d, n+1) = \alpha \cdot E(d, n) + GAP_k \cdot (1 - \alpha) \cdot M_D(d, n+1) \quad (1)$$

where $M_D(d, n+1)$ represents the median of previous days' values, and GAP_k represents the scaling of the current day against the previous days. The inclusion of both patterns from multiple previous days as well as the use of values from the current

Table 1 Solar power prediction algorithm comparison

Algorithm	Absolute mean error (%)		
	Consistent conditions	Inconsistent conditions	Severely inconsistent conditions
EWMA	12.7	32.5	46.8
eEWMA	4.9	23.4	58.7
WCMA	4	9.6	18.3

day itself help WCMA provide a better pattern for the performance of solar panels. The three algorithms discussed above are tested using real solar power traces from the UCSD Microgrid. Absolute mean error is calculated against the measured data, shown in Table 1. The optimal parameter values have been determined empirically for each algorithm. The results demonstrate the importance of incorporating recent data to reduce error. The WCMA algorithm provides a significant improvement over EWMA and extended EWMA algorithms due to its highly adaptive nature, and its effective use of the GAP factor to scale future intervals based on the deviation from previous days.

2.2 *Wind Energy Prediction*

Wind prediction algorithms may use physical or statistical models. Physical models use spatial equations for wind speed at the locations of each turbine, and then predict wind power with theoretical or measured power curves [25]. Statistical models aggregate measured or forecasted meteorological variables and develop a relationship between the variables and the output power.

Several data-mining models have been used to predict the wind speed based on the meteorological variables collected from SCADA data acquisition units at each wind turbine [26]. The heuristics developed for wind speed prediction are then applied to wind power prediction, demonstrating 19.8 % mean error for 10 min-ahead prediction. Nearest-neighbor tables (k-NN) algorithm reduces this error to 4.23 % by mapping wind speed to wind power [27]. However, when forecasted wind speed is used, the power prediction error grows to 27.83 %.

Power curves, which describe the output power of wind turbines mapped against wind speed, form the basis of many predictors [25, 26, 28]. The work presented in [29] analyzes power curves and demonstrates their inaccuracy. Instead, the paper uses a dynamic combination of several statistical predictors, most notably the Autoregressive Moving Average (ARMA) model with past wind power, speed and direction as inputs. The results show a 50 % reduction in power prediction error, with ability to reduce error levels between prediction horizons of 2–45 h.

The above algorithms pose difficulties in implementation: unlike the solar prediction algorithms, which only require past power data, the wind prediction algorithms require various types of high-overhead input. Instead, we dramatically lower the overhead with our new wind energy predictor: we construct weighted nearest-neighbor tables based on the two most correlated variables contributing to wind energy output: the wind speed and direction [7]. The weighted tables show preference to the most recent results and allow the algorithm to adapt to gradual changes, while the power curves, based on both wind speed and direction, provide versatility. The algorithm to add a new entry to the table is in Eq. 2, where $P_{new}(v, d)$ is the new power curve table entry for a given wind velocity v and direction d, $P_{old}(v, d)$ is the existing value, and $P_{obs}(v, d, t)$ is the observed value at time t. Future interval prediction is determined by Eq. 3.

$$P_{new}(v, d) = \alpha \cdot P_{obs}(v, d, t) + (1 - \alpha) \cdot P_{old}(v, d) \tag{2}$$

$$P_{pred}(v, d, t + k) = P(v(t + k), d(t + k)) \tag{3}$$

The algorithms described above have been tested using real power data from a Lake Benton, MN wind farm, and the meteorological data was provided by published reports from the National Renewable Energy Laboratory (NREL). For better comparison, we have all the predictors use the same inputs: wind speed and direction. We also include the commonly used baseline—persistence prediction, which assumes that the future interval is the same as the current one.

Persistence has a high error at 137 %, affected by the high variability of the wind farm power. The data-mining algorithm's error is at 84 %, despite using the two most-correlated variables. This can be attributed to the unreliability of using forecasted wind speed for a region as opposed to measured wind speeds at each turbine level. The ARMA model performed better, at 63 % error, but the accuracy is hampered by the limited input data available. The wind-speed-based nearest-neighbor predictor, the kNN algorithm, performed the best, with an error of 48 %, which can be attributed to the higher variance of the Lake Benton wind farm than the wind farm in the original work. Our custom nearest-neighbor predictor, which uses both wind speed and direction to develop a relationship with wind farm energy output, has only 21 % mean error as it is more adaptive to recent conditions and as a result is 25 % better than the next-best algorithm (Table 2).

3 Datacenter and WAN Models

Multiple datacenters increase the capacity, redundancy, and parallelism of computation, but a fast, reliable inter-datacenter network is essential to maintain performance. Since large datacenters consume a lot of power, they usually undergo a tiered power pricing. The tier level depends on the overprovisioned input power to avoid high

Table 2 Wind power prediction algorithm comparison

Algorithm	Mean error (%)	Std. dev. (%)
Persistence	137	340
Data-mining [26]	84	101
ARMA model [29]	63	12
kNN predictor [6]	48	32
Our NN predictor [7]	21	17

prices in-peak periods [21]. This can be seen as a power budget. In this work, we also study the effects of different power tier levels and how these levels can affect the job migration decisions and network utilization. To avoid power tier violations, datacenters may deploy several techniques: CPU capping, virtual CPU management and dynamic voltage and frequency scaling (DVFS), all of which incur performance overhead. If the power goes beyond the tier level, it is charged at higher rates, which can be 3×–10× higher than the base cost [5, 8]. One way to remedy this problem is to leverage a network of globally distributed datacenters along with renewable energy prediction for peak power reduction. In the next subsections, we present our datacenter and backbone network models, which we then use in our frameworks for managing renewable-powered globally distributed datacenters and related WAN.

3.1 Backbone Network Model

Our network topology is a subset of the LBNL Esnet [30], containing 5 datacenters and 12 routers in total, distributed over the USA (Fig. 1), where each link has a predefined capacity from 10 to 100 Gbps. A portion of this capacity is normally

Fig. 1 Network topology; *squares* datacenters, *circles* routers

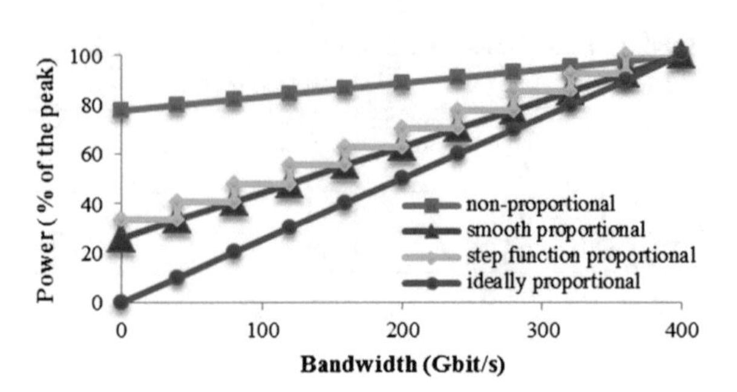

Fig. 2 Router energy proportionality schemes

reserved for background traffic (10 % in our experiments). When calculating the network energy cost, we consider the router power and add a fixed offset for the link power consumption. The power consumption of the router can be estimated using a simple linear model [31] based on bandwidth utilization.

Current router's idle power is very close to active power. In our previous work, we investigate methods for designing more energy-proportional routers and construct representative power proportionality models [30]. Figure 2 reflects our findings: the non-proportional curve represents an actual state-of-the-art router [32], and the step function depicts shutting down idle line cards. Smooth proportionality is a linear correlation of the step function, while ideal proportionality reflects an idle router with no idle power cost.

In our model, we account for the network transfer delay as an increase in the total response time of a job. The state-of-the-art is Dijkstra's Shortest Path Routing (SPR). For comparison, we use our own Green Energy Aware Routing algorithm (GEAR), which minimizes the amount of brown energy used [30]. GEAR is a power- and performance-constrained dynamic routing algorithm that performs online path computation and constructs virtual circuits [33]. We use different price models to calculate the network lease costs including, fixed bandwidth (BW) cost, where cost does not increase with utilization; and linear BW cost increase, which assumes that cost of operation and revenue are proportional to usage. These two options represent the different models that network operators might incorporate in their service level agreements (SLAs). In the results section, we show how the total cost is affected by these cost schemes.

3.2 Datacenter Model

In order to represent a multi-datacenter network more accurately, each datacenter is modeled separately based on actual measurements. Each includes computation models to represent servers and how they interact with each other and the workloads they execute. For each datacenter in Fig. 1, we implement a detailed model designed to faithfully represent a fully populated set of datacenter containers. Each container has

8 racks populated with 250W Intel Xeon Quad core machines [34]. We create and run multiple virtual machines using Xen VM [35], and measure run-time characteristics such as CPU, memory, and power consumption. We use these measurements to construct our datacenter model for simulation.

The workload is divided into two representative categories: service and batch jobs, both of which are run on our servers. We model the former with RUBiS [36] and the latter with MapReduce [37]. Short-running service jobs have tight latency constraints, and the longer-running batch jobs represent the background tasks (i.e. indexing results or correlating data) and need to maintain a certain level of throughput. Their inter-arrival time distributions are modeled with a lognormal distribution based on measured values [7]. We model service and batch job arrival to the system independently, and place them in separate local job queues. Each MapReduce job consists of multiple tasks and multiple tasks are dispatched from an active job and put in different servers.

Servers run the jobs assigned to them, and prioritize service jobs over batch jobs because of their response time requirements. These requirements are indicators for the quality of service (QoS) a datacenter has to maintain to ensure its profitability. We also measure the interference of running different types of jobs simultaneously on the same server. Since these jobs have different resource requirements, the interference of one on the other might lead to performance degradation for either job. In our experiments, we observe that service requests have negligible impact on MapReduce jobs, but MapReduce jobs are detrimental to both service jobs and other MapReduce jobs. In order to meet QoS of service jobs and maintain the throughput hit of batch jobs to less than 10 %, we limit the total number of simultaneous MapReduce jobs on a single server. The baseline of this study is established in [7].

We calculate the server power consumption with a linear CPU-utilization based equation [38] and scale the aggregate server power cost using power usage effectiveness (PUE) metric, which is set to 1.15 [39], to find the total power of the datacenter as a function of overheads related to cooling, powering, and other loads [40]. The deviation between our simulations and measurements is 3 % for power consumption, 6 % for service job QoS and 8 % for MapReduce job performance.

4 Relocating Jobs to Improve Efficiency

4.1 Background

Multi-datacenter networks offer advantages for improving both performance and energy. As each datacenter is in a different location, its peak hours and energy prices vary. A datacenter with high electricity prices may need to migrate work to another datacenter with a lower price, incurring some performance and power cost due to data migration. The live migration of virtual machines over high-speed WAN has made this idea feasible, as it offers fast transmission with limited performance hit [41].

However, the migration costs through WAN need to be considered. For example, WAN may be leased, with lease costs quantified per increment of data transferred, and thus might be too high to justify frequent migration of jobs between datacenters [42]. Furthermore, datacenters often undergo a tiered power pricing scheme. The energy under a specific level may cost a fixed amount and this fixed price changes depending on the location, so it is beneficial to run jobs in a datacenter at a lower fixed price. Data migration should not increase the total power consumption to more than the amount specified by the specific tier level. Otherwise, extra power costs are calculated using higher prices, generally much higher than the fixed price.

Table 3 summarizes and compares the key state of the art contributions for managing distributed datacenters in order to minimize an objective function, e.g. the overall cost of energy. Buchbinder et al. [5], Qureshi et al. [42] and Rao et al. [11] relocate jobs to where the energy is cheaper to minimize the energy cost. They do not model different energy types; perform detailed workload performance analysis and different routing options for both WAN providers and datacenters. Le et al. [44] solves a similar problem including green energy in their model but they assume a centralized dispatcher and do not analyze network latency or cost. Liu et al. [14] and Mohsenian-Rad et al. [43] minimize the brown energy usage or carbon footprint. They either do not consider the variability of green energy or do not have a network model. Aksanli et al. [30] solve a load-balancing problem by modeling network properties, but do not consider energy costs. As we can see from this analysis, previous studies do not consider all the important aspects of multi-datacenter networks simultaneously in their models. As we show in this chapter, this can lead to overestimated cost savings or overlooked performance implications due to not considering both the requirements of different types of applications and WAN characteristics.

In this chapter, we generalize the problem of migrating jobs among datacenters to minimize the cost of energy and analyze the effects of using WAN for the transfer. Our design considers both brown and locally generated green energy, and variable energy market pricing. We simultaneously investigate energy proportionality of routers and servers and tiered energy pricing, which are at best considered individually in previous works. Additionally, we account for the latency and cost of the WAN, the costs of leasing or owning the WAN, and the impact of different routing algorithms. Our work is also the first showing the potential of green energy to improve performance in addition to addressing environmental concerns or reducing energy costs.

4.2 *Cost Minimization and Performance Maximization Algorithms*

We now describe our cost minimization algorithm, which considers the properties of both the datacenters and the backbone network simultaneously. Our algorithm performs in discrete time steps of 30 minutes. Each datacenter has its own workload distributions that represent different types of applications in a datacenter

Table 3 Summary and comparison of the related work

	Buchbinder 2011 [5]	Qureshi 2009 [42]	Mohsenian-Rad 2010 [43]	Rao 2010 [11]
Goal	Minimize electricity bill	Minimize electricity bill	Minimize carbon footprint and job latency	Minimize electricity bill
How	Move jobs where energy is cheaper	Move jobs where energy is cheaper	Migrate jobs to different locations depending on the goal	Move jobs where energy is cheaper
Workload	No specification	No specification	Service requests only	Service requests only
Perf. constraints	X	X	Latency of service requests	Latency of service requests
Network cost model	Fixed cost per bandwidth	Fixed cost per bandwidth	X	X
Routing	X	Distance based routing	X	X
Green energy	X	X	Local green energy, carbon tax	X
Network delay	X	✓	✓	X
	Liu 2011 [14]	Le 2010 [44]	Aksanli 2011–2012 [7, 30]	
Goal	Minimize brown energy use	Minimize the total cost of energy	Maximize batch job performance and minimize brown energy use	
How	Move jobs to local green energy	Forward jobs to datacenters	Move jobs where utilization is low	
Workload	No specification	Different job types (not explicitly specified)	Mix of service and batch jobs	
Perf. constraints	X	SLA of service requests	Latency of service requests and throughput of batch jobs	
Network cost model	X	X	X	
Routing	X	X	Static routing versus energy aware	
Green energy	Local green energy	Grid green energy carbon tax	Local green energy with prediction	
Network delay	X	X	✓	

environment. The properties of these distributions are determined by applying statistical analysis on real datacenter traces (Sect. 3.2 outlines the distributions and Sect. 5 presents the real workloads we use).

The goal of our algorithm is to determine which workloads we need to transfer among different datacenters during each interval to minimize the energy cost. The current algorithm assumes a centralized implementation for control for job migration decisions, though each datacenter generates its own workloads. We assume that green energy is generated and used locally, and is prioritized over brown energy to minimize the total cost, as green energy is a fixed, amortized cost. Thus, we transfer workloads to datacenters which have available capacity and extra green energy. Because of datacenters' energy pricing scheme, energy in a particular location may have a fixed, low cost up to a specified amount of peak power capacity. After this level, energy becomes much more expensive. Therefore, our goals include maintaining utilization in datacenters such that we do not increase the power consumption further than the power tier levels.

Figure 3 illustrates our cost minimization algorithm. Each interval begins with the calculation of the amount of energy required by each datacenter, incorporating the previous and incoming load rates. The former represents the active jobs at a given time, and the latter is determined by the statistical distributions of real applications. We estimate the green energy availability using prediction (Sect. 2), obtain the current brown energy pricing, and check power restrictions. Based on the energy need and green energy availability, each datacenter determines if it has surplus green energy.

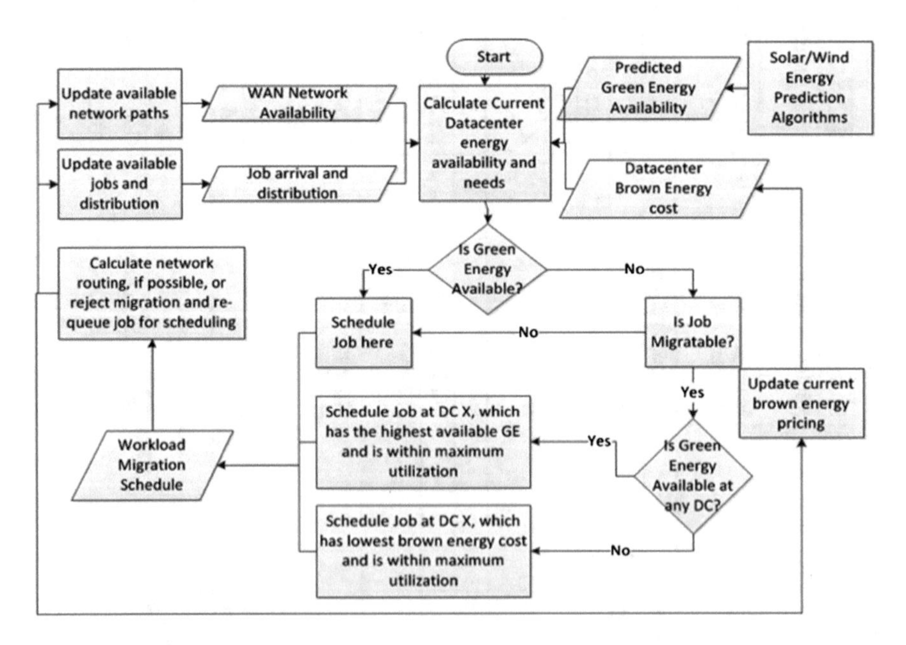

Fig. 3 High-level overview of the algorithm

The key assumption is that if brown energy has already been within the lower price region, it makes sense to use it for running jobs, while green energy can be used to both reduce power consumption and to run extra jobs which otherwise might not be scheduled.

Then workloads are transferred from the datacenters with the highest need to those with the highest available green energy. The workload that can be transferred from a datacenter is determined by what is migrateable, while the workload that can be transferred to a particular datacenter is limited by the amount of additional green energy and WAN availability. This process continues until every datacenter is analyzed. If there are workloads remaining in any datacenters at the end, the algorithm focuses on datacenters with the cheapest brown energy cost. It moves workloads from the datacenters with higher energy costs to those with the cheapest brown energy. The amount of data that can be transferred is limited by receiving datacenter's peak power constraints and tiered power levels. If there are still unscheduled jobs remaining at the end of this process, they are scheduled in datacenters where the market electricity prices are the lowest.

We can also modify this iterative part of our algorithm to maximize the performance of the workloads instead of minimizing the total cost of energy. In this case, we transfer the jobs that are actively waiting in the execution queue to datacenters with excess green energy availability. The iterative process of the cost minimization algorithm is also valid here, but the migration depends only on green energy availability, i.e. jobs are not migrated to datacenters with cheaper brown energy prices because extra brown energy would be required for these additional jobs. We denote this process as performance maximization as it runs additional jobs with surplus green energy.

At the end of this iterative process, we obtain a matrix representing workload transfers among datacenters. This transfer matrix is then provided to the networking algorithm, which calculates the paths to be used and the amount of bandwidth that needed by each selected path. In our study, we analyze different path selection algorithms, such as shortest path routing (SPR), green energy aware routing (GEAR), and network lease based routing. A detailed description of SPR and GEAR implementations is in [30]. Network lease based routing selects the path with the least per-bandwidth price in the case the WAN is leased. In our results, we analyze different network cost functions as well. If a selected path in the transfer matrix is unavailable due to network limitations, the job is rescheduled with a limitation on target datacenters.

Our algorithm is similar to those proposed in previous studies, but it minimizes the cost of energy more comprehensively. This is because it has a more complete view of datacenter energy costs, modeling both fixed energy costs under fixed amounts and variable, higher tier energy prices. This helps us to calculate the energy cost savings in a more accurate way. Second, it considers the side effects of the WAN, analyzing both the performance implications of different routing algorithms and additional leasing costs if necessary. This is key when multi-datacenter systems lease the WAN. Job migration may not be feasible for those systems if the cost of leasing the network is too high. Third, the green energy availability information is enhanced by using prediction which can provide information 30-min ahead and thus help us allocate the workloads across multiple datacenters in a more effective manner. Last but not the least; our algorithm is flexible in the sense that it can perform for both cost minimization and performance maximization purposes. Also, we are the first to show that green energy can be used to maximize the performance rather than just minimizing the total cost of energy of geographically distributed multi-datacenter systems.

5 Methodology

We use an event-based simulation framework to analyze and compare the results of our solution to the problems described above. The inputs to our simulator are derived from measurements performed on our datacenter container (Sect. 3.2) and data obtained from industrial deployments. This section discusses how we construct the simulation environment, including the datacenter loads, simulation parameters, green energy availability, and brown energy prices.

5.1 Datacenter Load

We analyze real datacenter workload traces to accurately capture the characteristics. We use a year of traffic data from Google Orkut and Search, reported in the Google Transparency Report [39], to represent latency-centric service jobs and reproduce the waveform in Fig. 3 of [45] to represent MapReduce workloads in order to model throughput-oriented batch jobs. In Fig. 4, we show a sample workload combination of these jobs. We use this data to find the parameters of the statistical workload models fed into our simulator (Sect. 3.2), listed in Table 4. We also only migrate batch jobs due to the tight response time constraints of service jobs.

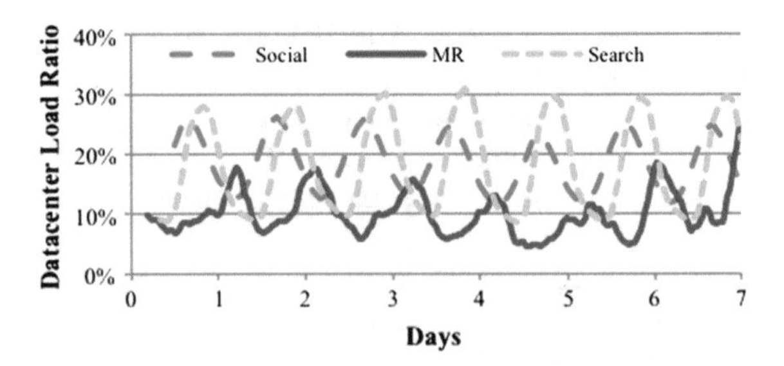

Fig. 4 Sample datacenter load ratio with different job types

Table 4 Simulation parameters

Parameter	Value	Parameter	Value
Mean web request inter-arrival time per client	5 ms	Average # tasks per MR job	70
Mean web request service time	20 ms	Avg. throughput level per MR job	0.35
Service request SLA	150 ms	Servers in a datacenter	1000
Mean MR Job inter-arrival time	2 min	Number of datacenters	5
Mean MR task service time	4 min	Number of routers	12
Idle server power	212.5 W	Idle router power	1381 W
Peak server power	312.5 W	Peak router power	1781 W
Single link capacity	100 Gbps	Average batch VM size	8 GB

5.2 Green Energy Availability

In Fig. 5, we show a subset of the green energy availability measurements. Solar data is gathered from the UCSD Microgrid and wind data is obtained from a wind farm in Lake Benton, MN, made available by the National Renewable Energy Laboratory. The representative outputs for the other various locations in our experiments (San Francisco, Chicago, etc.) are obtained by scaling and time-shifting the measured results from our available sources to published average availability data for the target areas [46, 47].

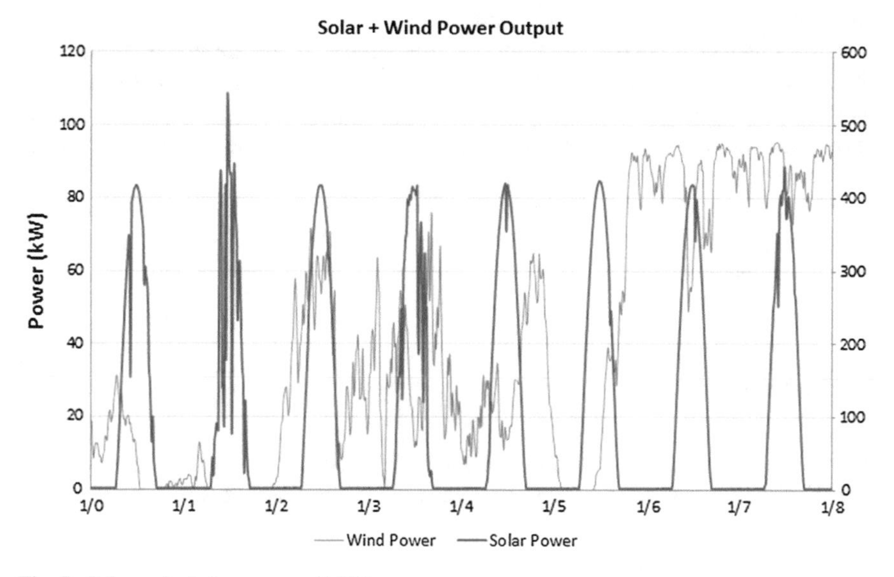

Fig. 5 Solar and wind energy availability

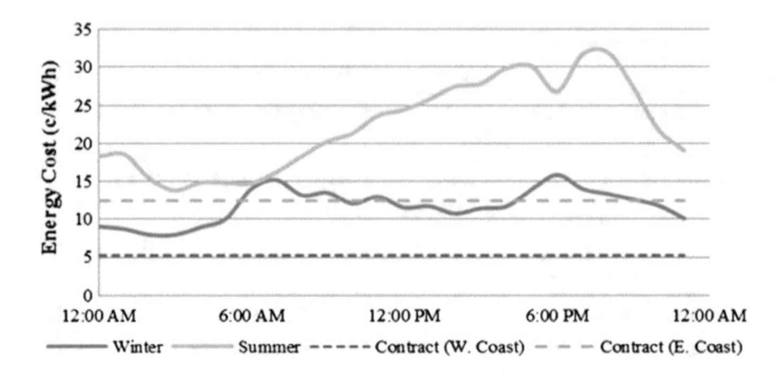

Fig. 6 Daily brown and amortized green energy cost (¢/kWh)

5.3 *Brown and Green Energy Costs*

Datacenters contract power from utilities to obtain competitive prices for their expected loads. This can be seen as a tiered pricing scheme. If a datacenter exceeds the tiered amount in an interval, it is relegated to higher prices, sometimes even market prices. We obtain sample fixed pricing for the midwest, the east and the west coasts [48]. Since market prices change over time, we use the California ISO [49] whole-sale pricing database to obtain brown energy prices for various California locations, and time-shift and scale those values for the other locations based on published averages [50]. Figure 6 shows daily pricing values for brown energy in comparison to fixed costs. The straight lines correspond to fixed, under-tier prices and the others show samples of variable, market prices which can be used to charge datacenters that go over their tiered amounts.

Local green energy costs are typically amortized over the lifetime of an installation, incorporating the capital and the maintenance costs. This is represented by a fixed offset to our cost model. We use data from [4] to obtain the capital and operational expenses of several solar and wind farms, amortized over their lifetimes, as representative solar and wind costs per interval.

We list our simulation parameters in Table 4 and present our network topology and green energy locations in Table 5. Green energy is scaled to 80 % of peak datacenter and router energy needs.

6 Results

This section presents the simulation results for the base case of *no migration*, and the workload migration policies for *performance maximization* and *cost minimization*.

Table 5 Available renewable energy type for each location

Location	Node	Type	Location	Node	Type
Chicago	DC + router	Wind	New York	DC + router	Wind
Atlanta	DC + router	Solar	San Diego	DC + router	Solar
Kansas	Router	–	El Paso	Router	Solar
Nashville	Router	Wind	Cleveland	Router	Wind
San Francisco	DC + router	Solar and wind	Houston	Router	Solar
Denver	Router	–	Washington DC	Router	–

6.1 No Migration

In this scenario, each datacenter runs its own workload using only locally available green energy. This is the baseline for our comparisons, as it represents the nominal brown energy need and quantifies the performance of batch jobs without the overhead of migration. A power tier level accounts for 85 % of datacenter's power needs, while the rest, when needed, is provided at variable market prices. We allow service and batch jobs to run on the same servers while ensuring that they meet quality of service (QoS) requirements (service job $QoS_{ratio} < 1$), and find that the average MapReduce job completion time is 22.8 min. Only 59 % of the total green energy supply is consumed by datacenters locally, motivating the distributed policies described previously. The next two sections quantify the impacts of *performance maximization* and *cost minimization* policies.

6.2 Performance Maximization Using Migration

In this algorithm, we leverage migration to complete more batch jobs than previously possible. Datacenters with high utilization transfer jobs to locations with low utilization or where there is excess green energy, effectively completing more work in the same amount of time.

Most MapReduce jobs (representative of batch jobs) complete within 30 min [7], which becomes the threshold for both the green energy prediction interval and the interval for checking datacenter utilization. At each interval, the controller retrieves the resource usage and green energy profiles of each datacenter and optimizes the system by initiating extra workloads in datacenters with green energy availability while still meeting under-tier power constraints. It calculates the available transfer slots between each end-point pair, and selects the tasks to be executed remotely from each datacenter's active batch jobs. Once the tasks finish execution in a remote datacenter, the results are sent back to the original center. The key to this policy is

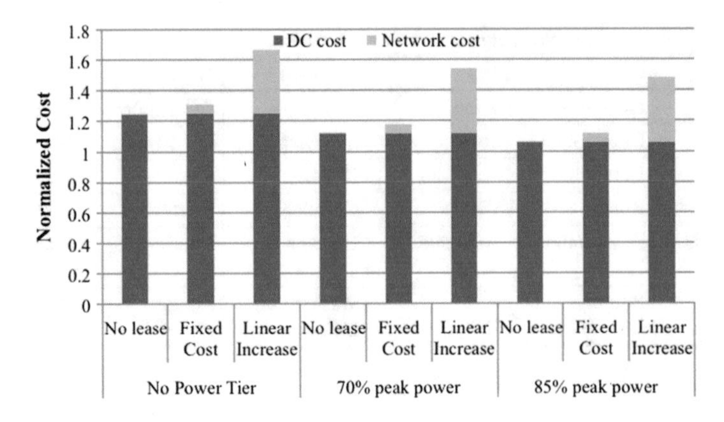

Fig. 7 Normalized performance maximization algorithm costs for datacenters and network

that waiting tasks are migrated, as opposed to active tasks, resulting in more jobs executed overall (Sect. 4).

Our simulation results show that the average completion time of MapReduce jobs is 16.8 min, 27 % faster than the baseline, with no performance hit for service requests. Furthermore, since we are leveraging all available green energy for extra workloads, the percentage of green energy used is 85 %, significantly higher than the baseline.

Figure 7 reports the total cost normalized against the no migration case with different tier levels specified as a percentage of the datacenter's peak power capacities and network lease options. Without tiered energy pricing (where all the consumption is charged using market prices), we demonstrate a 25 % increase in the total energy cost. However, when we do include tiered energy pricing, we see more accurate results, with a cost increase of only 12 % for a 70 % level, and a total cost increase of 6 % for an 85 % level.

Since the WAN may not be owned by a datacenter, we also analyze the case where the network is leased. In this case, a bandwidth-dependent cost is incurred. Figure 7 shows the results of this analysis over different cost functions that network providers use. For linear increase (Sect. 3.2), we see that the network cost can be up to 40 % of the datacenter cost. This ratio increases with tiered energy pricing from <1 % to 25 %, since this pricing scheme reduces datacenter power consumption and magnifies the network cost.

For this policy, we also calculate the profit of network providers based on the energy costs associated with the WAN. Table 6 shows the profit normalized against fixed bandwidth cost and non-energy-proportional routers. Energy proportionality of routers enables up to 37 % more profit for network providers with ideal power curves and 20 % with step proportionality WAN router power curve. We also observe that different power tier levels do not affect the savings of the network provider because the migration is based only on green energy availability in other locations.

Table 6 Profit of network providers for performance maximization with different router energy proportionality schemes

Network cost function	Profit			
	Non-prop	Step	Smooth	Ideal
Fixed cost	1×	1.2×	1.2×	1.4×
Linear increase	4.5×	6.7×	6.8×	6.9×

6.3 Cost Minimization Using Migration

The main goal of the cost minimization policy is to maximize green energy usage and then leverage as much as possible inexpensive brown energy. Also, we show the impact of energy-proportional servers to quantify the policy's benefit in future systems.

Unlike *performance maximization*, *cost minimization* does not transfer extra jobs, and thus, does not obtain any performance improvement. Furthermore, the overhead of network transfer decreases the performance of MapReduce jobs. We observe 23.8 min average job completion time for MapReduce jobs, 4.5 % worse than the no migration case with green energy efficiency of 66 %, a 7 % improvement over no migration, with no performance overhead for service jobs.

In Fig. 8, we show the impact of energy proportionality and tiered energy pricing to our model, normalized against the no migration case. We observe a 10 % decrease in total cost when tiered energy pricing is incorporated into the model. Cost reduction grows to 15 % when energy-proportional servers are used. This shows the potential of cost minimization method in the future when servers become more energy proportional.

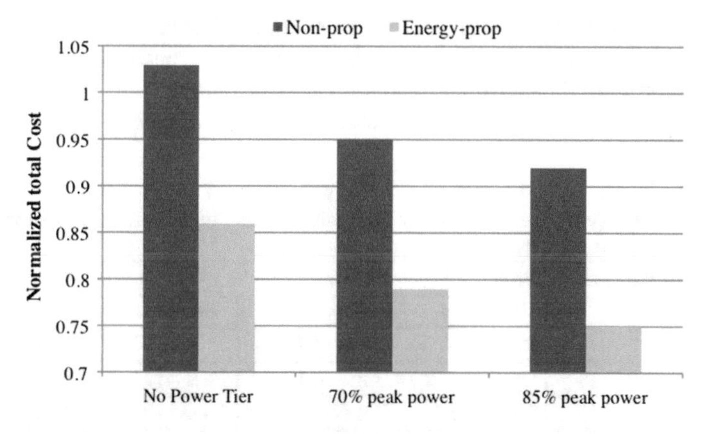

Fig. 8 Normalized cost minimization algorithm costs with different power tier levels and energy proportionality

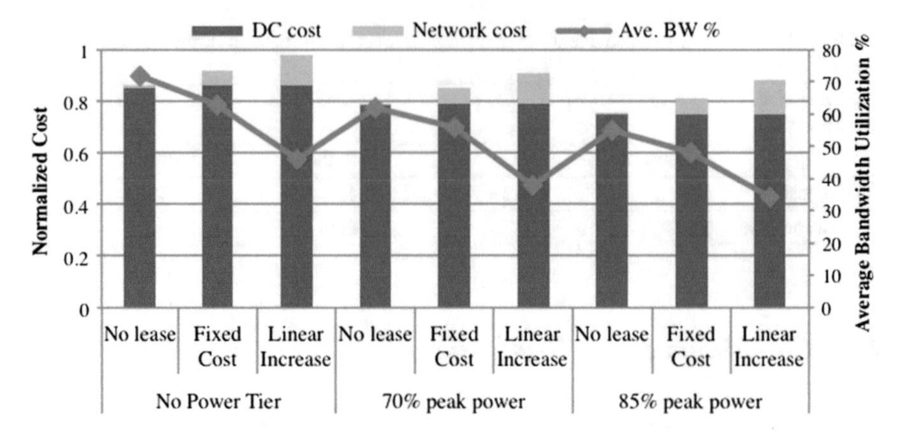

Fig. 9 Normalized total cost and utilization for cost min. with different power tier levels and network lease options using energy-proportional servers

We also analyze how the total cost of datacenters changes if the network is leased. Unlike the *performance maximization* policy, we prevent migration if the cost is higher than the potential savings. Figure 9 shows the results of this analysis, and additionally incorporates server energy proportionality. We use the same coefficients for the network cost functions as in the previous case. Neglecting the cost of network leasing can result in up to 15 % error. The network costs are up to 17 % of the datacenter cost, which is significantly less than results we saw with the performance maximization, where it is up to 40 %. This is mainly because this policy sacrifices a potential increase in performance if the cost of a data transfer outweighs the cost savings. Figure 9 also shows how bandwidth utilization changes with different power tier levels and network lease options. First, as network costs become more dominant, bandwidth utilization decreases due to a growth in unfeasible data transfers. As a result, if the lease cost is not modeled, the average bandwidth utilization has up to 60 % error. Introducing tiered power levels decreases network utilization because they create a more balanced energy cost scheme across datacenters. Table 7 shows the normalized profit of the network providers. The cost minimization policy inherently limits network profits, since it only allows financially profitable transfers.

Table 7 Profit of network providers for cost. min. with different router energy proportionality and with server energy proportionality

Network cost function	Profit							
	Non-prop		Step		Smooth		Ideal	
	85 %	70 %	85 %	70 %	85 %	70 %	85 %	70 %
Fixed cost	1×	1.2×	1.2×	1.4×	1.2×	1.4×	1.4×	1.6×
Linear increase	2.2×	2.45×	3.26×	3.6×	3.4×	3.8×	3.5×	3.9×

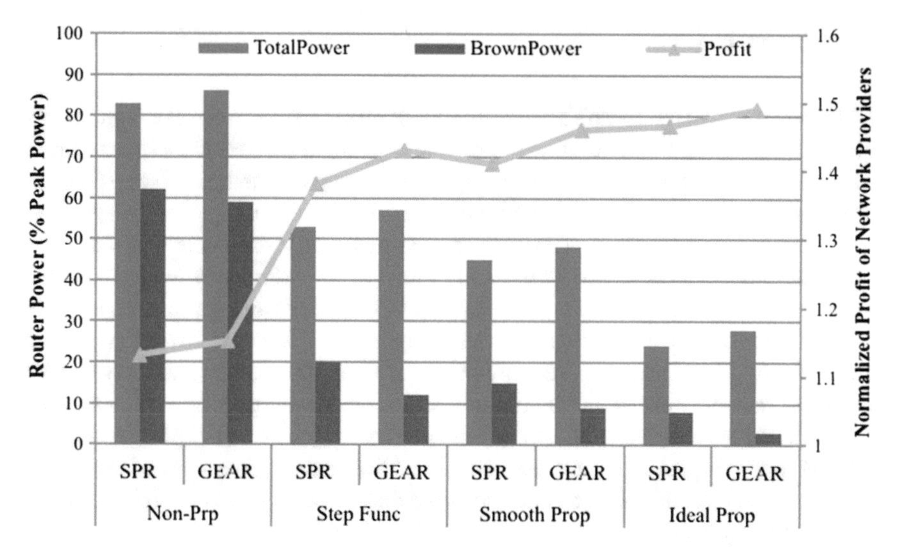

Fig. 10 Comparison between SPR and GEAR energy consumption of routers and network profit with different energy proportionality schemes

6.4 *Cost Minimization Using a Green Energy Aware Network*

We now investigate the cost minimization policy incorporating green energy aware routing (GEAR). Instead of simply selecting the shortest path between two datacenters, GEAR chooses the path with the least brown energy need. As we only change the network routing policy for this scenario, datacenter cost values are similar compared to the previous case. An energy-aware network provides several benefits. Reducing brown energy costs of the WAN improves overall networking costs for both providers and datacenters. It also provides a viable alternative for datacenters, opting for cheaper green energy at the cost of GEAR's slightly increased network latency. Also, as network elements become more energy proportional in the future, we expect the energy savings obtained by GEAR to be more prominent.

Figure 10 compares SPR and GEAR in terms of router energy consumption and network provider profit, using fixed cost per bandwidth. GEAR with energy proportionality increases profits by 50 % compared to the base case (non-proportional, SPR), and provides profit for all proportionality schemes. Without energy proportional routers, GEAR's brown energy consumption is slightly lower than SPR (62 vs. 65 % of SPR) with a 3 % increase in network delay as a result of occasionally choosing a longer path, though with negligible overall effect on the job completion time.

7 Discussion

In this section, we first recap the most important results of the above case studies. We then compare our methodology with previous work, and explore the lessons learned with our analysis. Table 8 shows the comparison among the methods discussed in the previous sections. Our performance maximization algorithm uniquely leverages both workload and green energy differences across distributed datacenters to maximize both throughput (27 % improvement) and green energy efficiency (44 % increase). We also demonstrate that the same variations in workloads and green energy can be leveraged for cost minimization, where our algorithm utilizes tiered energy pricing, and both migration and green-energy-aware routing. The results show up to 19 % reduction in energy cost and 7 % improvement in green energy usage while meeting QoS of latency sensitive applications, and increasing job completion time of batch jobs by only 4 %. Additionally, the comprehensive and novel aspects of our model provide a level of realistic simulation that previous models do not exhibit to make a complete analysis.

Green Energy Prediction and Workload Migration Green energy prediction mitigates the inefficiency caused by the variability of renewable sources. We further improve inefficiency by matching our prediction horizon to the long-running batch jobs. The result is better decision making, and as the results indicate, up to 26 % improvement in green energy efficiency. Previous work [43, 48] only uses green energy as a method to reduce carbon footprint, and deploy workload migration to improve performance considering load balancing and resource availability [14]. In contrast, we show green energy can also be used to improve performance. We initially propose the idea in [7] for a single datacenter, but now leverage prediction and availability across a network to run extra batch jobs in remote locations. We obtain 27 % better batch job completion time compared to no migration with only a 6–12 % increase in total energy cost. Our work is the first to demonstrate the potential of green energy not only as a resource for environmental concerns, but also a means of performance improvement. While cost minimization precludes all potential migrations due to network costs, it still has 7 % improvement in green energy usage.

WAN Ownership and Leasing Related work assumes that WAN is part of the datacenter network, or applies static bandwidth costs. However, the WAN may be leased or owned, typically with bandwidth-dependent pricing. Our work is the first to accurately consider such costs. Our first observation is that higher network cost reduces the bandwidth utilization. Secondly, despite increasing network costs with larger cost functions, datacenters can obtain 2–19 % cost savings by checking the financial feasibility of each potential migration. In contrast, when the datacenter owns the network, disregarding the initial WAN cost, it achieves up to 22 % cost savings.

Tiered Energy Pricing Previous work on minimizing total energy cost, [5, 11, 45] uses grid electricity pricing as either fixed or variable with load. Others [3] attempt to limit datacenter peak load but do not consider how different power levels can

Table 8 Comparison of different policies with respect to total cost, MapReduce performance and green energy usage

Policy	MapReduce job completion time (min)	Non-energy proportional servers			Energy proportional servers		
		Power tier (%)	Total energy cost	Green energy usage (%)	Power tier (%)	Total energy cost	Green energy usage (%)
No mig.	22.8	–	1.22×	59	–	0.99×	47
		85	1×		85	0.85×	
		70	1.10×		70	0.92×	
Perf. max.	16.8	–	1.25×	85	–	1.03×	80
		85	1.06×		85	1×	
		70	1.12×		70	1.05×	
Cost min.	23.8	–	1.03×	66	–	0.86×	60
		85	0.92×		85	0.75×	
		70	0.95×		70	0.79×	

affect overall energy cost. Not modeling different cost regions for data center energy consumption may not be correct due to large power consumption of the datacenters. We demonstrate that proposed improvements might be overestimated by up to 20 % when accurate pricing is taken into account. Both of our algorithms inherently attempt to remain below tiered power levels in order to avoid higher energy prices, and only exceed those limits when inevitable, i.e. when all datacenters are over-provisioned. Consequently, while our algorithms' performance and cost benefits are tempered by the incorporation of tiered energy pricing, we can still show up to 15 % cost savings.

Energy-Proportional Routing We investigate the future of datacenter communication, analyzing the impact of energy proportionality of routers on network provider profit, which has not been explored before. We show that dynamic, green energy aware routing (GEAR) policies can improve energy efficiency by reducing brown energy consumption up to 65 %. We quantify that energy proportionality can increase the profit of network providers up to 35 and 57 % with fixed and linear policies, respectively. The difference in profit between an implementable proportionality scheme (i.e. step-function) and the ideal case is between 5–17 % and decreases with increasing network lease costs. The key observation is that router energy-proportionality schemes can increase profits significantly if deployed, and that GEAR can decrease network brown energy use up to 3× with energy proportionality [30] with negligible performance impact.

Power-Proportional Computing for Future Systems Current datacenter hardware is highly non-energy proportional, resulting in power-inefficient systems. There has been recent work [51] on designing energy-proportional elements. Our work quantifies the benefits of this trend in both major aspects of a datacenter network: servers and network elements. We show the benefit of optimizing the components individually and together into an ideal energy-proportional system, with up to 30 % energy savings despite being limited by tiered energy pricing and network contracts. Table 8 quantifies both the impact of such systems, and the continued benefit of our algorithms in a power-proportional environment.

8 Conclusion

Energy efficiency and green energy usage in datacenters and their networks has gained importance as their energy consumption, carbon emissions, and costs have increased dramatically. Previous work leverages geographically-separated datacenters by migrating workloads over WAN, leveraging demand and price differences. However, the work neglects several key cost and energy contributions: the financial network, and consequently, data migration costs, focusing solely on latency and QoS costs. Additionally, these publications assume a simpler, and ultimately inaccurate, model for datacenter energy costs. To counteract this, we explore tiered energy pricing for datacenters, network cost models and the costs of owning/leasing a datacenter WAN. We then quantify the inaccuracy of conclusions (25–40 % error) drawn when these two are omitted. We solve the variability problem of green energy by using novel prediction algorithms, and subsequently develop algorithms for energy management, focusing on (1) performance maximization, and (2) cost minimization. With the performance maximization algorithm, we demonstrate the ability to leverage green energy to actually improve workload throughput, rather than simply reducing the operational costs. We explore and quantify up to 22 % cost savings when realistic WAN costs are incorporated, and up to 65 % reduction in network costs when deploying Green Energy Aware Routing (GEAR). We further explore the viability of our new algorithms in the face of emerging technologies in datacenter infrastructure, showing continued benefit of both the performance maximization and the cost minimization algorithms in the presence of energy-proportional computing and communication. Our results show that our performance maximization improves batch job performance by 27 % while meeting the QoS of services, and that our cost minimization policy decreases overall energy cost by 16 %, even when tempered by realistic energy pricing schemes and networking contracts. In future work, we will look to merge the two algorithms to create a balance between the performance gains and cost reduction, for an optimal cost-performance tradeoff for different datacenter configurations.

9 Acknowledgment

This work was sponsored in part by the Multiscale Systems Center (MuSyC), National Science Foundation (NSF) Project GreenLight, Energy Sciences Network (Esnet), NSF ERC CIAN, NSF Variability Expedition, NSF Flash Gordon, CNS, NSF IRNC TransLight/StarLight, Oracle, Microsoft and Google. We also want to thank Chin Guok for his guidance on path computation algorithms.

References

1. Koomey, J.: 2011. Growth in Data Center Electricity Use 2005 to 2010. Analytics Press, Oakland. http://www.analyticspress.com/datacenters.html
2. Mankoff, J., Kravets, R., Blevis, E.: Some computer science issues in creating a sustainable world. Computer **41**(8), 102–105 (2008). doi:10.1109/MC.2008.307
3. Gmach, D., Rolia, J., Bash, C., Chen, Y., Christian, T., Shah, A., Sharma, R., Wang, Z.: Capacity planning and power management to exploit sustainable energy. In: International Conference on Network and Service Management (CNSM), pp. 96, 103. 25–29 Oct. 2010. doi:10.1109/CNSM.2010.5691329
4. Miller, R.: Green Data Centers. Data Center Knowledge (2011) http://www.datacenterknowledge.com/archives/category/infrastructure/green-data-centers/
5. Buchbinder, N., Jain, N., Menache, I.: Online job-migration for reducing the electricity bill in the cloud. In: Domingo-Pascual, J., Manzoni, P., Pont, A., Palazzo, S., Scoglio, C. (eds.) Proceedings of the 10th International IFIP TC 6 Conference on Networking—Part I (NETWORKING'11), pp. 172–185. Springer, Heidelberg (2011)

6. Krioukov, A., Goebel, C., Alspaugh, S., Chen, Y., Culler, D., Katz, R. Integrating Renewable Energy Using Data Analytics Systems: Challenges and Opportunities (2011)
7. Aksanli, B., Venkatesh, J., Zhang, L., Rosing, T.: Utilizing green energy prediction to schedule mixed batch and service jobs in data centers. In: Proceedings of the 4th Workshop on Power-Aware Computing and Systems (HotPower '11). ACM, New York, NY, USA, Article 5 (2011). doi:10.1145/2039252.2039257
8. Sankaranarayanan, A.N., Sharangi, S., Fedorova, A.: 2011. Global cost diversity aware dispatch algorithm for heterogeneous data centers. In: Proceedings of the 2nd ACM/SPEC International Conference on Performance Engineering (ICPE '11), pp. 289–294. ACM, New York, NY, USA. doi:10.1145/1958746.1958787
9. Bergler, B., Preschern, C., Reiter, A., Kraxberger, S.: Cost-effective routing for a greener internet. In: Green Computing and Communications (GreenCom), 2010 IEEE/ACM Int'l Conference on & Internationall Conference on Cyber, Physical and Social Computing (CPSCom), pp. 276, 283. 18–20 Dec. 2010 doi:10.1109/GreenCom-CPSCom.2010.112
10. Le, K., Bianchini, R., Martonosi, M., Nguyen, T.D.: Cost-and energy-aware load distribution across data centers. In: Proceedings of HotPower, pp. 1–5 (2009)
11. Rao, L., Liu, X., Xie, L., Liu, W.: Minimizing electricity cost: optimization of distributed internet data centers in a multi-electricity-market environment. In: INFOCOM, 2010 Proceedings IEEE, pp. 1, 9. 14–19 March 2010. doi:10.1109/INFCOM.2010.5461933
12. Rao, L., Liu, X., Ilic, M., Liu, J.: MEC-IDC: joint load balancing and power control for distributed internet data centers. In: Proceedings of the 1st ACM/IEEE International Conference on Cyber-Physical Systems (ICCPS '10), pp. 188–197. ACM, New York, NY, USA. doi:10.1145/1795194.1795220 (2010)
13. Kutare, M., Eisenhauer, G., Wang, C., Schwan, K., Talwar, V., Wolf, M.: 2010. Monalytics: online monitoring and analytics for managing large scale data centers. In: Proceedings of the 7th International Conference on Autonomic Computing (ICAC '10), pp. 141–150. ACM, New York, NY, USA. doi:10.1145/1809049.1809073
14. Liu, Z., Lin, M., Wierman, A., Low, S.H., Andrew, L.L.H.: 2011 Greening geographical load balancing. In: Proceedings of the ACM SIGMETRICS Joint International Conference on Measurement and Modeling of Computer Systems (SIGMETRICS '11), pp. 233–244. ACM, New York, NY, USA. doi:10.1145/1993744.1993767
15. AT&T: (2007) http://www.att.com/gen/pressroom?pid=4800&cdvn=news&newsarticleid=24555
16. Fratto, M.: (2009) http://www.networkcomputing.com/data-center/229503323
17. Mahimkar, A., Chiu, A., Doverspike, R., Feuer, M.D., Magill, P., Mavrogiorgis, E., Pastor, J., Woodward, S.L., Yates, J.: Bandwidth on demand for inter-data center communication. In: Proceedings of the 10th ACM Workshop on Hot Topics in Networks (HotNets-X). ACM, New York, NY, USA, Article 24 (2011). doi:10.1145/2070562.2070586
18. Global Action Plan Report: An inefficient truth (2007) http://www.globalactionplan.org.uk/
19. Abts, D., Marty, M.R., Wells, P.M., Klausler, P., Liu, H.: Energy proportional datacenter networks. In: Proceedings of the 37th Annual International Symposium on Computer Architecture (ISCA '10), pp. 338–347. ACM, New York, NY, USA (2010). doi:10.1145/1815961.1816004
20. Zhang, Y., Wang, Y., Wang, X.: 2011 Capping the electricity cost of cloud-scale data centers with impacts on power markets. In: Proceedings of the 20th International Symposium on High Performance Distributed Computing (HPDC '11), pp. 271–272. ACM, New York, NY, USA. doi:10.1145/1996130.1996170
21. Kontorinis, V., Zhang, L.E., Aksanli, B., Sampson, J., Homayoun, H., Pettis, E., Tullsen, D.M., Simunic Rosing, T.: Managing distributed UPS energy for effective power capping in data centers. In: 39th Annual International Symposium on Computer Architecture (ISCA), pp. 488, 499. 9–13 June 2012. doi:10.1109/ISCA.2012.6237042
22. Holt, C.C.: Forecasting seasonals and trends by exponentially weighted moving averages. Int. J. Forecast. **20**(1), 5–10 (2004)
23. Dondi, D., Zappi, P., Rosing, T.: A Scheduling Algorithm for High-Performance Monitoring WSN with Hybrid Energy Harvester. ISLPED (2010)

24. Piorno, J.R., Bergonzini, C., Atienza, D., Rosing, T.S.: Prediction and management in energy harvested wireless sensor nodes. In: 1st International Conference on Wireless Communication, Vehicular Technology, Information Theory and Aerospace & Electronic Systems Technology, 2009. Wireless VITAE 2009, pp. 6, 10. 17–20 May 2009. doi:10.1109/WIRELESSVITAE.2009.5172412
25. Chow, C.W., Urquhart, B., Lave, M., Dominguez, A., Kleissl, J., Shields, J., Washom, B.: Intra-hour forecasting with a total sky imager at the UC San Diego solar energy testbed. Sol. Energy **85**(11), 2881–2893 (2011)
26. Kusiak, A., Zheng, H., Song, Z.: Wind farm power prediction: a data mining approach. Wind Energy **12**(3), 275–293 (2009)
27. Kusiak, A., Zheng, H., Song, Z.: Short-term prediction of wind farm power: a data mining approach. IEEE Trans. Energy Convers. **24**(1), 125–136 (2009). doi:10.1109/TEC.2008.2006552
28. Giebel, G., Brownsword, R., Kariniotakis, G., Denhard, M., Draxl, C.: The State-Of-The-Art in Short-Term Prediction of Wind Power: A Literature Overview, 2nd edn. ANEMOS.plus (2011)
29. Sanchez, I.: Short-term prediction of wind energy production. Int. J. Forecast. **22**(1), 43–56 (2006)
30. Aksanli, B., Rosing, T.S., Monga, I.: Benefits of green energy and proportionality in high speed wide area networks connecting data centers. In: Design, Automation and Test in Europe Conference and Exhibition (DATE), 2012, pp. 175, 180, 12–16 March 2012. doi:10.1109/DATE.2012.6176458
31. Mahadevan, P., Sharma, P., Banerjee, S., Ranganathan, P.: A power benchmarking framework for network devices. In: Fratta, L., Schulzrinne, H., Takahashi, Y., Spaniol, O. (eds.) Proceedings of the 8th international IFIP-TC 6 networking conference (NETWORKING '09), pp. 795–808. Springer, Heidelberg (2009)
32. Tucker, R.S., Baliga, J., Ayre, R.W.A., Hinton, K., Sorin, W.V.: Energy consumption in IP networks. In: 34th European Conference on Optical Communication, 2008. ECOC 2008, p. 1. 21–25 Sept. 2008. doi:10.1109/ECOC.2008.4729202
33. Guok, C.P., Robertson, D.W., Chaniotakisy, E., Thompson, M.R., Johnston, W., Tierney, B.: A user driven dynamic circuit network implementation. In: GLOBECOM Workshops, 2008 IEEE. pp. 1, 5. Nov. 30 2008–Dec. 4 2008. doi:10.1109/GLOCOMW.2008.ECP.14
34. Intel. (n.d.). www.intel.com/Xeon
35. Dhiman, G., Marchetti, G., Rosing, T.: vGreen: a system for energy efficient computing in virtualized environments. In: Proceedings of the 2009 ACM/IEEE International Symposium on Low Power Electronics and Design (ISLPED '09), pp. 243–248. ACM, New York, NY, USA (2009). doi:10.1145/1594233.1594292
36. RUBiS. (n.d.). http://rubis.ow2.org/
37. Hadoop. (n.d.). http://hadoop.apache.org/
38. Economou, D., Rivoire, S., Kozyrakis, C., Ranganathan, P.: Full-system power analysis and modeling for server environments. In: International Symposium on Computer Architecture-IEEE (2006)
39. Google. (n.d.). http://www.google.com/transparencyreport/traffic/
40. Barroso, L.A., Hlzle, U.: The datacenter as a computer: an introduction to the design of warehouse-scale machines. Synth. Lect. Comput. Archit. **4**(1), 1–108 (2009)
41. Travostino, F., Daspit, P., Gommans, L., Jog, C., de Laat, C., Mambretti, J., Monga, I., van Oudenaarde, B., Raghunath, S., Wang, P.Y.: Seamless live migration of virtual machines over the MAN/WAN. Future Gener. Comput. Syst. **22**(8), 901–907 (2006). doi:10.1016/j.future.2006.03.007
42. Qureshi, A., Weber, R., Balakrishnan, H., Guttag, J., Maggs, B.: 2009. Cutting the electric bill for internet-scale systems. In: Proceedings of the ACM SIGCOMM 2009 Conference on Data Communication (SIGCOMM '09), pp. 123–134. ACM, New York, NY, USA. doi:10.1145/1592568.1592584

43. Mohsenian-Rad, A.H., Leon-Garcia, A.: Energy-information transmission tradeoff in green cloud computing. Carbon **100**, 200 (2010)
44. Le, K., Bianchini, R., Nguyen, T.D., Bilgir, O., Martonosi, M.: Capping the brown energy consumption of Internet services at low cost. In: International Green Computing Conference, pp. 3, 14. 15–18 Aug. 2010. doi:10.1109/GREENCOMP.2010.5598305
45. Chen, Y., Ganapathi, A., Griffith, R., Katz, R.: The case for evaluating MapReduce performance using workload suites. In: IEEE 19th International Symposium on Modeling, Analysis and Simulation of Computer and Telecommunication Systems (MASCOTS). pp. 390, 399. 25–27 July 2011. doi:10.1109/MASCOTS.2011.12
46. NREL Solar Maps: (2012) http://www.nrel.gov/gis/solar.html
47. NREL Wind Maps: (2012) http://www.nrel.gov/gis/wind.html
48. Le, K., Bilgir, O., Bianchini, R., Martonosi, M., Nguyen, T.D.: Managing the cost, energy consumption, and carbon footprint of internet services. In: Proceedings of the ACM SIGMETRICS International Conference on Measurement and Modeling of Computer Systems (SIGMETRICS '10). ACM, New York, NY, USA (2010). doi:10.1145/1811039.1811085
49. California ISO: Retrieved from OASIS (2012) http://oasis.caiso.com/
50. U.S. Energy Information Administration. Electric Power Monthly. (n.d.). http://www.eia.gov/electricity/monthly/
51. Holzle, U.: The case for energy-proportional computing. Computer **40**(12), 33–37 (2007)

A Hybrid Machine Learning and Knowledge Based Approach to Limit Combinatorial Explosion in Biodegradation Prediction

Jörg Wicker, Kathrin Fenner and Stefan Kramer

Abstract One of the main tasks in chemical industry regarding the sustainability of a product is the prediction of its environmental fate, i.e., its degradation products and pathways. Current methods for the prediction of biodegradation products and pathways of organic environmental pollutants either do not take into account domain knowledge or do not provide probability estimates. In this chapter, we propose a hybrid knowledge-based and machine learning-based approach to overcome these limitations in the context of the University of Minnesota Pathway Prediction System (UM-PPS). The proposed solution performs relative reasoning in a machine learning framework, and obtains one probability estimate for each biotransformation rule of the system. Since the application of a rule then depends on a threshold for the probability estimate, the trade-off between recall (sensitivity) and precision (selectivity) can be addressed and leveraged in practice. Results from leave-one-out cross-validation show that a recall and precision of approximately 0.8 can be achieved for a subset of 13 transformation rules. The set of used rules is further extended using multi-label classification, where dependencies among the transformation rules are exploited to improve the predictions. While the results regarding recall and precision vary, the area under the ROC curve can be improved using multi-label classification. Therefore, it is possible to optimize precision without compromising recall. Recently, we integrated the presented approach into enviPath, a complete redesign and re-implementation of UM-PPS.

J. Wicker (✉) · S. Kramer
Institut für Informatik, Johannes Gutenberg-Universität Mainz,
Staudingerweg 9, 55128 Mainz, Germany
e-mail: wicker@uni-mainz.de

S. Kramer
e-mail: kramer@informatik.uni-mainz.de

K. Fenner
Eawag, Swiss Federal Institute for Aquatic Science and Technology,
CH-8600 Dübendorf, Switzerland
e-mail: kathrin.fenner@eawag.ch

J. Lässig et al. (eds.), *Computational Sustainability*,
Studies in Computational Intelligence 645,
DOI 10.1007/978-3-319-31858-5_5

1 Introduction

In chemical industry, one of the most challenging tasks concerning the sustainability of a new product, i.e., a new chemical, is the prediction of the environmental fate of the chemical. All chemicals eventually end up in the environment, e.g., drugs developed by pharmaceutical companies or cosmetics are released via canalization into the soil or rivers. Important for the sustainability of each chemical is the biodegradation of the chemical. Chemicals can degrade quickly into harmless compounds, but they may also be difficult to degrade or degrade from harmless compounds to toxic compounds.

In silico methods to predict products and pathways of microbial biotransformations of chemical substances are increasingly sought due to rapidly growing data requirements for regulatory chemical risk assessment at the European (cf. REACH [16]) and global level. Existing methods for the prediction of biotransformation products and pathways can be categorized as either knowledge-based or machine learning-based approaches. Each of the two approaches have strengths and weaknesses. Knowledge-based approaches such as METEOR for the prediction of mammalian metabolism [9] or the University of Minnesota Pathway Prediction System (UM-PPS) for microbial biodegradation [11] utilize expert knowledge on the basis of sets of transformation rules. However, they run the risk of including potentially overly-general, incomplete, or inconsistent rules. In contrast, machine learning approaches produce accurate probability estimates on the basis of empirical data, but often lack the ability to incorporate prior domain knowledge. Also, recent machine learning approaches for biotransformation prediction only predict quite general classes.

The goal of this chapter is to combine the two approaches: We assume a given set of biotransformation rules and learn the probability of transformation products proposed by the rules from known, experimentally elucidated biodegradation pathways. Only two comparable systems currently exist: META [14] which is similar in spirit, but uses a less advanced problem formulation and machine learning approach than the one presented here; and CATABOL [5], the only rule-based method explicitly aiming for probability estimates. However, the CATABOL system works with a fixed pathway structure for training, which is different from the approach presented here, working on the basis of individual rules.

Rule-based systems, such as UM-PPS, work on the basis of rules that are generalizations and abstractions of known reactions. In the case of UM-PPS, it is the underlying Biocatalysis/Biodegradation Database (UM-BBD) [6]. UM-BBD is a manually curated compilation of over 200, experimentally elucidated microbial biotransformation pathways encompassing enzymatic reactions for roughly 1,000 parent compounds and intermediates. If certain functional groups of a query substrate match with any of the biotransformation rules in UM-PPS, then its structure is transformed into one or several products according to the rule(s). These rules are typically relatively general, either to cover all known reactions or because there is not enough information known to restrict them. As a consequence, UM-PPS produces a large number of possible reaction products, especially when used to predict several subsequent generations of transformation products. This combinatorial explosion is a phenomenon also known from other rule-based systems and approaches. It is particularly aggravated for the structurally more complex contaminants of current concern, e.g., pesticides, biocides or pharmaceuticals. Potential users of such a system, e.g., environmental microbiologists, risk assessors, and analytical chemists, are overwhelmed by the number of possible products, and find it difficult to identify the most plausible products.

In an effort to restrict combinatorial explosion, some of the knowledge-based approaches to metabolic prediction employ what is called *relative reasoning* [2]. In relative reasoning, the possibility to apply a rule depends on the presence of other applicable rules. Practically, this requires additional rules for the prioritization of rules and the resolution of conflicts. These meta-rules, or relative reasoning rules, express that some reactions take priority over others, and vice versa—that some reactions only occur if others are not possible.

Relative reasoning rules have been derived automatically for the set of UM-PPS biotransformation rules and have been successfully implemented into the working UM-PPS [7]. However, although reductions in the number of predicted products in one prediction step of about 20 % were achieved, the selectivity (precision) of UM-PPS still remained rather low, at approximately 16–18 %. Thus, the question remains how we can further refine the process of selecting and accepting those rules that most likely lead to observed products, when a set of rules applies to the structure of a given substrate.

In this chapter, we propose a solution that transfers the idea of relative reasoning to a machine learning setting [25, 26] to further improve the system's selectivity. Rule probabilities are to be estimated such that they depend not only on all other rules that are applicable but also on the structure of the substrate. The priorities are learned statistically from data on known biodegradation pathways. In our solution, one classifier is learned for each rule, generalizing over the molecular substructures of the substrate and the "activation patterns" of the rules as given by the set of all other rules that are triggered by the same substrate.

The prediction of multiple transformation products at once can be considered from a multi-label perspective. Multi-label classifiers exploit the dependencies among multiple target values to improve the prediction. In the case of the prediction of biodegradation products, there clearly exists dependencies between the target values. When a compound is degraded, similar reactions will occur and other reactions will be prevented. Using these dependencies, we are able to increase the number of transformation rules on which the expert and machine learning-based approach performs well, simultaneously increasing the performance on all transformation rules.

Most multi-label classifiers predict probabilities for the predicted target values. Given the availability of such probabilistic classifiers, the decision to accept a product or not can be made dependent on a probability threshold: The application of individual rules can be tuned such that only transformations with a probability above a certain threshold are accepted. In this way, one can also control the generality of whole rule sets and the overall number of products. Thus, it is simple to address the fundamental trade-off between the completeness and the accuracy of predictions. In technical terms, we can analyze the performance of both individual rules and the entire system in recall-precision space, and visualize their performance in two-dimensional plots. Moreover, it is possible to explicitly choose a suitable point in recall-precision space by setting the probability threshold for accepting a rule to a certain level.

2 A Hybrid Knowledge and Machine Learning-Based Approach

To illustrate the problem and the proposed solution, we start with an example shown in Fig. 1a: Given a new compound c_{new}, several rules of the UM-PPS are applicable and suggest possible transformation products. In the example, a subset of rules r_1, r_2, r_5, etc. triggers for the given input structure. In the illustration, triggering rules are indicated by solid arrows and rules that are not triggering by dashed arrows. As mentioned above, the problem is that the rules of the system are overly-general, i.e., they suggest a wide range of possible products, many of them false positives.

To restrict the number of possible products, it would be desirable to score the proposed transformations by estimated probabilities. In this way, it would also be possible to tune the number of products depending on a user-defined threshold: If the estimated probability of a transformation exceeds a threshold, it is accepted, otherwise it is discarded. The probability for each rule r_i is estimated by a corresponding function f_i. Function f_i indicates how likely a transformation suggested by rule r_i is, depending on the structure of the input compound and all other triggering rules. This is illustrated in Fig. 1b: Function f_1 estimates that the probability of obtaining a correct product from applying r_1 to substrate c_{new} is 0.6, given the molecular structure of c_{new} and the other applicable rules $(r_2, r_5, \ldots, r_{179})$. The dependency of the decision on all other transformation options reflects that, under certain conditions, one reaction should be given priority over another. If the cut-off was set to 0.5 in the example, we would only accept the transformations proposed by r_1 and r_5.

The problem is to derive suitable probability scores. In this chapter, the solution is based on a training set of examples and machine learning. In Fig. 2a, a sample of three compounds from a hypothetical training database is shown. For the three training compounds, we assume that we not only know which rules are applicable, but also which rule applications lead to observed products. In the figure, the observed transformation products are indicated by a check mark, whereas the spurious products are indicated by a cross. Given this information, it is possible to learn under which conditions the suggested product of a transformation rule can actually be observed. As a classifier is only needed when a rule triggers, the training set for a rule also includes only those compounds for which the rule suggests a product. Figure 2b, c show two training sets constructed from the three training compounds c_1 to c_3, one for rule r_1 (upper table) and one for rule r_2 (lower table). The first group of features $(s_1, \ldots, s_m)$ is a fingerprint-based representation of the structure of the input compound. The second group of features (all $r_1, \ldots, r_{179}$ except the rule for which the classifier is built) indicates which other rules are applicable to the compound, for

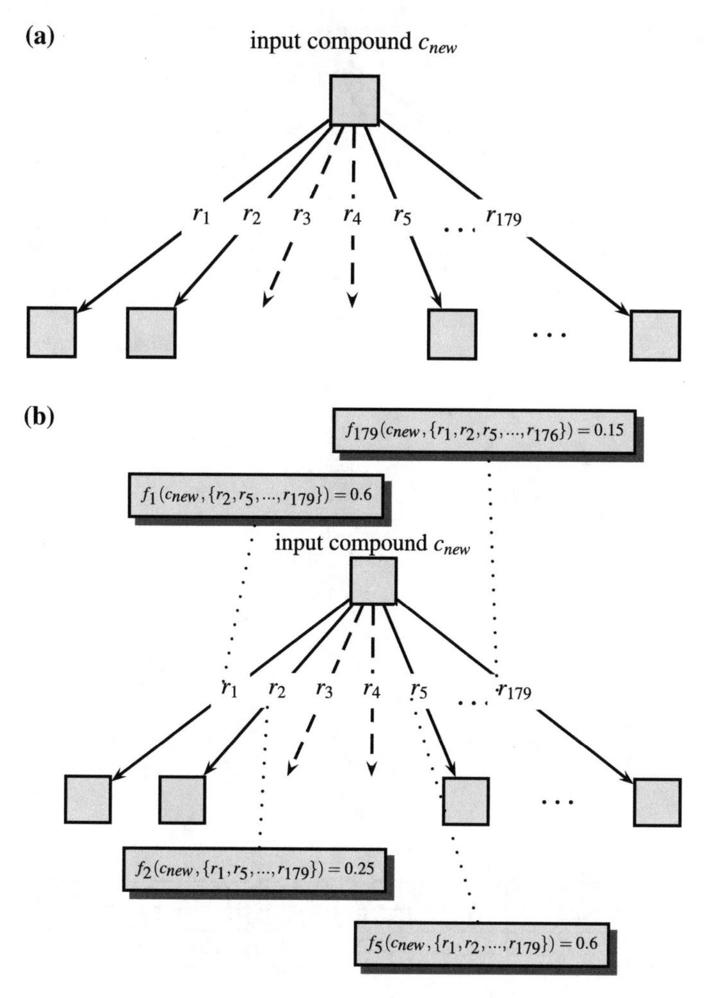

Fig. 1 **a** Indicates the rules applicable to an input compound c_{new} by *solid arrows*, rules not triggered are given by *dashed arrows*. **b** Illustrates the use of one classifier for each rule to determine the probability of obtaining a valid product, depending on the structure and other applicable rules

example a feature is set to $+1$, if the corresponding rule fires, and 0, otherwise. As explained above, the training set for f_1 does not contain an entry for c_2, because rule r_1 is not applicable to that compound. Similarly, c_1 is not listed in the training set for f_2, because rule r_2 cannot be applied. Also note that c_3 is a positive example for f_1, while it is a negative example for f_2. Given such training sets, any machine learning algorithm for classification can be applied to induce a mapping from the structural and rule descriptors to the target variable, i.e., whether a rule generates an observed product.

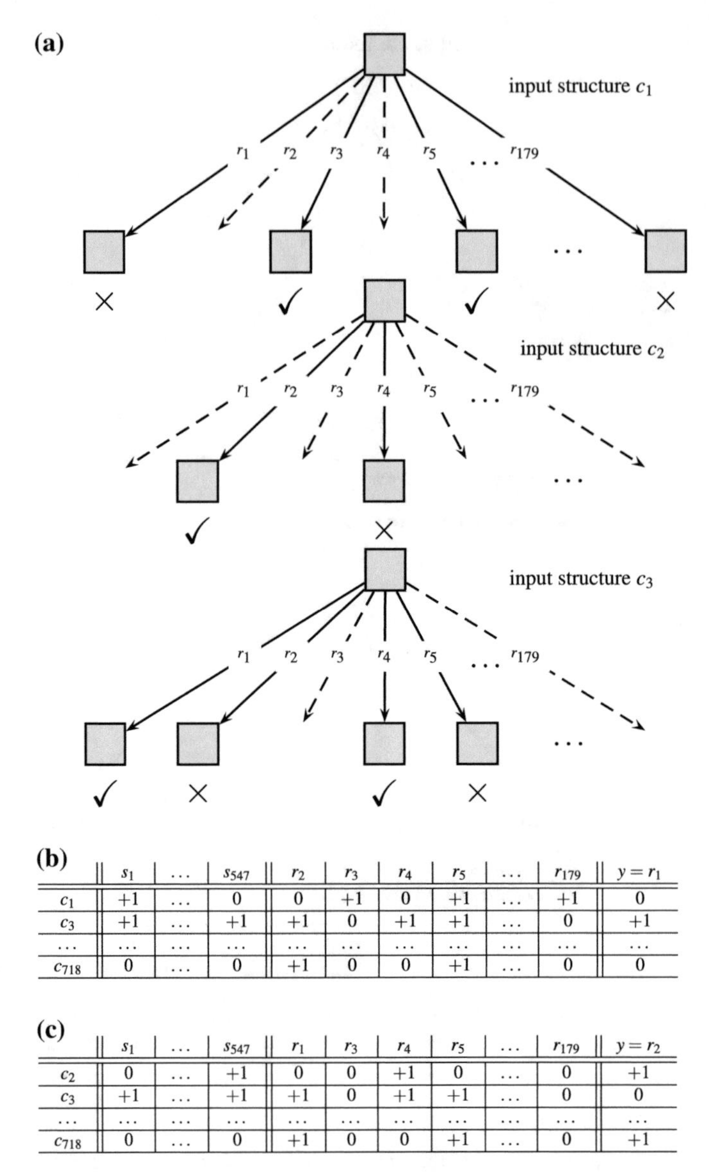

(b)

	s_1	...	s_{547}	r_2	r_3	r_4	r_5	...	r_{179}	$y = r_1$
c_1	+1	...	0	0	+1	0	+1	...	+1	0
c_3	+1	...	+1	+1	0	+1	+1	...	0	+1
...	...	...	...	...	...	...	...	...	...	...
c_{718}	0	...	0	+1	0	0	+1	...	0	0

(c)

	s_1	...	s_{547}	r_1	r_3	r_4	r_5	...	r_{179}	$y = r_2$
c_2	0	...	+1	0	0	+1	0	...	0	+1
c_3	+1	...	+1	+1	0	+1	+1	...	0	0
...	...	...	...	...	...	...	...	...	...	...
c_{718}	0	...	0	+1	0	0	+1	...	0	+1

Fig. 2 Figure (**a**) gives examples for the construction of two training sets, one for f_1 (**b**) and one for f_2 (**c**). Observed products are indicated by a check mark, spurious products are indicated by a cross

To be more precise (amongst others, to enable reproducibility) we introduce some notation: In the following, C denotes the set of compounds c_i, and R the set of rules r_j. $triggers(r_j, c_i)$ is a predicate indicating whether r_j triggers on compound c_i. $observed(r_j, c_i)$ is a predicate indicating that rule r_j fires and provides an observed degradation product. For instance, we have the following list of facts for c_1 and c_2, and r_1 to r_3 from Fig. 2a:

$$triggers(r_1, c_1). \tag{1}$$
$$triggers(r_3, c_1).\ observed(r_3, c_1). \tag{2}$$
$$triggers(r_5, c_1).\ observed(r_5, c_1). \tag{3}$$
$$triggers(r_2, c_2).\ observed(r_2, c_2). \tag{4}$$
$$triggers(r_4, c_2). \tag{5}$$

Finally, S denotes the set of molecular substructures s_l, and predicate $occurs(s_l, c_i)$ checks the occurrence of a substructure s_l in a compound c_i.

To prepare for training we need two transformation operators, one for the construction of individual examples, and one for the construction of whole training sets. The first one, $\tau_{instance}$, is defined as follows:

$$\begin{aligned} \tau_{instance}(c_i, k) = x_i,\ such\ that & \\ x_{i,j} &= occurs(s_j, c_i)\ for\ 1 \le j \le |S| \wedge \\ x_{i,j} &= triggers(r_j - |S|, c_i)\ for\ |S| < j \le |S| + k - 1 \wedge \\ x_{i,j} &= triggers(r_j - |S| + 1, c_i)\ for\ |S| + k - 1 < j \le |R| + |S| - 1 \end{aligned} \tag{6}$$

This means that operator $\tau_{instance}(c_i, k)$ constructs the description of an individual example without its class information. It takes a compound c_i and constructs a feature vector (see the example above), taking into account substructures and applicable rules. Parameter k is used to exclude the information for the kth rule, which is convenient for our purposes because it constitutes the target for training. With $\tau_{instance}(c_i, k)$ we are ready to define a transformation operator generating a training or test set for rule k from a given set of compounds C: τ_{set} takes a compound c_i from C and checks whether rule k triggers. Only if this is the case, a training example (x_i, y_i) is constructed:

$$\begin{aligned} \tau_{set}(C, k) = \{(x_i, y_i) | c_i \in C \wedge triggers(r_k, c_i) \wedge x_i = \tau_{instance}(c_i, k) \wedge \\ y_i = 1\ if\ observed(r_k, c_i),\ y_i = 0\ otherwise\} \end{aligned} \tag{7}$$

In the example above, $\tau_{set}(\{c_1, c_2, c_3..., c_{718}\}, 1)$ gives the training set for classifier f_1 shown in the table of Fig. 2b and $\tau_{set}(\{c_1, c_2, c_3..., c_{718}\}, 2)$ gives the training set for f_2 in the table of Fig. 2c. A training procedure $train$ returns the classifiers needed for the restriction of the rules based on such training sets. As already indicated above, classifiers are represented as functions f_j returning class probability estimates for given examples.

Given those preliminaries, we can explain how training and testing is performed and how it is embedded into the working system (see Algorithm 1 for the pseudocode).

Algorithm 1 Pseudocode for training and testing classifiers for biotransformation rules.

```
Input: Training data 𝒟_Trg, rules r, test instance c_new, threshold θ
Output: Predicted products of c_new
/* training one classifier per rule                                  */
1 foreach rule r_k do
2 |  𝒟^k_Trg ← τ_set(C_Trg, k)  f_k ← train(𝒟^k_Trg)
/* testing for a new test compound c_new                             */
/* the cut-off for acceptance is given by parameter θ                */
3 foreach rule r_k do
4 |  if triggers(r_k, c_new) then
5 |  |  if f_k(τ_instance(c_new, k)) > θ then
6 |  |  |  classify as "product of k"
7 |  |  else
8 |  |  |  classify as "no product of k"
```

In the training phase, a classifier is trained for each rule in turn. In the testing phase, we first obtain a list of rules applicable to each test compound using the UM-PPS. If a rule triggers, we apply the rule's classifier to the instance, where information from the molecular structure and all competing rules is taken into account to obtain a probability estimate. If this estimate exceeds a threshold θ, the product suggested by rule r_k is accepted, otherwise the proposed transformation is rejected.

3 Using Multi-label Classifiers and Extended Encoding

Clearly, there are dependencies between the rules. If one rule is correctly applied to a structure, other rules might not be applicable or supported by this transformation. Nevertheless, the method proposed above does not exploit the dependencies for the training. Multi-label classifiers can exploit the dependencies between multiple binary target values. This changes the method in a way that we do not learn one function per transformation rule, but one predicting a set of probabilities for all rules. The advantage of this is the potential improvement in the prediction quality due to the additional information gained from the other transformation rules.

There exists a wide range of multi-label classifiers.[1] Each classifier has certain qualities, we decided to use two main classifiers: (i) *ensembles of classifier chains* (ECC) [18] and (ii) MLC-BMaD [27]. ECC proved to be a fast and well-performing multi-label classifier which performs well on typical multi-label data sets and is flexible due to the possibility to select a base classifier. MLC-BMaD proved to be capable of training well-performing models on data sets with a large number of labels

[1]For an overview of multi-label classification see the paper by Tsoumakas et al. [23].

and a rather small number of instances. Compared to other multi-label data sets, the number of labels in this data set is rather large and the number of instances rather small, hence, MLC-BMaD should be capable of training well-performing models.

Preliminary experiments showed that multi-label classification could slightly improve the performance of the models. Nevertheless, analysis of the experiments showed that the selected data structure in this setting could be improved. The default approach is to use all triggered transformation rules as features, together with the structural features, and information if the rule is triggered correctly as labels. If a rule is not triggered at all, the corresponding label is set to be missing.

Hence, we changed the encoding of the target information to improve the classification and make the problem more suitable for multi-label classifiers. An overview of the mapping is given in Table 1, an example data set is given in Table 2. The first set of labels for the classifiers was set to binary values describing if the corresponding rule was triggered correctly or not. If a rule was not triggered at all, this value was set to be missing. Additionally, the second set of labels provided information about the activation (triggering) of rules: if the rule is incorrectly triggered or not triggered (negative label), or if triggered correctly (positive label). While this encoding of the information might seem redundant and not intuitive, multi-label algorithms seem to benefit from this redundant information. The second set of labels describes if a rule should be used for the compound or not, regardless whether they are triggered or not. The latter means that the environmental fate strictly is predicted on a machine learning basis, leaving the expert-based knowledge aside. The expert-based knowledge is given by the features indicating if a rule is triggered or not. For the final prediction, we used a combination of the two labels per rule. For each rule, we combined the predicted probability of each set using the mean of the predictions. For the evaluation process, we only used the label set that tells if a label is correctly triggered. Only from them, evaluation measures are calculated, which are the same values on which the single-label approach is evaluated.

Table 1 Extended encoding for multi-label classification

	Correctly triggered	Incorrectly triggered	Not triggered
Label 1 (λ_i)/correctly triggered	1	0	?
Label 2 (λ_i')/known product	1	0	0
Feature (x_i)/triggered	1	1	0

"?" indicates a missing value, which is ignored by the classifiers and evaluation. The data is translated into three new attributes, two are used as labels as they store the target information if a rule is triggered correctly. The third attribute is used as feature which is used for the predictions. For the final prediction, a combination of the probabilities of label 1 and label 2 is used. However, for the evaluation, only label 1 is used

Table 2 Example data set for the multi-label encoding

	s_1	...	s_{547}	x_1	x_2	...	x_{179}	λ'_1	λ'_2	...	λ'_{179}	λ_1	λ_2	...	λ_{179}
c_1	+1	...	0	1	0	...	0	1	0	...	0	1	?	...	?
c_2	+1	...	+1	1	1	...	0	0	1	...	0	0	1	...	?
c_3	0	...	+1	1	0	...	1	0	0	...	1	0	?	...	1
...	...	...	...	...	...	...	...	...	...	...	...	...	...	...	...
c_{718}	0	...	0	0	1	...	1	0	0	...	0	?	0	...	0

c_1 correctly triggers rule 1, rule 2 and 179 are not triggered. c_2 incorrectly triggers rule 1, rule 2 is triggered correctly, rule 179 is not triggered. c_3 incorrectly triggers rule 1, rule 2 is not triggered, and rule 179 is correctly triggered. c_{718} does not trigger rule 1, rule 2 and 179 are incorrectly triggered. Note that in this encoding, training and test use an identical schema. The only difference is that the prediction does not produce missing values

4 Data and Implementation

It is possible to validate the procedure described above by running a cross-validation over the compounds of the UM-BBD database. The running time can be optimized considerably if the predicates *triggered*, *observed*, and *occurs* are pre-computed once for all compounds and stored for later use. In our implementation, we pre-computed a $|C| \times |R|$ table indicating the rules' behavior on the UM-BBD compounds: The value $+1$ of an entry encodes that a rule is triggered and produces an observed product, 0 encodes that a rule is triggered but the product is not observed, and -1 encodes that the rule is not triggered for a given compound. Simple database and arithmetic operations can then be applied to extract training and test sets, e.g., for (*leave-one-out*) *cross-validation*. Additionally, we learned the classifiers on the complete data set and tested the proposed approach on an external validation set of 25 xenobiotics (pesticides) which was also used in previous work [7].[2] Pesticide biodegradation data is the largest cohesive data set available because these compounds are made to be put into the environment and are among the most heavily regulated class of chemicals.

The matrix encoding described above was applied to the UM-BBD/UM-PPS from July 24, 2007, containing 1,084 compounds and 204 biotransformation rules (btrules). Of these, 366 compounds did not trigger by any rule (terminal compounds of reported pathways, compounds containing metals or other compounds whose biodegradation should not be predicted) were removed. Likewise, 25 strictly anaerobic (unlikely or very unlikely) btrules and btrules not triggered by any compounds in the UM-BBD were removed. Finally, 48 transformation rules triggered by only one structure were removed from the set as well. The remaining 718 UM-BBD compounds were submitted to 131 UM-PPS btrules. The predicate *triggered* was then

[2]Those 25 pesticides were also tested in our previous experiments investigating the sensitivity and selectivity of the method (see Table 6 in [7]). In addition, 22 other xenobiotics (pharmaceuticals) were only used for determining the reduction of predictions (see Table 4) because their degradation products are not known.

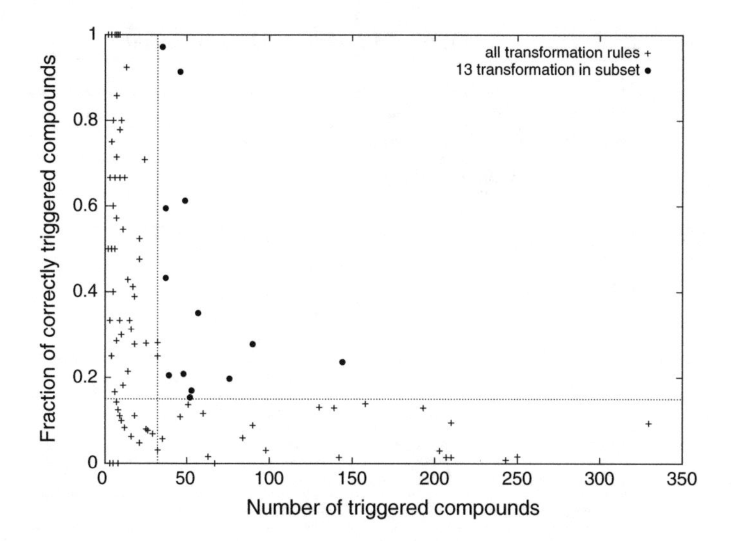

Fig. 3 Characteristics of data sets for rules: size (number of triggered compounds) and class distribution (fraction of correctly triggered compounds). The 13 transformation rules used in the subset are marked. The dotted lines are the cutoffs (Number = 35, Fraction = 0.15) used to select the subset

defined to be true if a rule applied to a compound, and *observed* was defined to be true if the product could be found in the database.

The class distribution in the data set is very diverse (see Fig. 3). There are few transformation rules with both a balanced class distribution and a sufficient number of structures triggering them. Thus, we decided to implement single-label classifiers for a subset of the transformation rules. We chose rules that provide at least a certain amount of information for the construction of the classifiers. The transformation rules needed to be triggered by at least 35 structures. On the other hand, for the ratio of "correct triggers", we set a minimum of at least 0.15. These parameters seem sufficient to cover a sufficient number of cases and exclude overly-skewed class distributions. Varying the parameters in further experiments did not lead to an improvement of the results. However, multi-label classifiers are able to exploit dependencies among the labels and, hence, use more information in the training process which leads to an improvement of the prediction in the remaining labels. Thus, multi-label classifiers are able to train well-performing models on all transformation rules.

Of the 131 transformation rules in the training set, this leaves 13 rules for the validation process (see Table 3). Considering the class distribution and number of examples of the remaining rules, it is not reasonable to learn classifiers for these transformation rules. However, to compare the results with previous work and to evaluate all transformation rules, we generated a *default classifier (DC)* for these rules which predicts the ratio of positive examples as the probability to produce a correct product. Thus, if the chosen threshold is below the ratio of positive

Table 3 List of the 13 transformation rules in the subset used for prediction

Rule	Description
bt0001	primary Alcohol → Aldehyde
bt0002	secondary Alcohol → Ketone
	secondary Alcohol → Ester
bt0003	Aldehyde → Carboxylate
bt0008	vic-Dihydroxybenzenoid → extradiol ring cleavage
bt0029	organoHalide → RH
bt0036	aromatic Methyl → primary Alcohol
bt0040	1-Aldo/keto-2,4-diene-5-ol → Carboxylate + 1-ene-4-one
bt0055	1-carboxy-2-unsubstituted Aromatic → Catechol derivative
bt0060	vic-Hydroxycarboxyaromatic → Catechol derivative
	vic-Aminocarboxyaromatic → Catechol derivative
bt0063	primary Amine → Aldehyde or Ketone
	secondary Amine → Amine + Aldehyde or Ketone
	tertiary Amine → secondary Amine + Aldehyde or Ketone
	Methylammonium derivative → Trimethylamine + Aldehyde or Ketone
bt0065	1-Amino-2-unsubstituted aromatic → vic-Dihydroxyaromatic + Amine
bt0254	vic-Dihydroxyaromatic → intradiol ring cleavage
	vic-Dihydroxypyridine → intradiol ring cleavage
bt0255	vic-Dihydrodihydroxyaromatic → vic-Dihydroxyaromatic

examples, all structures are predicted as positive, i.e., they are predicted to trigger this transformation rule correctly.

For the multi-label classification setting, only one classifier can be chosen for the base classification. Hence, we only used a setting where a learned classifier is applied to all 131 transformation and no default classifier is used.

For the computation of frequently occurring molecular fragments, we applied the FreeTreeMiner system [19] as it builds on a cheminformatics library to handle structures and substructures conveniently. It produces fingerprints by searching for frequently occurring acyclic substructures in the set of structures.

5 Performance Measures

Clearly, we are facing a fundamental trade-off also found in many other applications of machine learning and classification: If the rules are too general, we will not miss many positive examples, but we might also include too many false positives. If the rules are too specific, we probably have few false positives but we will potentially miss too many positives. It is convenient to think of this trade-off in terms of recall (sensitivity) and precision (selectivity). If the overall system predicts an observed product for a given substrate, we can count this as a true positive. If the system predicts a product that is not observed, we have a false positive. If a product is missing for a substrate, we have a false negative.[3] The number of true positives is denoted by TP, the number of false positives by FP, and the number of false negatives by FN. Then the standard definitions of recall (sensitivity) and precision (selectivity) can be applied:

$$Recall = Sensitivity = \frac{TP}{TP + FN} \quad (8)$$

$$Precision = Selectivity = \frac{TP}{TP + FP} \quad (9)$$

The overall number of products predicted by the system critically depends on the cut-off parameter θ. To evaluate the performance of the system, this parameter does not need to be fixed in advance. Instead, the parameter can be varied over the entire range from 0 to 1, and the resulting values for recall and precision can be plotted in two dimensions: recall is plotted on the x-axis, and precision on the y-axis. Precision-recall plots (see Figs. 4 and 5) offer a simple and intuitive visualization of the trade-offs involved in choosing a certain value of θ. Also the results of approaches without cut-off parameters (e.g., relative reasoning as discussed above) appear as single data points in precision-recall space.

Precision-recall analysis can be performed on the system level as well as on the level of individual rules. In principle, one could set the threshold individually for each rule, but this would introduce a large number of parameters. For simplicity, we chose to visualize the system's performance below by applying the same threshold for all rules. Also, as individual classifier schemes should be sensitive and adaptive to different class distributions, one parameter for all should work reasonably well in the first approximation.

In addition to precision-recall analysis, we measure the area under the receiver operating characteristic (ROC) curve, which indicates the capability of a classifier to rank the examples correctly [3].

[3]We count the false negatives in a slightly different way than in a previous paper [7], as we only consider products that are suggested by any of the biotransformation rules. In other words, we do not take into account products of reactions that are not subsumed by any of the rules. This is done because only for the products suggested by the UM-PPS, the method proposed here becomes effective—the classifiers can only restrict the rules, not extend them.

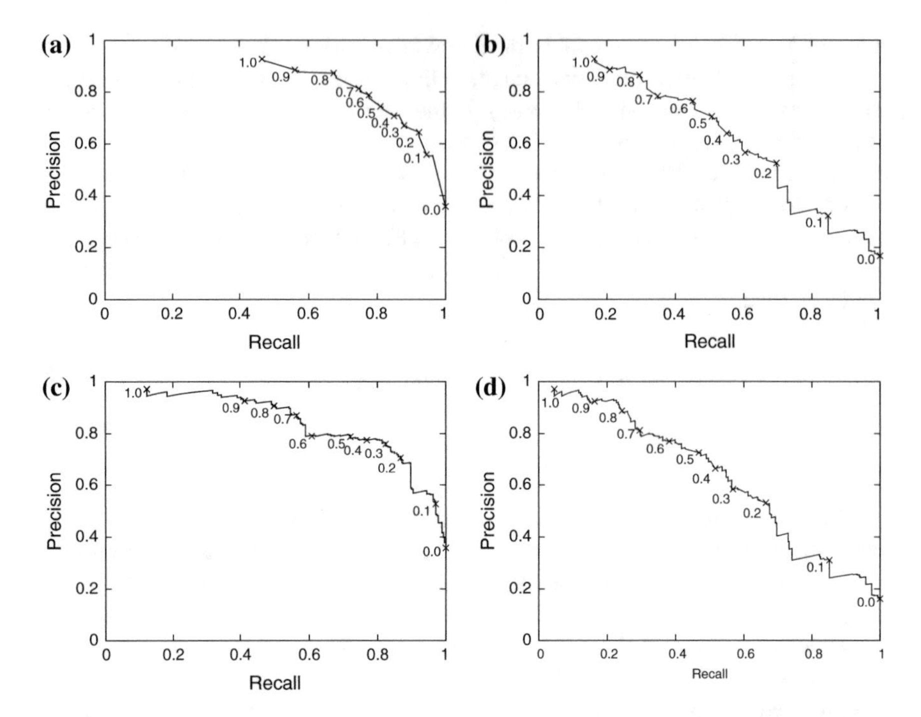

Fig. 4 Precision-recall plots from a *leave-one-out cross-validation* using the random forest classifier (**a** and **b**) and support vector machines (**c** and **d**). On the *left hand side* (**a** and **c**), only the results of classifiers on a subset of the rules are shown. On the *right hand side* (**b** and **d**), classifiers were generated for the same subset and a default classifier is used for the remaining rules. The subset was chosen by using transformation rules with at least 35 triggered examples and a minimum ratio of known products of 0.15. Using these parameters, 13 transformation rules were selected. The threshold θ is given in ten steps per plot. Note that the points in precision-recall space are connected by lines just to highlight their position. In contrast to ROC space, it is not possible to interpolate linearly. **a** Random forests on 13 rules. **b** Random forest on 13 rules, default classifier on remaining rules. **c** SVM on 13 rules. **d** SVM on 13 rules, default classifier on remaining rules

6 Experimental Results

In the following, we present the experimental results obtained with the proposed approach. After introducing various learning schemes and settings, we present the results on the xenobiotics test set and, more importantly, our main results from a *leave-one-out cross-validation* over the UM-BBD structures.

For the 13 transformation rules in the subset and 131 transformation rules in the case of multi-label classifiers, we applied the random forest algorithm [1] in the implementation of the Weka workbench [10], because it gave acceptable probability estimates in preliminary experiments. As a second classifier, we used support vector machines trained using sequential minimal optimization [15] in the implementation of the Weka workbench. We automatically adjusted the complexity constant of the

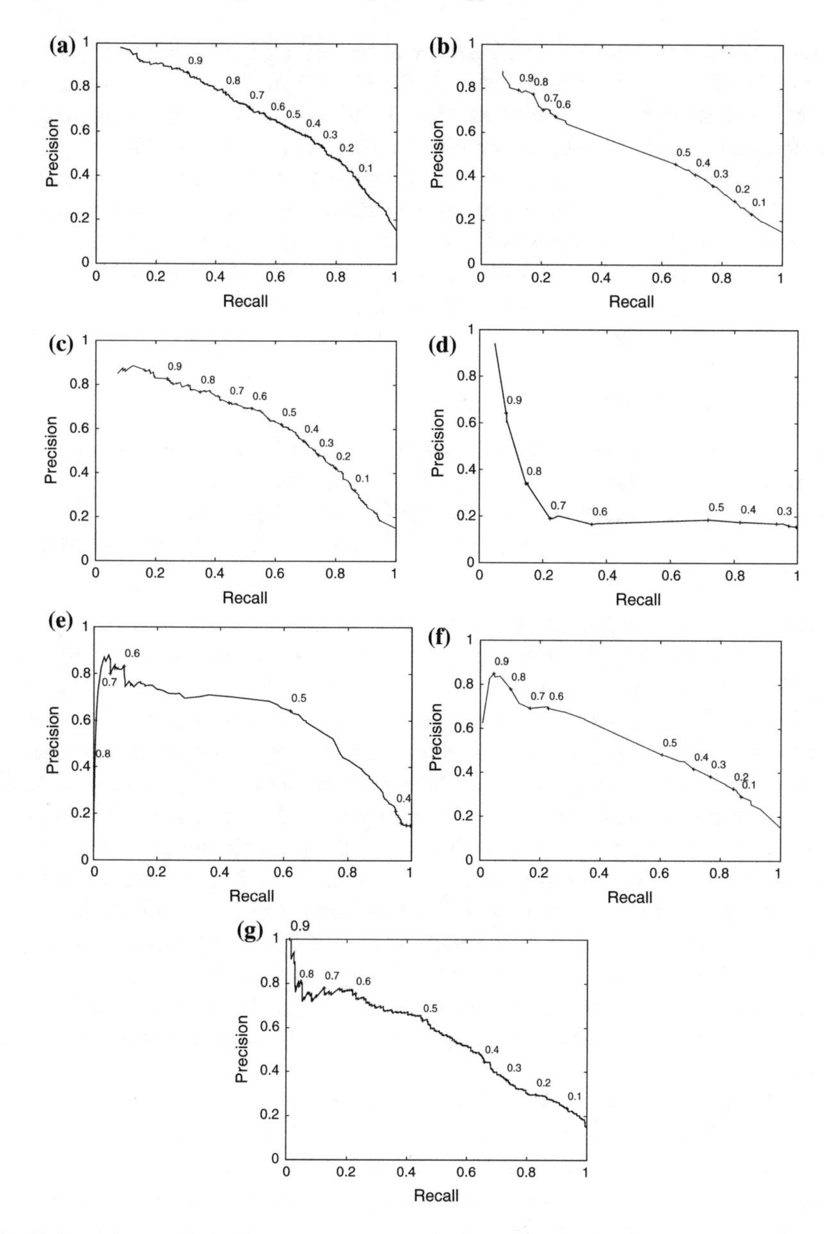

Fig. 5 Precision-recall plots from a *leave-one-out cross-validation* using different multi-label classifiers on all rules as labels. The threshold θ is given in ten steps per plot, except in plots (**d**) and (**e**) where the lower thresholds values are too close to each other to differentiate among them in the plots. Note that the points in precision-recall space are connected by lines just to highlight their position. In contrast to ROC space, it is not possible to interpolate linearly. **a** Ensembles of classifier chains. **b** Multi-label classification using Boolean matrix decomposition. **c** Include labels. **d** Label powerset. **e** Calibrated label ranking. **f** Binary Relevance2. **g** MLkNN

Support Vector Machine for each transformation rule separately. We used a *10-fold cross-validation* to generate the data for the logistic models [15] to obtain well-calibrated class probability estimates.[4] As a default classifier on the remaining 118 transformation rules, we used the ZeroR algorithm of the Weka workbench.

Multi-label classifiers were used in the implementation of the Mulan library [24]. We evaluated *ensembles of classifier chains* (ECC), *multi-label classification using Boolean matrix decomposition* (MLC-BMaD), *include labels* (IL) [22], *label powerset* (LP), *binary relevance 2* (2BR) [21], and *calibrated label ranking* (CLR) [8], which all use random forests (RF) as base classifiers. In addition, *multi-label k nearest neighbor* (MLkNN) [28] was evaluated, however it does not require any base classifier.

In total, we evaluated four variants:

(a) 13 learned classifiers (LC) (i.e., random forests or SVMs) only,
(b) 13 learned classifiers (LC) and 118 default classifiers (DC),
(c) 131 learned classifiers (LC) (without default classifiers), and
(d) one learned multi-label classifier on 131 labels, equivalent to 131 learned single label classifiers (LC).

The idea of (a) is to evaluate the performance of the machine learning component of the system only. In (b), the overall performance of the system is evaluated, where 13 classifiers are complemented by 118 default classifiers. The purpose of (c) is to reveal whether the default classifiers are sufficient, or whether learned classifiers should be used even when samples are very small and classes are unequally distributed. (d) gives the results of extended multi-label evaluation on the data set. Multi-label classifiers can increase the total performance by exploiting label dependencies. The goal is to use all the rules for learning and gain the maximum performance. All the results are shown in terms of manually chosen points in precision-recall space (e.g., before inflection points) as well as the area under the ROC curve (AUC). The possibility to choose thresholds manually is one of the advantages of working in precision-recall and ROC space; Instead of fixing the precise thresholds in advance, it is possible to inspect the behavior over a whole range of cost settings, and set the threshold accordingly. Finally, we compare the results to the performance of relative reasoning.[5]

For compatibility with a previous paper [7], we start with the results of training on all UM-BBD compounds and testing on the set of 25 xenobiotics. The results are given in the upper part of Table 4. Note that in this case the default classifiers were "trained" on the class distributions of the UM-BBD training data and subsequently applied to the external xenobiotics test set.

---

[4]Note that any other machine learning algorithm for classification and, similarly, any other method for the computation of substructural or other molecular descriptors could be applied to the problem.

[5]We cannot compare our results with those of CATABOL because the system is proprietary and cannot be trained to predict the probability of individual rules—the pathway structure has to be fixed for training (for details we refer to Sect. 7). This means that CATABOL addresses a different problem than the approach presented here.

Table 4 Recall and precision for one threshold (on the predicted probability of being in the positive class) of the machine learning approach and for relative reasoning

	Method	Variant	LC	DC	θ	Recall	Precision	AUC
Xenobiotics	RF	(a)	13	0	0.417	0.400	0.333	0.505
	RF	**(b)**	**13**	**118**	**0.296**	**0.525**	**0.447**	**0.676**
	RF	(c)	131	0	0.35	0.475	0.404	0.664
	SVM	(a)	13	0	0.023	0.800	0.235	0.389
	SVM	**(b)**	**13**	**118**	**0.296**	**0.475**	**0.463**	**0.674**
	SVM	(c)	131	0	0.157	0.410	0.390	0.599
	RR	–	–	–	–	0.950	0.242	–
UM-BBD	RF	(a)	13	0	0.600	0.777	0.788	0.902
	RF	**(b)**	**13**	**118**	**0.308**	**0.595**	**0.594**	**0.842**
	RF	(c)	131	0	0.485	0.653	0.632	0.857
	SVM	(a)	13	0	0.329	0.813	0.771	0.903
	SVM	**(b)**	**13**	**118**	**0.294**	**0.582**	**0.588**	**0.841**
	SVM	(c)	131	0	0.250	0.632	0.623	0.833
	ECC/RF	**(d)**	**131**	**0**	**0.510**	**0.630**	**0.628**	**0.894**
	MLC-BMaD/RF	**(d)**	**131**	**0**	**0.500**	**0.615**	**0.457**	**0.821**
	IL/RF	(d)	131	0	0.499	0.623	0.620	0.863
	MLkNN	(d)	131	0	0.477	0.557	0.553	0.844
	LP/RF	(d)	131	0	0.700	0.223	0.190	0.597
	2BR/RF	(d)	131	0	0.500	0.605	0.482	0.841
	CLR	**(d)**	**131**	**0**	**0.497**	**0.629**	**0.631**	**0.873**
	RR	–	–	–	–	0.942	0.267	–

The columns LC and DC indicate the number of transformation rules used for the learned classifiers (LC), SVMs or random forests, and the default classifiers (DC) ZeroR. The value of the threshold θ is determined manually considering the trade-off between recall and precision. We chose the threshold manually at an approximate optimum for recall and precision to provide a comparison to previous work [7]. Area under ROC Curve (AUC) is threshold-independent and only given for the new approach. The column with the variant refers to the assignment of rules to the different classifiers and is explained in the text. The multi-label classifiers are given with their used base classifier, in the case one was required. *Ensembles of classifier chains* (ECC), *Multi-Label Classification using Boolean Matrix Decomposition* (MLC-BMaD), *Include Labels* (IL) [22], *Label Powerset* (LP), *Binary Relevance2* (2BR), and Calibrated Label Ranking (CLR) all use random forests (RF) as base classifiers. *Multi-Label k Nearest Neighbor* (MLkNN) does not require any base classifier. RR refers to relative reasoning, which was used in previous work

First, we observe that ROC scores are on a fairly good level. The results in precision-recall space indicate that variant (b) is as good as variant (c). However, with an AUC of around 0.5 and having 13 learned classifiers, only variant (a) performs on the level of random guessing. An in-depth comparison of the two sets of structures (UM-BBD and xenobiotics) shows that this can be attributed to (i) the low structural similarity between the two sets, and (ii) the fact that a very limited set of rules trigger for the xenobiotics (a consequence of (i)). The average number of free tree substructures per compound is 48.76 in the xenobiotics data set, whereas it is 65.24 in the UM-BBD data set. Due to this structural dissimilarity, the transfer from one data set to the other is a difficult task. Therefore, we decided to perform a *leave-one-out cross-validation* over all UM-BBD compounds, where the structural similarity between test and training structures is higher than in the validation with the 25 xenobiotics as test structures.

Our main results from leave-one-out over the UM-BBD compounds are visualized in the precision-recall plots of Fig. 4 and shown quantitatively in Table 4. In Fig. 4, the top row (a and b) shows the plots of the random forest classifiers, the bottom row (c and d) displays the results of the Support Vector Machines. The left hand side (a and c) shows the results of the classifiers on the 13 transformation rules in the subset, the right hand side (b and d) uses the default classifier for the remaining rules. The plots tend to flatten when including predictions of the default classifier.

The overall performance does not differ much between random forest and Support Vector Machines. Using both classification methods, we can achieve recall and precision of slightly less than 0.8 (see Fig. 4 and also the values for variant (a) in the lower part of Table 4) for the LC only in a *leave-one-out cross-validation*. The quantitative results in Table 4 also show that the performance of random forests and support vector machines are on a similar level. Also, in this case, the performance of learned classifiers complemented by default classifiers is comparable to the performance of the learned classifiers for all rules, supporting the idea of having such a mixed (LC and DC) approach.[6] However, as expected, the machine learning component consisting of 13 learned classifiers only performs better on average than the overall system with 118 default classifiers added in this case. In summary, the AUC scores are satisfactory and the precision-recall scores of approximately 0.8 of the machine learning component indicate that improvements in precision are possible without heavily compromising recall. Therefore, the machine learning approach provides some added value compared to the relative reasoning approach developed previously [7].

To evaluate the enhancement of using both the structural information and the expert knowledge in the transformation rules, we applied the new method individually to the data set leaving out the structure and, in a second run, the transformation rules. As it tends to give smoother probability estimates, here and in the remainder of the section we focus on random forest classifiers. Using only the structural information gives an AUC of 0.895, whereas the transformation rules only give an AUC of 0.885. Taken together, we can observe an AUC of 0.902, which despite the apparent redundancy for the given data set, marks an improvement over the results of the individual feature sets.

In the multi-label experiments, we could improve the overall performance. We used seven multi-label classifiers to predict the probability of the biodegradation rules. Due to the higher complexity of the algorithms, the experiments are limited to random forests as base classifiers where needed. SVMs increase the running time of multi-label classifiers drastically. *Ensemble of classifier chains* (ECC) by Read

[6] In other words, it shows that informed classifiers do not pay for the rest of the rules.

et al. [17] showed the most promising results. Using all 131 rules in the prediction, the algorithm almost gained the same performance level regarding the area under the ROC curve as single-label SVMs on only the 13 selected rules. This provides the possibility to include a greater number of rules in the classification with some benefit. Single-label classifiers do not give better than random performance on the additional 118 rules.

The other multi-label classifiers all achieve a good performance with only the exception of *label powerset*. This algorithm does not perform that well on data sets with a large number of unique label combinations. Given the size of the data set and the rather large number of labels, this is the case for this data set. This explains the almost random performance of this classifier. The advantages of multi-label classifiers also can be seen regarding recall and precision. *ECC* raises both values compared to the single-label SVM case while simultaneously using more labels in the prediction. None are as biased towards one of the measures as Relative Reasoning. The precision-recall plots (see Fig. 5) give smoother plots than the single-label

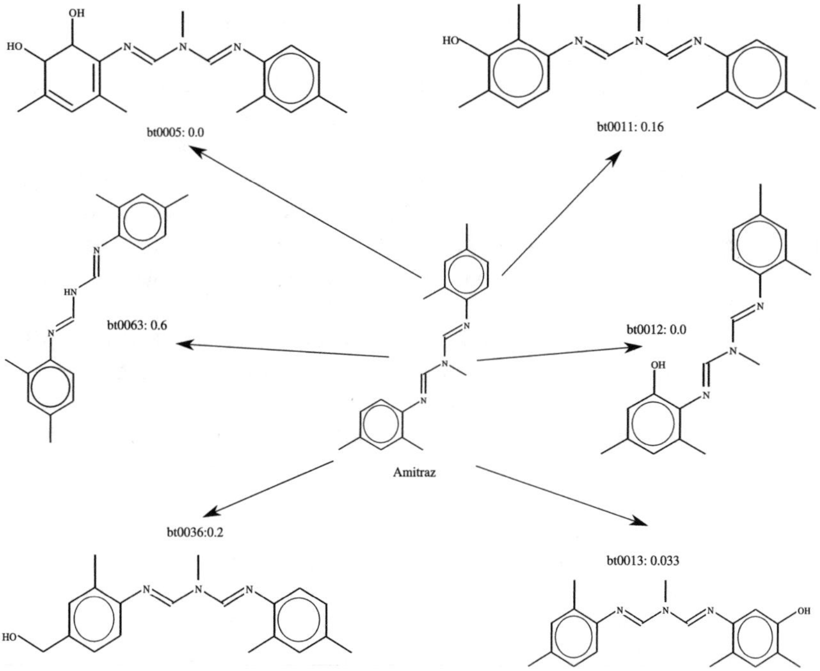

Fig. 6 Application of the new method to Amitraz, a compound from the xenobiotics data set. For each transformation rule triggered by this structure, an example product is given. Some of the transformation rules can produce more than one product from this structure. We applied random forest classifiers to the structure. The numbers indicate the predicted probability that the corresponding transformation rule produces a known product. From the transformations predicted by the UM-PPS, only bt0063 produces a known product. As shown in the figure, this is the only transformation rule with a relatively high predicted probability

predictions. In all but one case a threshold for addressing the trade-off between the measures is apparent and can be selected. The LP method again shows that it cannot perform well on this data set.

An example prediction of the biotransformation of a structure is given in Fig. 6. We applied the single label approach to Amitraz, a pesticide from the xenobiotics data set. The incorrectly triggered transformation rules all obtain a rather low probability whereas bt0063, a correctly triggered rule, is the only transformation rule being predicted with a probability higher than 0.53. As the xenobiotics data set is small, we generated random forest classifiers for every transformation rule triggered by this structure for the purpose of the example.

7 Discussion and Conclusion

In this chapter, we presented a combined knowledge-based and machine learning-based approach to the prediction of biodegradation products and pathways, which performs relative reasoning in a machine learning framework. One of the advantages of the approach is that probability estimates are obtained for each biotransformation rule. Thus, the results are tunable and can be analyzed in precision-recall space. In making the trade-off between recall (sensitivity) and precision (selectivity) explicit, one can choose which is more important. This obviously depends on the application scenario. In certain cases, it is more important to identify as many potential degradation products as possible with only discarding obvious wrong products, for example when trying to find potential toxic products. On the other hand, one might only be interested in certain products, e.g., predicting the most probable outcome and risk of one compound.

This provides a tool for the chemical industry and public agencies to identify the risk of chemicals being released in the environment. A produced chemical might be harmless and no threat to the environment, yet its biodegradation products can cause problems due to toxicity. By providing a probability for degradation products, the combinatorial explosion can be limited and more probable degradation products can be taken into account.

In contrast to CATABOL, the approach works on the level of rules and not on the level of pathways. In CATABOL, the structure of pathways must be laid out in advance in order to solve the equations based on the training data. To make the computations more stable, reactions must be grouped using expert knowledge. In contrast, we apply the rules to the training structures to extract a matrix, which is the basis for the creation of the training sets for each rule.

Our results illustrate that using multi-label classifiers and, hence, assigning a probability to each transformation rule, we can achieve a high precision for selecting correctly triggered transformation rules. By exploiting dependencies between the target values with multi-label classification, we can cover a broad range of transformation rules and incorporate rules with skewed class distribution and few examples triggering the rule. Consequently, we increase the predictive performance on a broad range of transformation rules and provide a method to limit the combinatorial explosion of the expert-based systems while keeping correct transformation products in the predicted biodegradation pathway.

Initial experiments indicated that the standard multi-label approach did not result in any significant improvement compared to the single-label case. Nevertheless, including a pseudo label λ_i' for each transformation rule λ_i, helped improve the multi-label predictions. The use of a pseudo label λ_i' splits the learning task into two problems: (i) The prediction is that a transformation rule produces a known product, and (ii) if that transformation rule is correctly triggered. The improvement in this case is mainly related to the missing values in the target label λ_i. In the standard multi-label approach, λ_i is missing if the transformation rule is not triggered. Hence, if $x_i = 0$ (i.e., rule i is not triggered), no model is learned. Hence, if, in the test case, a rule is not triggered and missing, the learned model can predict any probability independent of the information that it is triggered or not. While this is not a problem for this particular label λ_i, as it is ignored in the evaluation due to the missing value, the classifiers using this as an input might be confused by the prediction. In the case of using an additional label λ_i', the predicted label $\hat{\lambda}_i$ is in most cases set to 0, as there is a strong correlation between $\lambda_i' = 0$ and $\lambda_i = 0$. This is actually closer to the reality than the first case, as this is definitely not a correctly triggered rule. Furthermore, classifiers can reconstruct the information of a missing value from λ_i' and x_i. The training task for the models is still the same. In the case of using a default multi-label approach, it is to predict λ_i from x_i and all other labels; in the case of using pseudo labels, it is to predict λ_i' from x_i. The prediction of λ_i from λ_i' and x_i is trivial. This procedure can be understood as an imputation step. It levels out the effects of having missing values in the training data, which results in predictions that distort the input data for other classifiers.

In this domain, the run times of the algorithms are not that important. The data sets rarely change, hence the training process does not have to be repeated too often. The application of the models, of course, needs to be done fast. But this step is fast, independent of which classifier is chosen. Regarding the training of the models, in the case of the basic, single-label algorithm, the complexity depends linearly on the complexity of the base classifiers, in our experiments, Random Forests and Support Vector Machines. In case of the multi-label classification, it additionally depends on the multi-label classifiers. E.g. MLC-BMaD has a complexity of $O(kn^2)$ depending on the number of labels (n) and latent labels (k). Ensembles of Classifier Chains behave similarly to the basic algorithm, depending only on the complexity of the base classifiers.

CATABOL learns parameters for a fixed pathway structure, whereas the approach proposed here learns classifiers for (individual) transformation rules. During testing, only the pathways laid out for training can be used for making predictions in CATABOL. In contrast, the approach presented here predicts one transformation after the other according to the rule's applicability and priority determined by the classifiers. Overall, the training of CATABOL requires more human intervention than our approach, e.g., for grouping and defining hierarchies of rules (see Dimitrov et al. [4]).

One might speculate (i) which other methods could be used to address this problem, and (ii) where the proposed solution could be applied elsewhere. Regarding (i), it appears unlikely that human domain experts would be able and willing to write complex relative reasoning rules as the ones derived in this work. Alternatively, other machine learning schemes could be used to solve the problem, for instance, methods for the prediction of structured output [13]. These methods should be expected to require a large number of observations to make meaningful predictions. Also, with the availability of transformation rules, the output space is already structured and apparently much easier to handle than the typically less constrained problem of structured output. Regarding (ii), the approach could be used wherever expert-provided over-general transformation rules need to be restricted and knowledge about transformation products is available. It would be tempting to use the same kind of approach for other pathway databases like KEGG, if they were extended towards pathway prediction systems like the UM-BBD. Our extended pathway prediction system could also be used as a tool in combination with toxicity prediction, as the toxicity of transformation products often exceeds the toxicity of their parent compounds [20]. The procedure would be first to predict the degradation products and then use some (Q)SAR model to predict their toxicity.

Recently, we integrated the presented approach into enviPath [12], a complete redesign and re-implementation of UM-PPS. In the future, it may become necessary to adapt the method to more complex rule sets, e.g., (super-)rules composed of other (sub-)rules. Such complex rule sets should be useful for the representation of cascades of reactions.

References

1. Breiman, L.: Random forests. Mach. Learn. **45**(1), 5–32 (2001)
2. Button, W.G., Judson, P.N., Long, A., Vessey, J.D.: Using absolute and relative reasoning in the prediction of the potential metabolism of xenobiotics. J. Chem. Inf. Comput. Sci. **43**(5), 1371–1377 (2003)
3. Cortes, C., Mohri, M.: AUC optimization vs. error rate minimization. In: Proceedings of the 2003 Conference on Advances in Neural Information Processing Systems, vol. 16, pp. 313–320 (2004)
4. Dimitrov, S., Kamenska, V., Walker, J., Windle, W., Purdy, R., Lewis, M., Mekenyan, O.: Predicting the biodegradation products of perfluorinated chemicals using CATABOL. SAR QSAR Environ. Res. **15**(1), 69–82 (2004)
5. Dimitrov, S., Pavlov, T., Nedelcheva, D., Reuschenbach, P., Silvani, M., Bias, R., Comber, M., Low, L., Lee, C., Parkerton, T., et al.: A kinetic model for predicting biodegradation. SAR QSAR Environ. Res. **18**(5–6), 443–457 (2007)
6. Ellis, L.B., Roe, D., Wackett, L.P.: The University of Minnesota biocatalysis/biodegradation database: the first decade. Nucleic Acids Res. **34**(Database issue), D517–D521 (2006)
7. Fenner, K., Gao, J., Kramer, S., Ellis, L., Wackett, L.: Data-driven extraction of relative reasoning rules to limit combinatorial explosion in biodegradation pathway prediction. Bioinformatics **24**(18), 2079–2085 (2008)
8. Fürnkranz, J., Hüllermeier, E., Mencía, E.L., Brinker, K.: Multilabel classification via calibrated label ranking. Mach. Learn. **73**(2), 133–153 (2008)

9. Greene, N., Judson, P., Langowski, J., Marchant, C.: Knowledge-based expert systems for toxicity and metabolism prediction: DEREK, StAR and METEOR. SAR QSAR Environ. Res. **10**(2–3), 299–314 (1999)
10. Hall, M., Frank, E., Holmes, G., Pfahringer, B., Reutemann, P., Witten, I.H.: The WEKA data mining software: an update. ACM SIGKDD Explor. Newsl. **11**(1), 10–18 (2009)
11. Hou, B.K., Ellis, L.B., Wackett, L.P.: Encoding microbial metabolic logic: predicting biodegradation. J. Ind. Microbiol. Biotechnol. **31**(6), 261–272 (2004)
12. http://nar.oxfordjournals.org/content/44/D1/D502
13. Joachims, T., Hofmann, T., Yue, Y., Yu, C.N.: Predicting structured objects with support vector machines. Commun. ACM **52**(11), 97–104 (2009)
14. Klopman, G., Tu, M., Talafous, J.: META 3 a genetic algorithm for metabolic transform priorities optimization. J. Chem. Inf. Comput. Sci. **37**(2), 329–334 (1997)
15. Platt, J.C.: Fast training of support vector machines using sequential minimal optimization. In: Advances in Kernel Methods: Support Vector Learning, pp. 185–208. MIT Press (1999)
16. REACH: Regulation (EC) no 1907/2006 of the European Parliament and of the council of 18 December 2006 concerning the registration, evaluation, authorisation and restriction of chemicals (REACH). Off. J. Eur. Union **49**, L396 (2006)
17. Read, J., Pfahringer, B., Holmes, G.: Multi-label classification using ensembles of pruned sets. In: 8th IEEE International Conference on Data Mining, pp. 995–1000. IEEE (2008)
18. Read, J., Pfahringer, B., Holmes, G., Frank, E.: Classifier chains for multi-label classification. In: Machine Learning and Knowledge Discovery in Databases, pp. 254–269. Springer (2009)
19. Rückert, U., Kramer, S.: Frequent free tree discovery in graph data. In: Proceedings of the 2004 ACM Symposium on Applied Computing, pp. 564–570. ACM (2004)
20. Sinclair, C.J., Boxall, A.B.: Assessing the ecotoxicity of pesticide transformation products. Environ. Sci. Technol. **37**(20), 4617–4625 (2003)
21. Tsoumakas, G., Dimou, A., Spyromitros, E., Mezaris, V., Kompatsiaris, I., Vlahavas, I.: Correlation-based pruning of stacked binary relevance models for multi-label learning. In: Tsoumakas, G., Zhang, M.L., Zhou, Z.H. (eds.) Proceeding of ECML/PKDD 2009 Workshop on Learning from Multi-Label Data, pp. 101–116 (2009)
22. Tsoumakas, G., Katakis, I., Vlahavas, I.: A review of multi-label classification methods. In: Proceedings of the 2nd ADBIS Workshop on Data Mining and Knowledge Discovery, pp. 99–109 (2006)
23. Tsoumakas, G., Katakis, I., Vlahavas, I.: Mining multi-label data. In: Data Mining and Knowledge Discovery Handbook, pp. 667–685. Springer (2010)
24. Tsoumakas, G., Spyromitros-Xioufis, E., Vilcek, J., Vlahavas, I.: Mulan: a java library for multi-label learning. J. Mach. Learn. Res. **12**, 2411–2414 (2011)
25. Wicker, J., Fenner, K., Ellis, L., Wackett, L., Kramer, S.: Machine learning and data mining approaches to biodegradation pathway prediction. In: Bridewell, W., Calders, T., de Medeiros, A.K., Kramer, S., Pechenizkiy, M., Todorovski, L. (eds.) Proceedings of the 2nd International Workshop on the Induction of Process Models at ECML PKDD 2008 (2008)
26. Wicker, J., Fenner, K., Ellis, L., Wackett, L., Kramer, S.: Predicting biodegradation products and pathways: a hybrid knowledge- and machine learning-based approach. Bioinformatics **26**(6), 814–821 (2010)
27. Wicker, J., Pfahringer, B., Kramer, S.: Multi-label classification using Boolean matrix decomposition. In: Proceedings of the 27th Annual ACM Symposium on Applied Computing, pp. 179–186. ACM (2012)
28. Zhang, M.L., Zhou, Z.H.: A k-nearest neighbor based algorithm for multi-label classification. In: IEEE International Conference on Granular Computing, vol. 2, pp. 718–721. IEEE (2005)

Feeding the World with Big Data: Uncovering Spectral Characteristics and Dynamics of Stressed Plants

Kristian Kersting, Christian Bauckhage, Mirwaes Wahabzada, Anne-Kathrin Mahlein, Ulrike Steiner, Erich-Christian Oerke, Christoph Römer and Lutz Plümer

Abstract Modern communication, sensing, and actuator technologies as well as methods from signal processing, pattern recognition, and data mining are increasingly applied in agriculture, ultimately helping to meet the challenge of "How to feed a hungry world?" Developments such as increased mobility, wireless networks, new environmental sensors, robots, and the computational cloud put the vision of a sustainable agriculture for anybody, anytime, and anywhere within reach. Unfortunately, data-driven agriculture also presents unique computational problems in scale and interpretability: (1) Data is gathered often at massive scale, and (2) researchers and experts of complementary skills have to cooperate in order to develop models

K. Kersting (✉)
Computer Science Department, TU Dortmund University,
Otto-Hahn-Strasse 14, 44221 Dortmund, Germany
e-mail: kristian.kersting@cs.tu-dortmund.de

C. Bauckhage
B-IT, University of Bonn, and Fraunhofer IAIS, Schloss Birlinghoven,
53757 Sankt Augustin, Germany
e-mail: christian.bauckhage@iais.fraunhofer.de

M. Wahabzada · A.-K. Mahlein · U. Steiner · E.-C. Oerke
Institute for Crop Science and Resource Conservation (INRES)-Phytomedicine,
University of Bonn, Meckenheimer Allee 166a, 53115 Bonn, Germany
e-mail: mirwaes@uni-bonn.de

A.-K. Mahlein
e-mail: amahlein@uni-bonn.de

U. Steiner
e-mail: u-steiner@uni-bonn.de

E.-C. Oerke
e-mail: ec-oerke@uni-bonn.de

C. Römer · L. Plümer
Institute of Geodesy and Geoinformation, University of Bonn,
Meckenheimer Allee 172, Bonn, Germany
e-mail: roemer@igg.uni-bonn.de

L. Plümer
e-mail: pluemer@igg.uni-bonn.de

J. Lässig et al. (eds.), *Computational Sustainability*,
Studies in Computational Intelligence 645,
DOI 10.1007/978-3-319-31858-5_6

and tools for data intensive discovery that yield easy-to-interpret insights for users that are not necessarily trained computer scientists. On the problem of mining hyperspectral images to uncover spectral characteristic and dynamics of drought stressed plants, we showcase that both challenges can be met and that big data mining can—and should—play a key role for feeding the world, while enriching and transforming data mining.

1 Introduction

Facing a rapidly growing world population, answers to the daunting question of "How to feed a hungry world?" are in dire need. Challenges include climate change, water scarcity, labor shortage due to aging populations, as well as concerns as to animal welfare, food safety, changes in food consumption behavior, and environmental impact. Water scarcity is a principle global problem that causes aridity and serious crop losses in agriculture. It has been estimated that drought can cause a depreciation of crop yield up to 70 % in conjunction with other abiotic stresses [11, 48]. Climate changes and a growing human population in parallel thus call for a sincere attention to advance research on understanding of plant adaptation under drought. A deep knowledge of the adaptation process is essential in improving management practices, breeding strategies as well as engineering viable crops for a sustainable agriculture in the coming decades. Accordingly, there is a dire need for crop cultivars with high yield and strong resistance against biotic and abiotic stresses. Addressing this issue and the other ones mentioned above, agriculture –arguably the oldest economic endeavor of humankind– is receiving a technological makeover and information technology makes its appearance in the fields.

Agricultural information is gathered and distributed by means of smartphones, portable computers, GPS devices, RFID tags, and other environmental sensors. Farming companies are working on automation technologies such as GPS steering to operate tractors and other agricultural machines [5]. Aiming at increased food safety, RFID technologies are used to track animals in livestock; for example, since 2010, European sheep farmers are required to tag their flocks and the European Commission has suggested to extend this to cattle. RFID technologies also provide new possibilities for harvest asset management. For instance, by adding RFID tags, bales can be associated with measured properties such as weight and moisture level [5]. In general, mobile communication networks and technologies which are now commonly deployed in many areas around the world have become a backbone of pervasive computing in agriculture. Researchers and practitioners apply them to gather and disseminate information as well as to market products or to do business [12, 69]. As Farmers need to obtain and process financial, climatic, technical and regulatory information to manage their businesses, public and private institutions cater to their needs and provide corresponding data. For example, the U.S. Department of Agriculture, supplies information as to prices, market conditions, or newest production practices. Internet communities such as e-Agriculture allow users to exchange information, ideas, or procedures related to communication technologies in sustainable agriculture and rural development [5].

Agriculture is thus rapidly becoming a knowledge and data intensive industry. So far, however, much of the research and development in this regard has focused on sensing and networking rather than on computation. In this chapter, we survey our recent efforts on big data mining in agriculture [31, 32, 55, 65, 68]. We point out specific research challenges and opportunities—big data and reification—and hope to increase awareness of this new and exciting application domain.

2 Computational Sustainability in Agriculture

Looking at the scientific literature on precision farming, it appears that, most efforts so far were focused on the development and deployment of sensor technologies rather than on methods for data analysis tailored to agricultural measurements. In other words, up to now, contributions to computational intelligence in agriculture mainly applied off-the-shelf techniques available in software packages or libraries but did not develop specific frameworks or algorithms. Yet, efforts in this direction are noticeably increasing and in this section we survey some recent work on data mining and pattern recognition in agriculture.

Computational sustainability in agriculture involves different areas of computer- and information science. Here, we focus on key areas such as knowledge and information management, geo-information systems, and signal processing. Vernon et al. [67] highlight the importance of information systems for sustainable agriculture. While early work in this direction was focused on the design of (relational) databases, more recent approaches consider semantic web technologies for instance for pest control [42], farm management [61], or the integration of molecular and phenotypic information for breeding [7]. Others consider recommender systems and collaborative filtering to retrieve personalized agricultural information from the web [34] or the use of web mining, for instance, in localized climate prediction [15]. Geo-information processing plays a particular role in computational agriculture and precision farming. Research in this area considers mobile access to geographically aggregated crop information [35], region specific yield prediction [57], or environmental impact analysis [22]. It is clear that, in addition to information infrastructures, applications like these require advanced remote sensing or modern sensor networks. Distributed networks of temperature and moisture sensors are deployed in fields, orchards, and grazing land to monitor growth conditions or the state of pasture [12, 69]. Space- or airborne solutions make use of technologies such as Thermal Emission and Reflection Radiometers or Advanced Synthetic Aperture Radar to track land degradation [8] or to measure and predict levels of soil moisture [39]. Other agricultural applications include plant growth monitoring [38] and automated map building [59]. A particularly interesting sensing modality consists in airborne or tractor-mounted hyper-spectral imaging which records spectra of several hundred wavelengths per pixel. With respect to plant monitoring this allows, for instance, for assessing changes of pigment and chemical composition (water, starch, ligning et ist auch dabei) and information about plant architecture and leaf structure. This in turn allows for remotely measuring phenotypic and physiological reactions of plants due to biotic or abiotic stress [43]. Recently, hyper-spectral imaging is being increasingly used for near range plant monitoring in agricultural research. It enables basic research, for example regarding the molecular mechanisms of photosynthesis [50, 51], but is also used in plant phenotyping, for instance as an approach towards understanding phenotypic expressions of drought stress [4, 33, 55]. Classical image analysis and computer vision techniques are being used in agriculture, too. Examples include automated inspection and sorting in agricultural production facilities [36, 53], the detection of the activity of pests in greenhouses [6], or the recognition of plant diseases [45, 58]. Finally, artificial intelligence techniques are increasingly applied to address questions of *computational sustainability* [24]. Work in this area considers algorithmic approaches towards maximizing the utility of land [23], enabling sustainable water resource management [47], or learning of timber harvesting policies [17].

3 Plant Phenotyping: A Big Data and Reification Challenge

A common theme of the work just reviewed is that they require algorithms and architectures that can cope with massive amounts of data. Owing to the increased use of modern sensors, corresponding solutions have to cope with exploding amounts data recorded in dynamic and uncertain environments where there typically are many interacting components [24]. However, it appears that most work in this area so far did not involve specifically trained data scientists and that, from the point of view of computational intelligence, more efficient and accurate methods seem available. Yet, computer scientists entering the field must be aware that methods they bring have to benefit researchers and practitioners in agriculture. Practitioners “out in the fields” are in need of methods and tools that yield results with concrete connection to running mobile devices, i.e. under constrained computational resources, and in real time in order to help them in their daily work. From the perspective of farming professionals, problems are natural and real phenomena that may be addressed using scientific methods and advanced computing. To them purely theoretic concepts or mathematical abstractions are of little use. The world’s food producers are highly technology-oriented people but with a purpose. Even if they may not be adequately trained in information technology, they actually do not need to be. They know their business and if a new technology does not fit into their work flows they will either ignore it or wait until it meets their needs. In the next sections, we present our approaches to meet these big data and reification challenges in plant phenotyping. We shall use these examples to underline the above challenges and to illustrate practical solutions for large-scale data. More precisely, we present results from an ongoing efforts on recognizing and predicting levels of drought stress in plants based on the analysis of hyper-spectral images. There are estimates that drought in conjunction with other abiotic stresses causes a depreciation of crop yields of up to 70 % [11, 48]. Because of global warming this trend is expected to increase, so that an improved understanding of how plants adapt to drought is called for to be able to breed more resistant varieties. Yet, mechanisms of stress resistance are characterized by complex interactions between the genotype and the environment which lead to different phenotypic expressions [46]. Progress has been made towards the genetic basis of drought related traits [37, 41] and modern data analysis has lead to molecular insights into drought tolerance [1, 27, 49]. However, as genetic and biochemical research are time consuming and only moderately successful in predicting the performance of new lines in the field, there are increased efforts on phenomic approaches.

Hyper-spectral imaging provides an auspicious approach to plant phenotyping [50, 51]. In contrast to conventional cameras, which record only 3 wavelengths per pixel, hyper-spectral cameras record a spectrum of several hundred wavelengths ranging from approximately 400 nm to 2500 nm (see Fig. 1). These spectra contain information as to changes of the pigment composition of leafs which are the result of metabolic processes involved in plant responses to stress. Supervised classification of hyper-spectral signatures can thus be used to predict biotic stress before symptoms become visible to the human eye [54, 56].

However, scale poses a significant challenge in hyper-spectral image analysis, since the amount of phenotyping data easily grows into TeraBytes if several plants are monitored over time. For instance, each individual hypers-spectral recording considered below consists of a total of about 2 (resp. 5.8) Billion matrix entries. Manually labeling such data as well as running established supervised classification algorithms therefore quickly becomes infeasible. Thus, the main goal of plant phenotyping—*the identification of phenotypic features and complex traits which are relevant for stress resistance and to understand the underlying causal networks in the interaction between genotype and environment*—poses important and challenging problems for big data mining: *easy-to-interpret, (un)supervised data mining solutions for massive and high dimensional data over time that scale at most linearly with the amount of data*. This requirement makes it difficult—if not impossible—to use prominent

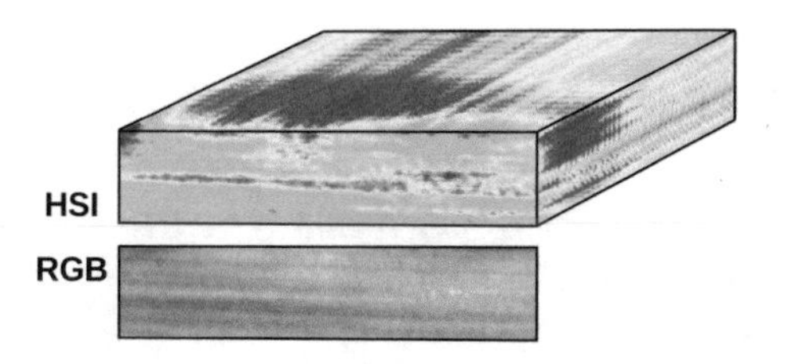

Fig. 1 While conventional RGB images record only three color values (*red*, *green*, and *blue*) per pixel, each pixel of a hyper-spectral image records how a whole spectrum of visible or invisible light waves is reflected from a scene. (best viewed in color)

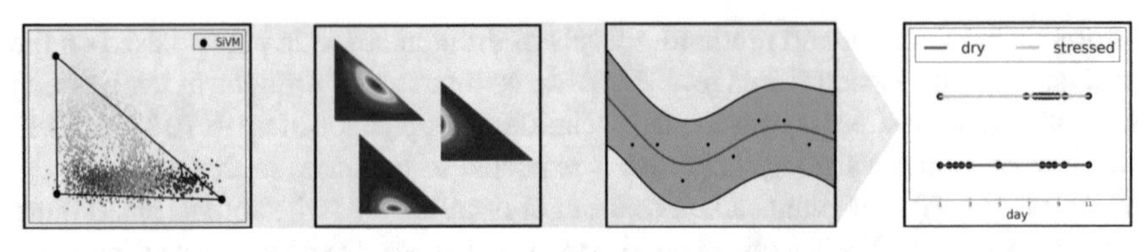

Fig. 2 Pipeline for generating interpretable summaries from hyper-spectral imaging data. (Best viewed in color). **a** Simplex Volume Maximization. **b** Distribution on the Simplex. **c** Dirichlet Aggregation Regression. **d** Summary of Dynamics

statistical classification or clustering techniques such as SVD, kMeans, (convex) NMF, NMF with volume constraints (see e.g. [3, 44, 60] and references in there), and SVMs, which typically scale at least quadratically with the amount of data if no form of approximation is used that is often accompanied by information loss, may require label information, and/or do not provide easy-to-interpret features/models; they typically produce features that are mathematical abstractions computable for any data matrix. As Mahoney and Drineas argue, "they are not 'things' with a 'physical' reality" [40] and consequently it might be difficult—if not impossible—to provide the plant physiological meaning of, say, an eigenvector.

Addressing these challenges, we have developed novel data mining methods for plant phenotyping that we will review in this chapter. They make only weak assumption on the generating distribution of observed signatures. For instance, one key ingredient is a recent linear time, data-driven matrix factorization approach to represent hyper-spectral signatures by means of convex combinations of only few extreme data samples. Practical results show that the resulting pipeline can predict the level of drought stress of plants well and in turn stress before it even becomes visible to the human eye. Moreover, it can provide an abstract and interpretable view on drought stress progression. In the following we will go through the main ingredients of our pipeline as illustrated in Fig. 2: (1) detecting informative hyper-spectral signatures, (2) modeling distributions over these signatures, (3) smoothing and predicting the evolution of these distributions over time, and (4) summarizing these dynamics using graphical sketches.

4 Interpretable Factorization of Hyper-Spectral Images

Scientists working on plant phenotyping regularly need to find meaningful patterns in massive, high dimensional and temporally diverse observations. For instance, in one of our projects hyper-spectral data of resolution $640 \times 640 \times 69$ were taken of 10 (resp. 12) plants l at 7 (resp. 20) days t. Each record can thus be viewed as a data matrix $\mathbf{X}^{t,l} \in \mathbb{R}^{m \times n}$ with $m = 640 \times 640$ and $n = 69$. Horizontally stacking the data

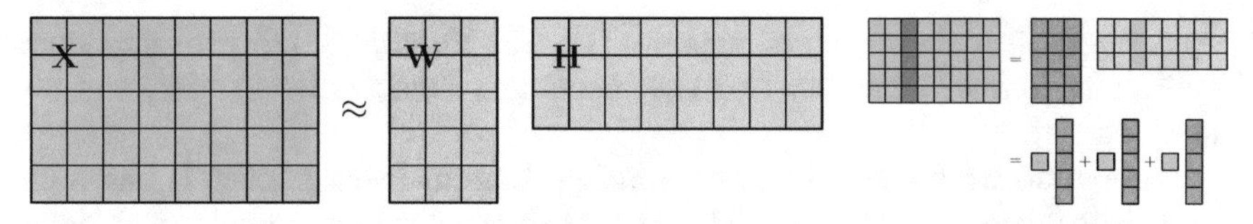

Fig. 3 Data analysis using matrix factorization. (*Left*) Given an integer $k \leq \min\{m, n\}$, two factor matrices $\mathbf{W} \in \mathbb{R}^{m \times k}$ and $\mathbf{H} \in \mathbb{R}^{k \times n}$ are determined such that $\mathbf{X} \approx \mathbf{WH}$. (*Right*) Doing so can be viewed as latent factor analysis respectively dimensionality reduction. For each $\mathbf{x}_j \in \mathbb{R}^m$, there is a $\mathbf{h}_j \in \mathbb{R}^k$ expressing $\mathbf{x}_j$ in terms of the found latent factors $\mathbf{W} = [\mathbf{w}_1, \mathbf{x}_2, \ldots, \mathbf{w}_k]$. (Best viewed in color)

Algorithm 2 Interpretable matrix factorization

Input: Matrix $\mathbf{X} \in \mathrm{R}^{m \times n}$, integer c

1 Select c columns from $\mathbf{X}$ and construct $\mathbf{W} \in \mathbb{R}^{m \times c}$ Compute reconstruction matrix $\mathbf{H} \in \mathbb{R}^{c \times n}$ such that the Frobenius norm $||\mathbf{X} - \mathbf{HW}||$ is minimized with respect to $\mathbf{1}^t \mathbf{h}_j = 1$, i.e., all rows of $\mathbf{H}$ sum to one **Return W** and **H** with $\mathbf{X} \approx \mathbf{WH}$

matrices recorded in all experiment then results in a single matrix $\mathbf{X}$ with about 2 (resp. 5.8) Billion entries. Matrix factorization is commonly used to analyze such data. As illustrated in Fig. 3, it factorizes of a matrix $\mathbf{X}$ into a product of (usually) two matrices. That is, $\mathbf{X}$ is approximated as $\mathbf{X} \approx \mathbf{WH}$ where the matrix of basis elements $\mathbf{W} \in \mathbb{R}^{m \times k}$, the coefficient matrix $\mathbf{H} \in \mathbb{R}^{k \times n}$, and $k \ll \min\{m, n\}$. It's useful to think of each column vector in $\mathbf{W}$ as a kind of discoveredthe original data matrix $\mathbf{X}$. A column in $\mathbf{H}$ represents an original data point in terms of the discovered features. This allows for mapping high dimension al data $\mathbf{X}$ to a lower dimensional representation $\mathbf{H}$ and can thus mitigate effects due to noise, uncover latent relations, or facilitate further processing and ultimately help finding patterns in data.

A well known low-rank approximation approach consists in truncating the Singular Value Decomposition (SVD), which expresses the data in terms of linear combinations of the top singular vectors. While these basis vectors are optimal in a statistical sense, the SVD has been criticized for it less faithful to the nature of the data at hand. For instance, if the data are sparse the compressed representations are usually dense which leads to inefficient representations. Or, if the data consists entirely of non-negative vectors, there is no guarantee for an SVD-based low-dimensional embedding to maintain non-negativity. However, the data mining practitioner—as in our application—often tends to assign a "physical" meaning to the resulting singular components.

A way of circumventing these problems, which also hold for other classical techniques such as NMF, kMeans, and sub-space clustering, consists in computing low-rank approximations from selected columns of a data matrix [26] as sketched in the vanilla "interpretable" matrix factorization Algorithm 2. Corresponding approaches yield naturally interpretable results, since they embed the data in lower dimensional spaces whose basis vectors correspond to actual data points. They are guaranteed to preserve properties such as sparseness or non-negativity and enjoy increasing popularity in the data mining community [19, 20, 29, 40, 63, 66] with important applications to fraud detection, fMRI segmentation, collaborative filtering, and co-clustering.

But how do we select columns in Line 1? A prominent approach is based on the statistical leverage score [40, 63]. We first compute the top-k right/left singular vectors $\mathbf{V}^{k\times n}$ of $\mathbf{X}$. Then, the statistical leverage score π_i for a particular column i is computed by summing over the rows of the singular vectors, i.e. $\pi_i = \frac{1}{k}\sum_{j=1}^{k} v_{j,i}^2$. The scores π_i form a probability distribution over the columns, and we essentially select columns using that score as an importance sampling probability distribution. Thereby, columns which capture the dominant part of the spectrum of $\mathbf{X}$ are preferred and assigned a higher importance score/probability, cf. [40]. As an alternative, it was shown that a *good* subset of columns maximize their volume [13, 25]. That is we maximize the volume of the parallelepiped (the value of the determinant $\det \mathbf{W}$) spanned by the columns of $\mathbf{W}$. Given a matrix $\mathbf{X}^{m\times n}$, we select c of its columns s.t. the volume $Vol(\mathbf{W}^{m\times c}) = |\det \mathbf{W}|$ is maximized, where $\mathbf{W}^{m\times c}$ contains the selected columns. The criterion, however, is provably NP-hard [13]. Thurau et al. [66] introduced recently an approximation, called Simplex Volume Maximization and illustrated in Fig. 4, that was empirically proven to be quite successful. For a subset $\mathbf{W}$ of c columns from $\mathbf{X}$, let $\Delta(\mathbf{W})$ denote the $c-1$-dimensional simplex formed by the columns in $\mathbf{W}$. Now, the volume of the c-simplex $\text{Vol}(\Delta(\mathbf{W}))$ is $\text{Vol}(\Delta(\mathbf{W}))_c^2 = \theta \mathbf{det} \mathbf{A}$ where $\theta = \frac{-1^{c+1}}{2^c (c!)^2}$ and $\det \mathbf{A}$ is the so-called *Cayley-Menger* determinant [9] that essentially only involves the squared distance $\text{d}_{i,j}^2$ between the vertices i and j (or columns i and j of $\mathbf{W}$), see [9, 66] for details. As Thurau et al. have shown, since the distance geometric formulations is entirely based on vector norms and edge lengths, it allows for the development of an efficient greedy algorithm.

Specifically, finding a globally optimal subset $\mathbf{W}$ that maximizes the volume requires the computation of all pairwise distances among the columns in $\mathbf{W}$. For large data sets, this is ill-advised as it scales quadratically with the number n of data points. To arrive at an iterative, approximative $O(cn)$ procedure, Thurau et al. proceed greedily: Given a simplex S consisting of $k-1$ vertices, we seek a new

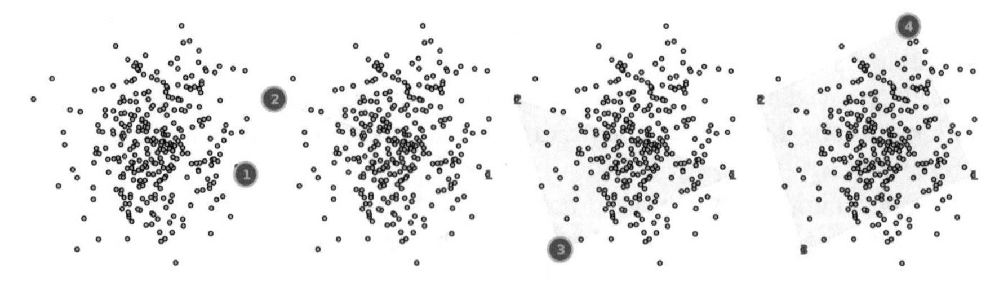

Fig. 4 (From *left* to *right*) Didactic example of how Simplex Volume Maximization iteratively determines basis vectors for representation of a data sample by means of convex combinations. (Best viewed in color)

Algorithm 3 Simplex Volume Maximization (SiVM) as introduced in [65].

Input: Matrix $\mathbf{X} \in \mathbb{R}^{m \times n}$, integer c, interger l
2 Randomly select r from $1, 2, \ldots, n$ $\mathbf{z} = \arg\max_j d(\mathbf{X}_{*,r}, \mathbf{X}_{*,j})$ **for** $j = 1 \ldots n$ **do**
3 | $p_j \leftarrow \log(d(\mathbf{z}, \mathbf{X}_{*,j}))$; $\Phi_{0,j} \leftarrow n_j$; $\Lambda_{0,j} \leftarrow n_j^2$; $\Psi_{0,j} \leftarrow 0$
4 $\mathrm{a} = \max_j(p_j)$ **for** $i = 2 \ldots c$ **do**
5 | **for** $j = 1 \ldots n$ **do**
6 | | $p_j \leftarrow \log(d(\mathbf{w}_{i-1}, \mathbf{X}_{*,j}))$; $\Phi_{i,j} \leftarrow \Phi_{i-1,j} + p_j$; $\Lambda_{i,j} \leftarrow \Lambda_{i-1,j} + p_j^2$ $\Psi_{i,j} \leftarrow \Psi_{i-1,j} + p_j * \Phi_{i-1}$; $p_j \leftarrow \mathrm{a} * \Phi_{i,j} + \Psi_{i,j} - \frac{(i-1)}{2} \Lambda_{i,j}$
7 | select $= \arg\max_j\{p_j\}$ $\mathbf{w}_i = \mathbf{X}_{*,\mathrm{select}}$ $\mathbf{W}_{*,i} = \mathbf{X}_{*,\mathrm{select}}$
8 **Return** $\mathbf{W} \in \mathbb{R}^{m \times k}$

vertex $\mathbf{x}_\pi \in \mathbf{X}$ such that $\mathbf{x}_\pi = \arg\max_k \; \mathrm{Vol}(S \cup \mathbf{x}_k)^2$. Thurau et al. have shown that this leads to the following heuristic

$$\mathbf{v}_\pi = \arg\max_k \Bigg(\log(a) \sum_{i=1}^{n} \log(\mathrm{d}_{i,k}) + \sum_{\substack{i=1 \\ j=i+1}}^{n} \log(\mathrm{d}_{i,k}) \log(\mathrm{d}_{j,k}) - \frac{n-1}{2} \sum_{i=1}^{n} \log^2(\mathrm{d}_{i,k})\Bigg).$$

that locally increases the volume of the simplex in each iteration. Simplex Volume Maximization (SiVM) is illustrated in Fig. 4 and summarized in Algorithm 3. For the first data point to select, we simply take the two points, which are most likely furthest away from each other. In later iterations, we select points in lines 9-18. Pairwise distances computed in one iterations can be reused in later iterations so that, for retrieving c columns, we need to compute distances from the last selected column to all other data points exactly $c + 1$ times. As c is constant, we have an overall running time of $\mathcal{O}(n)$. Finally, we note that SiVM is more efficient than other deterministic methods as it supersedes the need for expensive projections of the data. Nevertheless, it aims for solutions that are similar to a greedy algorithm due to Civril and Magdon-Ismail [14] as the projection and orthogonality constraint is implicitly part of the distance geometric objective function.

To summarize, SiVM can be used for selecting columns in Line 1 of Algorithm 2 in $\mathcal{O}(cn)$ time. Then we compute the coefficients $\mathbf{H}$ in Line 2 by solving constrained quadratic programs [10]. This is $\mathcal{O}(cn) = \mathcal{O}(n)$ since c is a constant. Thus, applied to a data matrix $\mathbf{X}$ resulting from a stacked set of hyper-spectral images, SiVM can be used to detect few representative hyper-spectral signatures $\mathbf{W}$ in time linear of the number of hyper-spectral signatures. This fast plant phenotyping using SiVM is illustrated in Fig. 5. Next to exhaustive lab experiments, field experiments have shown that it can distinguish subtle differences of crop traits in the field [55]. The key to this success is that SiVM paves the way to statistical machine learning.

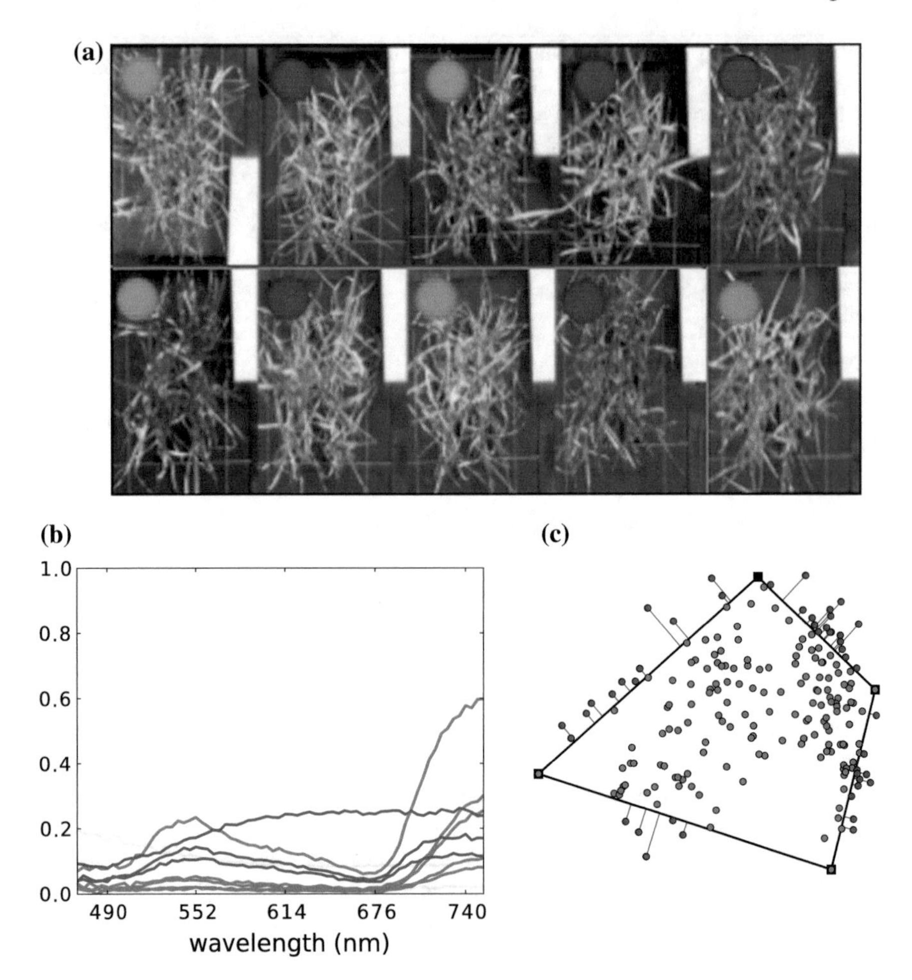

Fig. 5 Fast plant phenotyping using SiVM. From *left* to *right*: (**a**) actual examples of RGB images of plants on the fourth measurement day; corresponding hyper-spectral images were recorded from the same point of view and under the same conditions; (**b**) actual examples of different extreme high-dimensional spectra determined within the hyper-spectral recordings; each of these spectra corresponds to a hyper-spectral pixel and shows the fraction of light reflected at different wavelength; the automatically determined extreme spectra belong to images of "dry" (*red*) and "healthy" (*green*) leafs; it is noticeable that dry and healthy plants are not necessarily distinguishable from looking at the RGB images in (**a**); (**c**) didactic example of how any sample point can be expressed as a convex combination of selected extremes (see previous figure); while points inside of the convex hull of selected basis elements can be reconstructed exactly, points on the outside are approximated by their projection onto its closest facet; (best viewed in color)

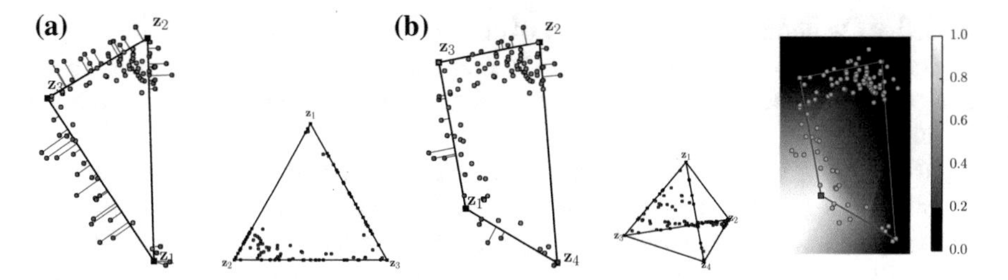

Fig. 6 Bridging geometry and probability. The coefficient vectors $\mathbf{h}_j$ are stochastic, i.e., they sum to one. Hence the coefficient h_{ij} can be thought of as $p(\mathbf{x}_j|\mathbf{w}_i)$ as shown for $k = 4$ (the darker the less probable). (best viewed in color). **a** $k = 3$. **b** $k = 4$

5 From Geometry to Probability: Densities over Signatures

From a geometric point of view, as illustrated in Fig. 6, the columns $\mathbf{h}_1, \ldots, \mathbf{h}_n$ of $\mathbf{H}$ are data points (signatures) that reside in a simplex spanned by the extreme elements in $\mathbf{W}$. On this simplex spanned by the extremes, there are natural parametric distributions to characterize the density of the $\mathbf{h}_i$. The best known one is the Dirichlet

$$\mathscr{D}(\mathbf{h}_i|\boldsymbol{\alpha}) = B(\boldsymbol{\alpha}) \prod\nolimits_{j=1}^{c} h_{ij}^{\alpha_j - 1} \tag{1}$$

where $\boldsymbol{\alpha} = (\alpha_1, \alpha_2, \ldots, \alpha_c)$. The normalization constant $B(\boldsymbol{\alpha}) = \Gamma(S(\boldsymbol{\alpha}))/\prod_{j=1}^{c} \Gamma(\alpha_j)$ where $\Gamma(\cdot)$ is the gamma function and $S(\boldsymbol{\alpha}) = \sum_{j=1}^{c} \alpha_j$. This distributional view on hyper-spectral data provides an intuitive measure of e.g. the drought stress: the expected probability of observing a healthy spot, which we call the "drought stress level" of a plant. This "drought stress level" is a distribution. To see this, given $\boldsymbol{\alpha}$, we note that the marginal distribution of the j-th reconstruction dimension follows a Beta distribution $\mathscr{D}(\alpha_j, S(\boldsymbol{\alpha}) - \alpha_j)$ and the expected value of the j-th reconstruction dimension is $\mu_j = \alpha_j / S(\boldsymbol{\alpha})$. Thus, each α_j controls "aggregation" of mass of reconstructions near the corresponding column c_j which explains the term *Dirichlet aggregation*. Now assume that each dimension was labeled either "background", "healthy", or "dry". Averaging the expected values of "healthy" or "dry" dimensions and treating them as parameters of a Beta distribution yields the drought stress level of a plant. As shown in [55], this can be used to detect drought stress up to 1.5 weeks earlier than by the naked eye.

An alternative to the Dirichlet is the log-normal distribution as e.g. proposed by Aitchison [2] or so-called logratio transformations also due to Aitchison. The latter transform the reconstructions from the simplex sample space to the Euclidean space. In the transformed space, we can then use any standard multivariate method such as estimating multivariate Gaussian distributions Normal $(\boldsymbol{\mu}, \boldsymbol{\Sigma})$ with mean $\boldsymbol{\mu}$ and covariance matrix $\boldsymbol{\Sigma}$. Doing soalternative. Under a Dirichlet, the components of the proportions vector are nearly independent. This leads to the strong and unrealistic modeling assumption that the presence of one extreme point is not correlated with the presence of another. A logratio transformation together with Gaussian distributions overcome this problem and may provide a richer view on the interactions of selected columns.

In any case, what do we gain by having distributions on the simplex induced by SiVM? In general, it opens the door to statistical data mining at massive scale. For instance, one could embed the hyper-spectral images into a low-dimensional Euclidean space. This yields easy-to-interpret representations of the relationships among the images and in turn among the plants (over time). Probably the most classical examples are multidimensional scaling (MDS) [16] and IsoMap [64]. Whereas MDS uses pairwise distances $D_{i,j}$ only to find an embedding that preserves the inter-point distances, IsoMap first creates a graph G by connecting each object to l of its neighbors, and then uses distances of paths in the graph for embedding using MDS. For plant phenotyping with images over time, one can strike a middle ground taking the temporal relations among plant images into account. The main step, however, is to compute distance metric among the densities Dirichlets. To do so, one could for instance employ the Bhattacharyya distance that is commonly used in data

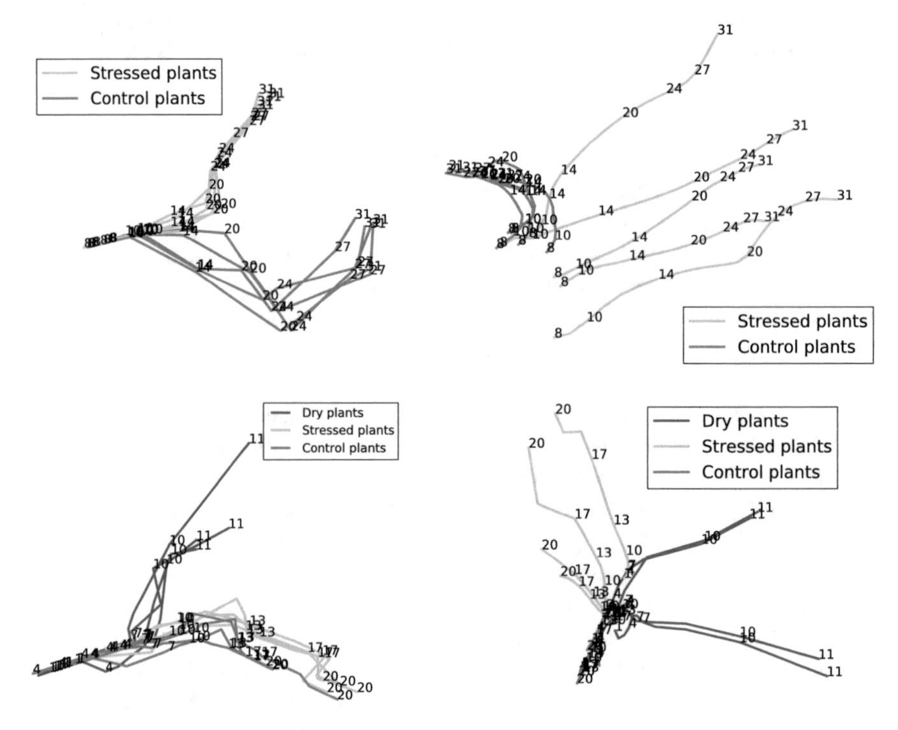

Fig. 7 Improved drought stress detection using DAR. From *left* to *right*: (1) Dirichlet traces for 2010 (two groups of measurements) without and (2) with DAR smoothing. (3) Dirichlet traces for 2011 (three groups of measurements) without and (4) with DAR smoothing. Colors indicate controlled/stressed plants; numbers denote the measurement days

mining to measure the similarity of two probability distributions [30]. It is computed by integrating the square root of the product of two distributions. Figure 7 shows Euclidean embeddings of hyper-spectral images computed using the Bhattacharyya distance between the induced Dirichlet distributions. Since the images were taken over time, we call this Dirichlet traces. Moreover, as we will show next, we can use more advanced machine learning methods to smooth and even predict the drought levels over time.

6 Pre-symptomatic Prediction of Plant Drought Stress

In order to track drought levels over time, we apply Dirichlet-aggregation regression (DAR) as proposed in [31]. We first select extreme columns from the overall data matrix **X**. This captures global dependencies as we represent the complete data by means of convex combinations extreme data points selected across all time steps. Then, on the simplex spanned by the extreme points, we estimate Dirichlet distributions specified by $\boldsymbol{\alpha}^{t,l}$ over all reconstructions per day t and plant l. This captures local dependencies. Finally, DAR puts a Gaussian process prior on these local Dirichlet distributions. The prior can be a function of any arbitrary types of observed continuous, discrete and categorical features such as time, location, fertilization, and plant species with no additional coding, yet inference remains relatively simple. More precisely, DAR iterates the following steps until convergence:

1. Optimize the logarithm of the complete likelihood w.r.t. the hidden Dirichlet aggregations $\boldsymbol{\alpha}^{t,l}$ for each plant l.
2. Optimize the log-likelihood of all plants w.r.t. the hyper-parameters ϑ of a common Gaussian process prior.

For more technical details, we refer to [32].

This non-parametric Bayesian approach can be used for smoothing the estimated drought level. Figure 8 shows drought levels estimated by DAR averaged over groups of plants in two data sets considered in our project. As one can see, DAR nicely smoothes SiVM's "hard" drought level (shown as dots). Having a Bayesian regression model at hand, however, we can also move on to make predictions. To do so, we iteratively obtain predictions by making repeated one-step ahead predictions, up to the desired horizon. For the one-step ahead prediction at time t^*, we apply standard Gaussian process regression [52]. For the multiple-step ahead prediction task we follow the method proposed in [21]. That is, we predict the next time step using the estimate of the output of the current prediction as well as previous outputs (up to some lag U) as input, until the prediction k steps ahead is made. Thus, the prediction k steps ahead is a random vector with mean formed by the predicted means of the lagged outputs.

To summarize, based on hyper-spectral images, drought stress levels of plants can be predicted as follows: (1) Using SiVM, we compute few extreme signatures, say 50, and label them accordingly. (2) On the simplex spanned by these extremes, we

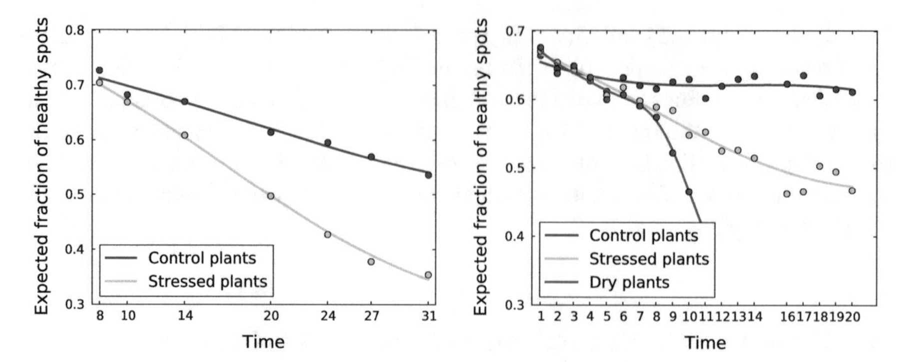

Fig. 8 Dirichlet-aggregation regression (DAR) of drought levels over several days in 2010 (*left*) and in 2011 (*right*) using all hyper-spectral images available. Colors indicate controlled/stressed plants. While the x-axis indicates measurement days, the y-axis indicates the fraction of pixels in the analyzed hyper-spectral images that show healthy parts or spots of a plant predicted from our DAR model. Note that experiments in agricultural research cannot seamlessly be repeated at any time but have to adhere to seasonal growth cycles of plants. Accordingly, data not recorded in an experiment may not be available until a year later. In this example, in the experiments in 2010, plants were watered or stressed but not deliberately dried out. In the experiments in 2011, a third set of data was recorded from dry plants (cf. the experimental procedure in Sect. 9)

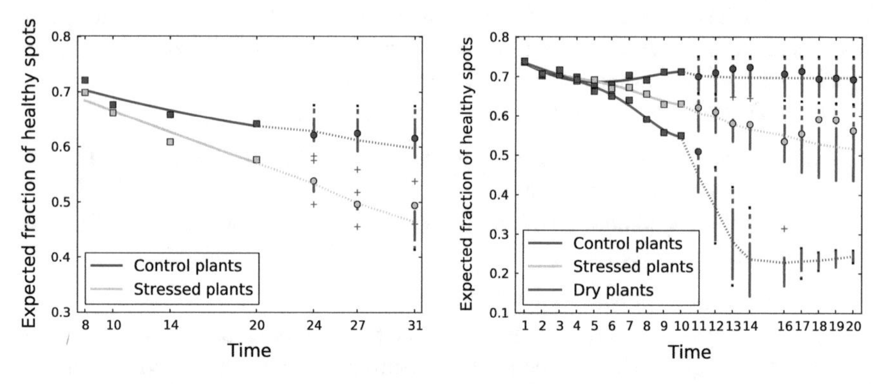

Fig. 9 Bayesian drought level predictions (over time indicated in days) for 2010 (*left*) and 2011 (*right*). While the x-axis indicates measurement days, the y-axis indicates the fraction of pixels in the analyzed hyper-spectral images that show healthy parts or spots of a plant. In both experiments, the drought levels of the second half of measurement days were predicted based on a DAR model (including the extraction of extreme spectra) obtained from the data gathered in the first half of measurement days. Colors indicate controlled/stressed plants. Again, for the measurements carried out in 2010, a control group of dry plants was not available

estimate the latent Dirichlet aggregation values per plant and time step using DAR. (3) Using the Gaussian process over the latent Dirichlet aggregation values, we compute the drought levels of each plant and time step using the labels of extreme spectra, i.e. "background", "healthy", and "dry". (4) Finally, we predict drought levels multiple steps ahead in time using the above Gaussian process approach. Figure 9 illustrates that this prediction can work well.

7 Sketching Drought Stress Progression

Indeed, one may argue that using non-parametric Bayesian machine learning contradicts the reification challenge. Practitioners are not necessarily trained statisticians or data scientists and hence may not be comfortable dealing them. However, as we will show now, one can compute easy-to-interpret summaries of them.

To create a single sketch describing the hyperspectral dynamics of stressed plants, we advocate the "Equally-Variance Bin Packing" (EBP) decomposition, where we are essentially motivated by the sequential bin-packing problem, a version of the classical bin-packing problem in which the objects are received one by one [28]. However, since we are in a batch[1] setting, we actually face a much simpler instance of the problem, actually with a linear time complexity: *we are looking for a segmentation of ordered objects in B equally weighted bins which preserves the original ordering of the objects.*

Given a matrix $\mathbf{X} \in \mathbb{R}^{K \times N}$ where the columns denote the hyperspectral signatures representing different stages of diseases progression, we can achieve and "Equally Variance Bin Packing" decomposition in B bins as follows: First, we compute the distances of consecutive spectra (columns) using Euclidean distance and compute the average bin size as

$$\delta = \frac{1}{B} \sum\nolimits_{i=1}^{N-1} d_{i(i+1)}(\mathbf{X}) \text{ where } d_{ij}(\mathbf{X}) = \sqrt{\sum\nolimits_{k=1}^{K} (x_i^k - x_j^k)^2} \tag{2}$$

is the Euclidean distance. Then we fill the $B - 1$ bins successively with the objects according to the bin size δ. The last bin is filled with the remaining objects.

This decomposition can be used to draw a single sketch, where the nodes denote the begin (resp. end) of an period and the length of the edge e_b between two consecutive nodes v_b and v_{b+1} is set relatively to the length of the period covered by the objects in bin b. An example of the resulting single sketches are shown in Fig. 10 (left) for drought stressed and actually drying out plants. Each sketch highlight the interesting periods, where a small edge (between two points) denotes a period of high impact (change in hyperspectral signature).

To illustrate the generality of this sketching approach, we additionally computed sketches on a financial dataset. More precisely, inspired by Doyle and Elkan [18], we applied our approach to financial data to obtain a alternative view of economic networks than that supplied by traditional economic statistics. The financial crisis 2008/2009 illustrates a critical need for new and fundamental understandings of the structure and dynamics of economic networks [62]. We computed sketches for the stock price changes of industrial sectors from the S&P 500 as listed on Yahoo! Financial. Specifically, our dataset consists of about 10 years worth of trading data from January 2000 to January 2011. The price of a stock may rise or fall by some percent on each day. We recorded the daily ups and downs for about 3 consecutive

[1] In the long run, when plants are monitored over month, the online setting will be relevant.

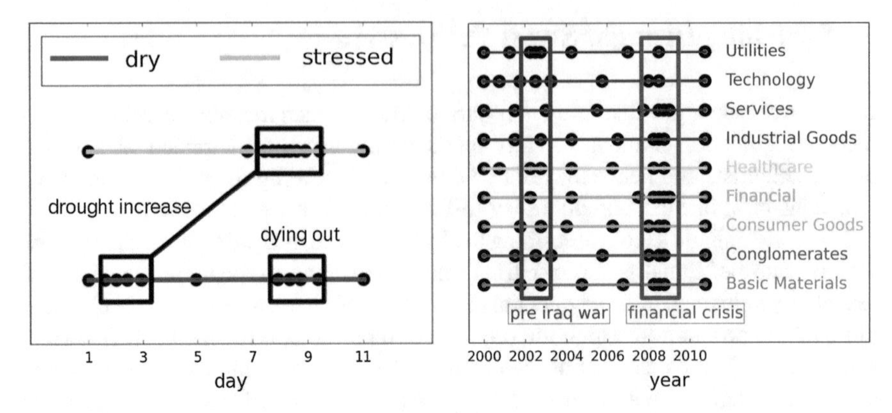

Fig. 10 (*Left*) Sketching the progression of drought stress over time. Each sketch highlight the interesting periods, where a small edge (between two points) denotes a period of high impact (change in hyperspectral signature). (*Right*) *Financial sketches* during the financial crises in 2008/2009 we have small periods of high impact on many sectors indicating a huge change in stock market prices. (Best viewed in color)

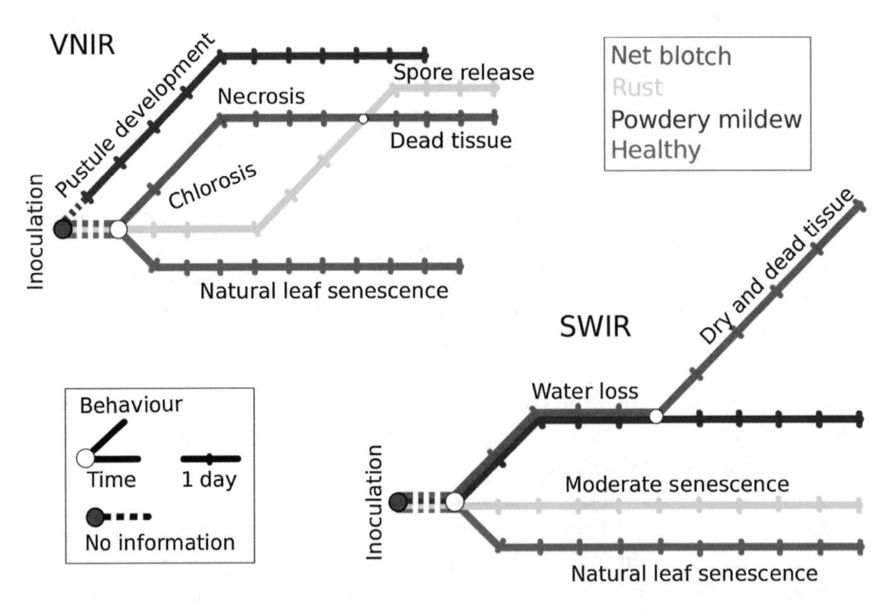

Fig. 11 Collective disease progression via Metro Maps of hyperspectral dynamics of diseased plants for visible-near infrared (VNIR) (*top*) and short-wave infrared (SWIR) wavelengths (bottom) taken from [68]. Each disease track from hyperspectral images exhibits a specific route in the metro map, the direction and the dynamic steps are in correspondence to biophysical and biochemical processes during disease development. The beginning of all routes is at the same time point/train station (day of inoculation, *gray circle*). (Best viewed in color)

months (columns) for all stocks (rows) into a single data matrix per sector. Then we proceed as for the drought stress data. The sketches produced are shown in Fig. 10 (right). As one can see, for the financial crises in 2008/2009 we have small periods of high impact for many of the sectors, indicating a huge change in stock market prices.

Finally, as demonstrated in [68], the sketches can be extended to compute structured summaries of collective phenomena that are inspired by metro maps, i.e. schematic diagrams of public transport networks. Applied on a data set of barley leaves (Hordeum vulgare) diseased with foliar plant pathogens Pyrenophora teres, Puccinia hordei and Blumeria graminis hordei, the resulting metro maps of plant disease dynamics as shown in Fig. 11 conform to plant physiological knowledge and explicitly illustrate the interaction between diseases and plants. Most importantly, they provide an abstract and interpretable view on plant disease progression.

8 Conclusion

Agriculture, the oldest economic venture in the history of humankind, is currently undergoing yet another technological revolution. Sparked by issues pertaining to sustainability, climate change, and growing populations, solutions for precision farming are increasingly sought for and deployed in agricultural research and practice. From the point of view of pattern recognition and data mining, the major challenges in agricultural applications appear to be the following:

1. The widespread deployment and ease of use of modern, (mobile) sensor technologies leads to exploding amounts of data. This poses problems throughputcomputation. Algorithms and frameworks for data management and analysis need to be developed that can easily cope with TeraBytes of data.
2. Since agriculture is a truly interdisciplinary venture whose practitioners are not necessarily trained statisticians or data scientists, techniques for data analysis need to deliver interpretable and understandable results.
3. Mobile computing for applications "out in the fields" has to cope with resource constraints such as restricted battery life, low computational power, or limited bandwidths for data transfer. Algorithms intended for mobile signal processing and analysis need to address these constraints.

In this chapter, we illustrated the first two challenges. More precisely, we considered the problems of drought stress recognition, prediction and summarization for plant phenotyping from hyper-spectral imaging. We presented algorithmic solutions that cope with TeraBytes of sensor recording and deliver useful, i.e. biologically plausible, and interpretable results. In particular, our approach was based on a distributional view of hyper-spectral signatures which we used for Bayesian prediction of the development of drought stress levels. Prediction models of this kind have great potential as they provide better insights into early stress reactions and to identify the most relevant moment when biologists or farmers have to gather samples for invasive, molecular examinations. Moreover, as we have illustrated, even the complex statistical machine learning models can be summarized into easy-to-understand sketches.

In conclusion, the problem of high-throughput phenotyping shows that methods from the broad field of artificial intelligence, in particular from data mining and pattern recognition, can contribute to solving problems due to water shortage or pests. Together with other contributions in the growing field of computational sustainability [24], it thus appears that developments such as mobile, wireless and positioning networks, new environmental sensors, and novel computational intelligence methods do have the potential of contributing to the vision of a sustainable agriculture for the 21st century: the dream of big data feeding a hungry world seems not to be insurmountable.

9 Experimental Setup

Data sets: In the experimental results reported here (that are actually taken from the corresponding references), we considered two sets of hyper-spectral images. Both data sets were recorded under semi-natural conditions in rain-out shelters at the experimental station of the University of Bonn. For the controlled water stress, three barley summer cultivars Scarlett, Wiebke, and Barke were chosen. The seeds were sown in 11.5 liter pots filled with 17.5 kg of substrate Terrasoil. In 2010 (first data set) the genotype Scarlett was used in *two* treatments (well-watered and with reduced water) with 6 pots per treatment. In 2011 (second data set) the genotypes Wiebke and Barke were used in pot experiments arranged in a randomized complete block design with *three* treatments (well-watered and two drought stressed) with 4 pots per genotype and treatment. The drought stress was induced either by reducing the total amount of water or by completely withholding water. In both cases, the stress was started at developmental stage BBCH31. By reducing the irrigation, the water potential of the substrate remained at the same level as in the well-watered pots for the first seven days but decreased rapidly in the following 10 days down to 40 % of the control. For the measurements, the plants were transferred to the laboratory and illumination was provided by 6 halogen lamps fixed at a distance of 1.6 meters from the support where the pots where placed to record hyper-spectral pictures. These were obtained using the Surface Optics Corp. SOC-700 which records images of 640 pixels x 640 pixels with a spectral resolution of approximately 4 nm with up to 120 equally distributed bands in the range between 400 and 900 nm. In 2010, images were taken at 10 time-points, twice per week starting from day four of water-stress. This provided 70 data cubes of resolution $640 \times 640 \times 69$. We transformed each cube into a dense pixel by spectrum matrix. Stacking them horizontally resulted in a dense data matrix with about 2 Billion entries. In 2011 images were taken every consecutive day starting at the second day of watering reduction. Images were taken at 11 time-points for the non-irrigated plants and at 20 time-points for plants with reduced water amount. Applying the same procedure as for the data from 2010 resulted in a matrix of about 5.8 Billion entries.

Analysis Setup: Where required, we split the data from 2011 (resp. 2010) into a first half, denoted 2011.A (resp. 2010.A) and a second half, denoted as 2011.B (resp. 2010.B). Then, we extracted 50 extreme signatures from 2011.A (resp. 2010.A) and determined a DAR regression model on 2011.A (resp. 2010.A). We labeled the extreme signatures as "healthy", "dry", and "background", computed drought levels for 2011.A (resp. 2010.A) based on the DAR model, and used them to predict the drought levels for 2011.B (resp. 2010.B). We also considered the complete 2010 (resp. 2011) data to determine a corresponding DAR model and computed Euclidean embeddings as described in [31] using the smoothed $\boldsymbol{\alpha}$s. Finally, we note that SiVM can be parallelized so that plant phenotyping from the given data required only about 30 minutes. Estimating DAR models and making predictions happened within minutes.

Acknowledgments Parts of this work were supported by the Fraunhofer ATTRACT fellowship "Statistical Relational Activity Mining" and by the German Federal Ministry of Education and Research (BMBF) within the scope of the competitive grants program "Networks of excellence in agricultural and nutrition research - CROP.SENSe.net", funding code: 0315529).

References

1. Abdeen, A., Schnell, J., Miki, B.: Transcriptome analysis reveals absence of unintended effects in drought-tolerant transgenic plants overexpressing the transcription factor ABF3. BMC Genomics **11**(69) (2010)
2. Aitchison, J.: The Statistical Analysis of Compositional Data. Chapman and Hall, London (1986)
3. Arngren, M., Schmidt, M.N., Larsen, J.: Bayesian nonnegative matrix factorization with volume prior for unmixing of hyperspectral images. In: Proceedings of the IEEE Workshop on Machine Learning for Signal Processing (MLSP), pp. 1–6 (2009)
4. Ballvora, A., Römer, C., Wahabzada, M., Rascher, U., Thurau, C., Bauckhage, C., Kersting, K., Plümer, L., Leon, J.: Deep phenotyping of early plant response to abiotic stress using non-invasive approaches in barley. In: Zhang, G., Li, C., Liu, X. (eds.) Advance in Barley Sciences, chapter 26, pp. 301–316. Springer (2013)
5. Bauckhage, C., Kersting, K., Schmidt, A.: Agriculture's technological makeover. IEEE Pervasive Comput. **11**(2), 4–7 (2012)
6. Bechar, I., Moisan, S., Thonnat, M., Bremond, F.: On-line video recognition and counting of harmful insects. In: Proceedings of the ICPR (2010)
7. Bergamaschi, S., Sala, A.: Creating and querying an integrated ontology for molecular and phenotypic cereals data. In: Sicilia, M.A., Lytras, M.D. (eds.) Metadata and Semantics, pp. 445–445. Springer (2009)
8. Blanco, P.D., Metternicht, G.I., Del Valle, H.F.: Improving the discrimination of vegetation and landform patterns in sandy rangelands: a synergistic approach. Int. J. Remote Sens. **30**(10), 2579–2605 (2009)
9. Blumenthal, L.M.: Theory and Applications of Distance Geometry. Oxford University Press (1953)
10. Boyd, S., Vandenberghe, L.: Convex Optimization. Cambridge University Press (2004)

11. Boyer, J.S.: Plant productivity and environment. Science **218**, 443–448 (1982)
12. Burrell, J., Brooke, T., Beckwith, R.: Vineyard computing: sensor networks in agricultural production. IEEE Pervasive Comput. **3**(1), 38–45 (2004)
13. Çivril, A., Magdon-Ismail, M.: On selecting a maximum volume sub-matrix of a matrix and related problems. Theoret. Comput. Sci. **410**(47–49), 4801–4811 (2009)
14. Çivril, A., Magdon-Ismail, M.: Column subset selection via sparse approximation of SVD. Theoret. Comput. Sci. (2011). (In Press). http://dx.doi.org/10.1016/j.tcs.2011.11.019
15. Chakraborty, S., Subramanian, L.: Location specific summarization of climatic and agricultural trends. In: Proceedings of the WWW (2011)
16. Cox, T., Cox, M.: Multidimensional Scaling. Chapman and Hall, London (1984)
17. Crowley, M., Poole, D.: Policy gradient planning for environmental decision making with existing simulators. In: Proceedings of the AAAI (2011)
18. Doyle, G., Elkan, C.: Financial topic models. In: Working Notes of the NIPS-2009 Workshop on Applications for Topic Models: Text and Beyond Workshop (2009)
19. Feng, P., Xiang, Z., Wei, W.: CRD: fast co-clustering on large datasets utilizing sampling based matrix decomposition. In: Proceedings of the ACM SIGMOD (2008)
20. Frieze, A., Kannan, R., Vempala, S.: Fast monte-carlo algorithms for finding lowrank approximations. J. ACM **51**(6), 1025–1041 (2004)
21. Girard, A., Rasmussen, C.E., Quinonero Candela, J., Murray-Smith, R.: Gaussian process priors with uncertain inputs—application to multiple-step ahead time series forecasting. In: Proceedings of the NIPS (2002)
22. Gocht, A., Roder, N.: Salvage the treasure of geographic information in farm census data. In: Proceedings of the International Congress European Association of Agricultural Economists (2011)
23. Golovin, D., Krause, A., Gardner, B., Converse, S.J., Morey, S.: Dynamic resource allocation in conservation planning. In: Proceedings of the AAAI (2011)
24. Gomes, C.P.: Computational sustainability: computational methods for a sustainable environment, economy, and society. Bridge **39**(4), 5–13 (2009)
25. Goreinov, S.A., Tyrtyshnikov, E.E.: The maximum-volume concept in approximation by low-rank matrices. In: DeTurck, D., Blass, A., Magid, A.R., Vogelius, M. (eds.) Contemporary Mathematics, vol. 280, pp. 47–51. AMS (2001)
26. Goreinov, S.A., Tyrtyshnikov, E.E., Zamarashkin, N.L.: A theory of pseudoskeleton approximations. Linear Algebra Appl. **261**(1–3), 1–21 (1997)
27. Guo, P., Baum, M., Grando, S., Ceccarelli, S., Bai, G., Li, R., von Korff, M., Varshney, R.K., Graner, A., Valkoun, J.: Differentially expressed genes between drought-tolerant and drought-sensitive barley genotypes in response to drought stress during the reproductive stage. J. Exp. Bot. **60**(12), 3531–3544 (2010)
28. György, A., Lugosi, G., Ottucsák, G.: On-line sequential bin packing. J. Mach. Learn. Res. **11**, 89–109 (2010)
29. Hyvönen, S., Miettinen, P., Terzi, E.: Interpretable nonnegative matrix decompositions. In: ACM SIGKDD (2008)
30. Kailath, T.: The divergence and Bhattacharyya distance measures in signal selection. IEEE Trans. Commun. **15**(1), 52–60 (1967)
31. Kersting, K., Wahabzada, M., Römer, C., Thurau, C., Ballvora, A., Rascher, U., Leon, J., Bauckhage, C., Plümer, L.: Simplex distributions for embedding data matrices over time. In: Proceedings of the SDM (2012)
32. Kersting, K., Xu, Z., Wahabzada, M., Bauckhage, C., Thurau, C., Römer, C., Ballvora, A., Rascher, U., Leon, J., Plümer, L.: Pre–symptomatic prediction of plant drought stress using dirichlet–aggregation regression on hyperspectral images. In: AAAI—Computational Sustainability and AI Track (2012)
33. Kersting, K., Xu, Z., Wahabzada, M., Bauckhage, C., Thurau, C., Römer, C., Ballvora, A., Rascher, U., Leon, J., Plümer, L.: Pre-symptomatic prediction of plant drought stress using dirichlet-aggregation regression on hyperspectral images. In: Proceedings of the AAAI (2012)

34. Kui, F., Juan, W., Weiqiong, B.: Research of optimized agricultural information collaborative filtering recommendation systems. In: Proceedings of the ICICIS (2011)
35. Kumar, V., Dave, V., Bhadauriya, R., Chaudhary, S.: Krishimantra: agricultural recommendation system. In: Proceedings of the ACM Symposium on Computing for Development (2013)
36. Laykin, S., Alchanatis, V., Edan, Y.: On-line multi-stage sorting algorithm for agriculture products. Pattern Recogn. **45**(7), 2843–2853 (2012)
37. Lebreton, C., Lazic-Jancic, V., Steed, A., Pekic, S., Quarrie, S.A.: Identification of QTL for drought responses in maize and their use in testing causal relationships between traits. J. Exp. Bot. **46**(7), 853–865 (1995)
38. Lin, H., Cheng, J., Pei, Z., Zhang, S., Hu, Z.: Monitoring sugarcane growth using envisat asar data. IEEE Trans. Geosci. Remote Sens. **47**(8), 2572–2899 (2009)
39. Loew, A., Ludwig, R., Mauser, W.: Derivation of surface soil moisture from ENVISAT ASAR wide swath and image mode data in agricultural areas. IEEE Trans. Geosci. Remote Sens. **44**(4), 889–899 (2006)
40. Mahoney, M.W., Drineas, P.: CUR matrix decompositions for improved data analysis. PNAS **106**(3), 697–702 (2009)
41. McKay, J.K., Richards, J.H., Sen, S., Mitchell-Olds, T., Boles, S., Stahl, E.A., Wayne, T., Juenger, T.E.: Genetics of drought adaptation in Arabidopsis thaliana II. QTL analysis of a new mapping population, KAS-1 x TSU-1. Evolution **62**(12), 3014–3026 (2008)
42. Medjahed, B., Gosky, W.: A notification infrastructure for semantic agricultural web services. In: Sicilia, M.A., Lytras, M.D. (eds.) Metadata and Semantics, pp. 455–462. Springer (2009)
43. Mewes, T., Franke, J., Menz, G.: Data reduction of hyperspectral remote sensing data for crop stress detection using different band selection methods. In: Proceedings of the IEEE International Geoscience and Remote Sensing Symposium (2009)
44. Miao, L., Qi, H.: Endmember extraction from highly mixed data using minimum volume constrained nonnegative matrix factorization. IEEE Trans. Geosci. Remote Sens. **45**(3), 765–777 (2007)
45. Neumann, M., Hallau, L., Klatt, B., Kersting, K., Bauckhage, C.: Erosion band features for cell phone image based plant disease classification. In: Proceedings of the 22nd International Conference on Pattern Recognition (ICPR–2014), pp. 3315–3320 (2014)
46. Passioura, J.B.: Environmental biology and crop improvement. Funct. Plant Biol. **29**, 537–554 (2002)
47. Petrik, M., Zilberstein, S.: Linear dynamic programs for resource management. In: Proceedings of the AAAI (2011)
48. Pinnisi, E.: The blue revolution, drop by drop, gene by gene. Science **320**(5873), 171–173 (2008)
49. Rabbani, M.A., Maruyama, K., Abe, H., Khan, M.A., Katsura, K., Ito, Y., Yoshiwara, K., Seki, M., Shinozaki, K., Yamaguchi-Shinozaki, K.: Monitoring expression profiles of rice genes under cold, drought, and high-salinity stresses and abscisic acid application using cDNA microarray and RNA gel-blot analyses. Plant Physiol. **133**(4), 1755–1767 (2010)
50. Rascher, U., Nichol, C., Small, C., Hendricks, L.: Monitoring spatio-temporal dynamics of photosynthesis with a portable hyperspectral imaging system. Photogram. Eng. Remote Sens. **73**(1), 45–56 (2007)
51. Rascher, U., Pieruschka, R.: Spatio-temporal variations of photosynthesis: the potential of optical remote sensing to better understand and scale light use efficiency and stresses of plant ecosystems. Precision Agric. **9**(6), 355–366 (2008)
52. Rasmussen, C.E., Williams, C.K.I.: Gaussian Processes for Machine Learning. The MIT Press (2006)
53. Rocha, A., Hauagge, D.C., Wainer, J., Goldenstein, S.: Automatic fruit and vegetable classification from images. Comput. Electron. Agric. **70**(1), 96–104 (2010)
54. Römer, C., Bürling, K., Rumpf, T., Hunsche, M., Noga, G., Plümer, L.: Robust fitting of fluorescence sprectra for presymptomatic wheat leaf rust detection with support vector machines. Comput. Electron. Agric. **74**(1), 180–188 (2010)

55. Römer, C., Wahabzada, M., Ballvora, A., Pinto, F., Rossini, M., Panigada, C., Behmann, J., Leon, J., Thurau, C., Bauckhage, C., Kersting, K., Rascher, U., Plümer, L.: Early drought stress detection in cereals: simplex volume maximization for hyperspectral image analysis. Funct. Plant Biol. **39**, 878–890 (2012)
56. Rumpf, T., Mahlein, A.-K., Steiner, U., Oerke, E.-C., Plümer, L.: Early detection and classification of plant diseases with support vector machines based on hyperspectral reflectance. Comput. Electron. Agric. **74**(1), 91–99 (2010)
57. Ruß, G., Brenning, A.: Data mining in precision agriculture: management of spatial information. In: Proceedings of the IPMU (2010)
58. Sankaran, S., Mishra, A., Ehsani, R., Davis, C.: A review of advanced techniques for detecting plant diseases. Comput. Electron. Agric. **72**(1), 1–13 (2010)
59. Satalino, G., Mattia, F., Le Toan, T., Rinaldi, M.: Wheat crop mapping by using ASAR AP data. IEEE Trans. Geosci. Remote Sens. **47**(2), 527–530 (2009)
60. Schachtner, R., Pöppel, G., Tome, A.M., Lang, E.W.: Minimum determinant constraint for non-negative matrix factorization. In: ICA, pp. 106–113 (2009)
61. Schmitz, M., Martini, D., Kunisch, M., Mosinger, H.-J.: agroxml: enabling standardized, platform-independent internet data exchange in farm management information systems. In: Sicilia, M.A., Lytras, M.D. (eds.) Metadata and Semantics, pp. 463–467. Springer (2009)
62. Schweitzer, F., Fagiolo, G., Sornette, D., Vega-Redondo, F., Vespignani, A., White, D.R.: Economic networks: the new challenges. Science **5939**(325), 422–425 (2009)
63. Sun, J., Xie, Y., Zhang, H., Faloutsos, C.: Less is more: compact matrix decomposition for large sparse graphs. In: SDM (2007)
64. Tenenbaum, J.B., De Silva, V., Langford, J.C.: A global geometric framework for nonlinear dimensionality reduction. Science **5500**(390), 2319–2323 (2000)
65. Thurau, C., Kersting, K., Wahabzada, M., Bauckhage, C.: Descriptive matrix factorization for sustainability: adopting the principle of opposites. DAMI **24**(2), 325–354 (2012)
66. Thurau, C., Kersting, K., Wahabzada, M., Bauckhage, C.: Descriptive matrix factorization for sustainability: adopting the principle of opposites. J. Data Min. Knowl. Disc. **24**(2), 325–354 (2012)
67. Vernon, R. (ed.): Knowing where you're going: information systems for agricultural research management. Int. Serv. Agric. Res. (ISNAR) (2001)
68. Wahabzada, M., Mahlein, A.-K., Bauckhage, C., Steiner, U., Oerke, E.-C., Kersting, K.: Metro maps of plant disease dynamics—automated mining of differences using hyperspectral images. PLoS ONE **10**(1) (2015)
69. Wark, T., Corke, P., Klingbeil, L., Guo, Y., Crossman, C., Valencia, P., Swain, D., Bishop-Hurley, G.: Transforming agriculture through pervasive wireless sensor networks. IEEE Pervasive Comput. **6**(2), 50–57 (2007)

Global Monitoring of Inland Water Dynamics: State-of-the-Art, Challenges, and Opportunities

Anuj Karpatne, Ankush Khandelwal, Xi Chen, Varun Mithal, James Faghmous and Vipin Kumar

Abstract Inland water is an important natural resource that is critical for sustaining marine and terrestrial ecosystems as well as supporting a variety of human needs. Monitoring the dynamics of inland water bodies at a global scale is important for: (a) devising effective water management strategies, (b) assessing the impact of human actions on water security, (c) understanding the interplay between the spatio-temporal dynamics of surface water and climate change, and (d) near-real time mitigation and management of disaster events such as floods. Remote sensing datasets provide opportunities for global-scale monitoring of the extent or surface area of inland water bodies over time. We present a survey of existing remote sensing based approaches for monitoring the extent of inland water bodies and discuss their strengths and limitations. We further present an outline of the major challenges that need to be addressed for monitoring the extent and dynamics of water bodies at a global scale. Potential opportunities for overcoming some of these challenges are discussed using illustrative examples, laying the foundations for promising directions of future research in global monitoring of water dynamics.

A. Karpatne · A. Khandelwal · X. Chen · V. Mithal · J. Faghmous · V. Kumar (✉)
Department of Computer Science & Engineering, University of Minnesota,
Minneapolis, USA
e-mail: kumar@cs.umn.edu

A. Karpatne
e-mail: anuj@cs.umn.edu

A. Khandelwal
e-mail: ankush@cs.umn.edu

X. Chen
e-mail: chen@cs.umn.edu

V. Mithal
e-mail: mithal@cs.umn.edu

J. Faghmous
e-mail: jfagh@cs.umn.edu

J. Lässig et al. (eds.), *Computational Sustainability*,
Studies in Computational Intelligence 645,
DOI 10.1007/978-3-319-31858-5_7

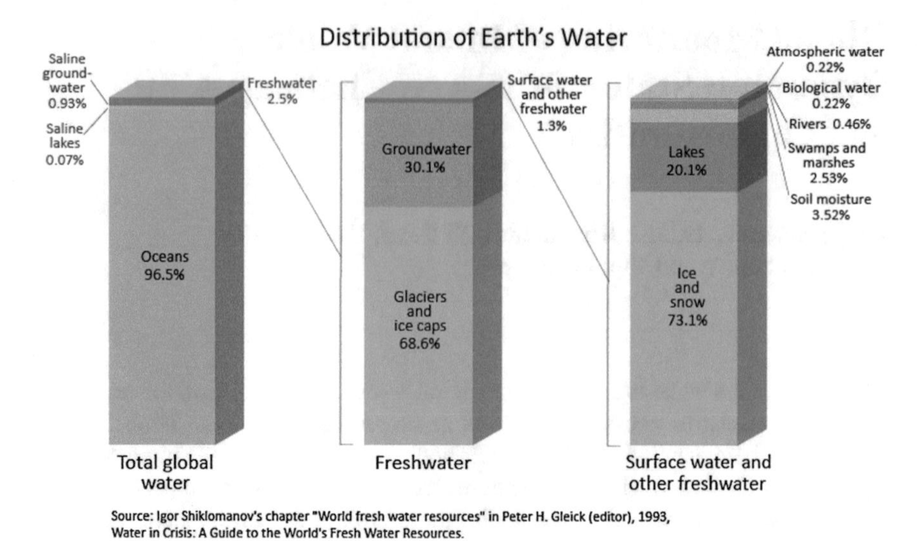

Fig. 1 An illustration of the distribution of Earth's water resources

1 Introduction and Motivation

The abundant availability of liquid water, which is unique to our planet Earth, has been of paramount importance in sustaining all forms of life. Water helps maintain the hydrological cycle which influences all major atmospheric processes of the Earth. Water also helps in preserving the biodiversity in water ecosystems by providing habitats to a plethora of flora and fauna [60, 92, 95].

The presence of inland water reserves plays an equally important role in maintaining the biogeochemical cycles, since lakes and reservoirs act as important sinks of nitrogen [34] and their annual storage and uptake of organic carbon is similar in magnitude to that of the oceans [7, 14, 87]. The availability of freshwater further plays an essential role in supporting a variety of human needs, such as drinking, agriculture, and industrial needs [91]. Even though approximately two-thirds of the Earth's surface is covered by water, only a small fraction of it is freshwater that is accessible for human consumption, as highlighted in Fig. 1. Further, available freshwater reserves are unevenly distributed across the Earth's surface, leading to scarcity of water in some regions and excessive abundance in others, as shown in Fig. 2.

The increasing growth of human population in the recent decades has additionally put immense pressure on the scarce and unevenly distributed freshwater resources. It has resulted in a lack of access to safe water and sanitation especially in the developing countries, which has led to the increase in the severity and frequency of water-related diseases as reported by the World Health Organization [31]. The growing demands for water are projected to further increase and together with the increasing trends of urbanization and water pollution, will lead to severe degradation

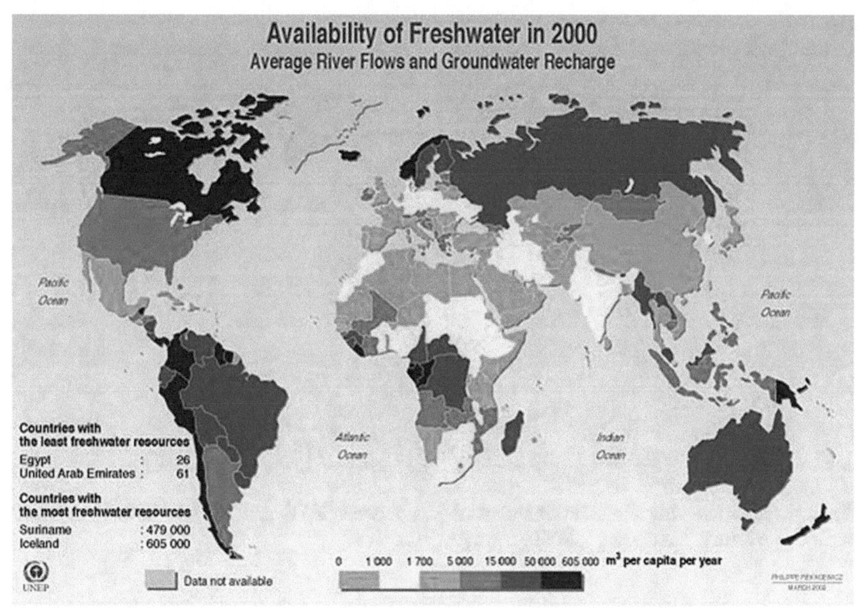

Source: *World Resources 2000-2001, People and Ecosystems: The Fraying Web of Life*, World Resources Institute (WRI), Washington DC, 2000

Fig. 2 Map showing the uneven distribution of freshwater at a global scale

of freshwater resources, which will have significant impacts on human and ecosystem sustainability. This calls for a need to take immediate preventive and restorative actions for conserving inland water resources as well as designing and promoting efficient strategies for water management. A global-scale initiative for managing water resources is further important since information about the available water resources is often not freely shared among national and international water monitoring agencies [4, 29, 33], and is thus often accessible only at local scales. This calls for a global surface water monitoring system that can quantify the available water stocks across the world and their dynamics over time for effectively managing water risks at a global scale.

Inland water bodies are dynamic in nature as they shrink, expand, or change their appearance or course of flow with time, owing to a number of natural and human-induced factors. Changes in water bodies have been known to have significant impacts on other natural resources and human assets, and further influence climate change. As an example, the Aral Sea has been steadily shrinking since the 1960s due to the undertaking of irrigation projects, which has resulted in the collapse of fisheries and other communities that were once supported by the lake, and has further altered the local climatic conditions [56, 71]. Figure 3 shows images of the Aral Sea in 2000 and 2014 available from NASA Earth Observatory, showing the magnitude of change encountered by the lake in the last fifty years that has brought the lake on the verge of extinction. Global mapping and monitoring of the extent and growth of inland water is thus important for assessing the impact of human actions on water resources, as

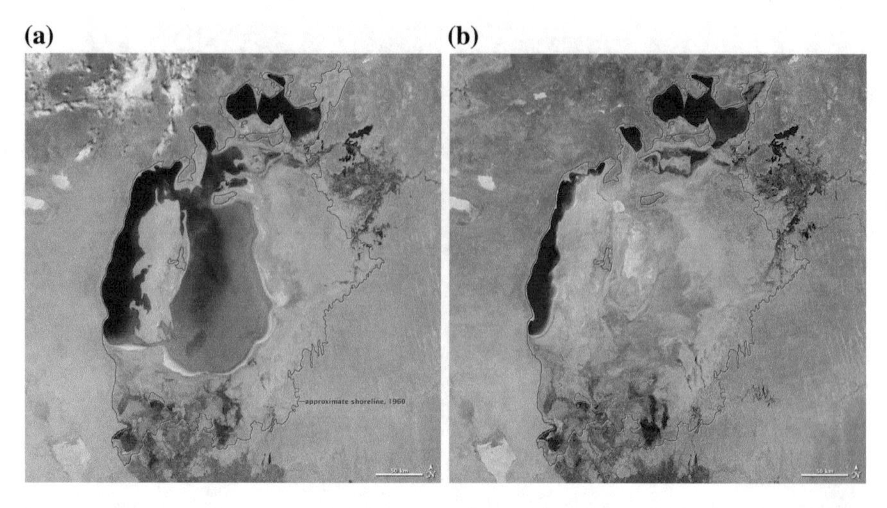

Fig. 3 Images showing the shrinking of Aral Sea between 2000 and 2014, obtained via NASA Earth Observatory. **a** August 25, 2000. **b** August 19, 2014

well as for conducting research that studies the interplay between water dynamics and global climate change.

An example of the impact of climate change on the health of inland water bodies includes the melting and expansion of several glacial lakes in Tibet due to the rapidly increasing temperatures in these regions [62, 74, 94, 98]. It was further mentioned in [96] that the black soot aerosols deposited on Tibetan glaciers due to the increasing rate of urbanization and atmospheric pollution in the Tibetan plateau have been one of the major contributing factors for the rapid glacier retreats. Figure 4 provides an illustrative example of the expansion of Cedo Caka Lake in Tibet from the year 1984 to 2011, using satellite images available via the Google Earth Engine. Monitoring and understanding the dynamics of such lakes is important as they influence the water availability in several major Asian rivers such as Yellow, Yangtze, Indus, Ganges, Brahmaputra, Irrawaddy, Salween and Mekong [5, 39], and are one of the largest sources of ice mass reserves, next only to the North and South poles.

Monitoring inland water dynamics is also helpful in obtaining descriptive insights about the natural processes that shape the landscape of water resources, thus advancing our understanding of the Earth's water system. As an example, [30] recently proposed an advancement in the classical theory of river hydraulics, termed as "at-many-stations hydraulic geometry", that enables an accurate estimation of a river's discharge solely using remotely sensed information about the river's width over time. Monitoring global water dynamics can help stimulate similar scientific advancements in our understanding of physical processes that are related to water resources.

Devising actionable plans for monitoring global water dynamics is also important for projecting the needs and demands for water resources [60, 75, 83], which can help in forecasting the water stress and security requirements in the future. It can help equip the policy-makers with the required information necessary in devising

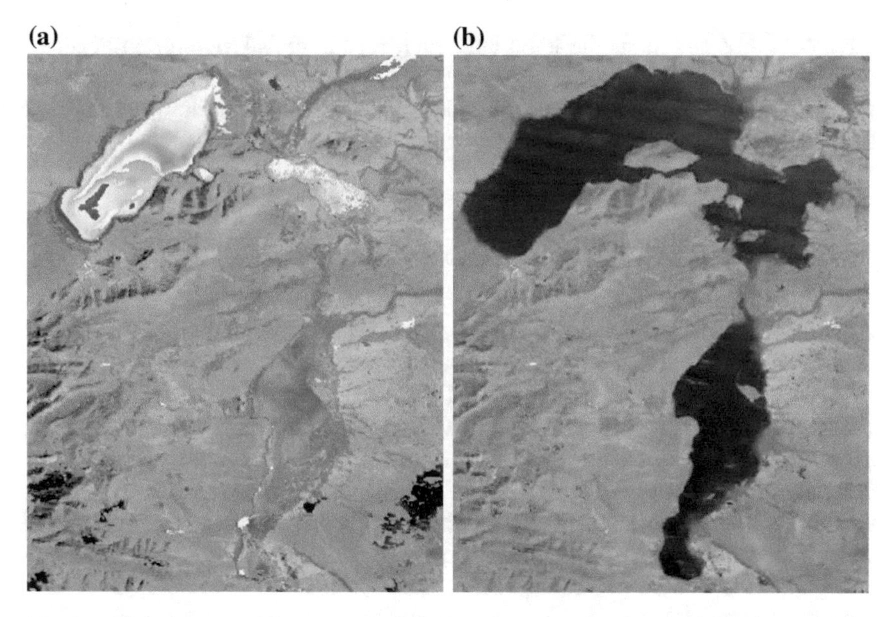

Fig. 4 Satellite images of Cedo Caka Lake in Tibet, obtained via the Google Earth Engine, showing the melting of the ice mass and expansion of the lake from 1984 to 2011. **a** Cedo Caka Lake in 1984. **b** Cedo Caka Lake in 2011

strategies for the effective planning and management of water resources at regional and global scales. Monitoring global inland water dynamics in near-real time is further important for mitigating and managing disaster events such as floods or harmful chemical discharges from coal extraction processes [73], which lead to large losses of human life and property.

The availability of vast and ever-growing volumes of remote sensing data from satellite instruments at a global scale and at regular time intervals offers a unique opportunity for the global monitoring of the extent and dynamics of inland water bodies. Remote sensing datasets provide spatially explicit and temporally frequent observational data of a number of physical attributes about the Earth's surface that can be appropriately leveraged for mapping the extent of water bodies at a global scale and monitoring their dynamics at regular and frequent time intervals. In this chapter, we present a survey of the existing approaches for monitoring water dynamics that vary in their choice of input datasets, type of algorithm used, and the scale of applicability of their results. We discuss their strengths and limitations and summarize the major challenges in monitoring the extent of water bodies at a global scale. We further present opportunities for overcoming some of these challenges using illustrative examples, which can be helpful in designing the road-map for future research in global monitoring of inland water dynamics.

The remainder of this chapter is organized as follows. Section 2 provides a survey of the existing approaches in monitoring the extent of inland water bodies based on their type of input data, type of application domain, and the type of algorithm used. Section 3 highlights the important challenges in global monitoring of inland water dynamics using remote sensing datasets, and presents opportunities in overcoming some of these challenges using illustrative examples. Section 4 summarizes the contributions of this chapter and provides concluding remarks.

2 Survey of Water Monitoring Approaches

Existing approaches for water monitoring have focused on monitoring a variety of physical attributes about water bodies, such as their water level or height, quality of water, or the extent or surface area of water. Techniques that have focused on monitoring changes in the water level in a given water body have made use of radar altimetry datasets, measured via satellite instruments [8–11, 62, 94, 98]. However, radar altimetry datasets are expensive and have limited coverage in space and time, making them unsuitable for monitoring the dynamics in small to moderately-sized water bodies. Techniques for monitoring the quality of water have focused on monitoring a number of attributes about water bodies, such the clarity of water [2, 44, 45], suspended particulate matter [19], soil wetness [32, 47, 81], and the trophic status [13, 69].

Water monitoring techniques that have focused on monitoring the extent or surface area of water bodies have made use of optical or radar remote sensing datasets [1, 27, 55, 80, 90]. These datasets contain discriminatory information about the land and water classes that can be used for estimating the spatial extent of water bodies. Due to the availability of optical remote sensing datasets at a global scale and at frequent time intervals, these methods have the potential of mapping the global surface area dynamics of water bodies. We primarily focus on techniques for monitoring the extent or surface area of water bodies, and provide a survey of existing approaches in this area.

Existing approaches for monitoring the extent of inland water dynamics can be categorized on the basis of (a) the type of input dataset used, e.g. LANDSAT, ASTER, MODIS, or SAR, (b) their type of application, e.g. type of water body being monitored and the region or scale of evaluation, and (c) their type of algorithm used, e.g. supervised, unsupervised, hybrid, or manual. We present a survey of existing approaches on the basis of each of these three categorizations in the following subsections.

2.1 *Based on Type of Input Data*

The use of optical remote sensing datasets has been extensively explored for mapping the extent of water bodies, since water and land bodies show distinguishing characteristics in the optical remote sensing signals, such as visible, infrared and thermal parts of the electromagnetic spectrum. A number of optical remote sensing datasets

Table 1 Table of references categorizing the existing water monitoring approaches based on the type of input data used

Input dataset	References
IKONOS (1–4 m resolution, commercially available)	[18, 22]
SPOT (1.5–20 m resolution, commercially available)	[12, 84–86]
ASTER (15–90 m resolution, via NASA satellites)	[41, 67, 73]
LANDSAT (30 m resolution, via USGS)	[21, 25, 26, 36, 38, 42, 46, 49, 53, 57–59, 61, 65–67, 73, 74, 80, 97, 100]
MODIS (250 m resolution, via NASA satellites)	[40, 50, 51, 68, 78, 79, 99]
SAR (via JERS, ScanSAR, TerraSAR-X, RadarSat, and PolSAR)	[1, 6, 35, 37, 47, 54, 64, 70, 81, 82, 84–86]

have been used for monitoring the extent of water bodies at varying degrees of spatial resolution, such as the IKONOS, SPOT, ASTER, and LANDSAT at finer spatial resolutions and the MODIS datasets at relatively coarser resolutions. Furthermore, radar datasets have been used for monitoring glacial lakes, floods, and wetlands, since they are free from cloud contaminations and have the ability to distinguish between water and snow or ice. The synthetically aperture radar (SAR) dataset, obtained via a number of satellite instruments such as JERS, ScanSAR, TerraSAR-X, RadarSat, and PolSAR, have been commonly used for monitoring water bodies using radar datasets. However, the high costs associated with obtaining radar datasets at a global scale and in a timely fashion limits their usefulness in monitoring the global surface area dynamics of inland water bodies.

Table 1 provides a summary of the various types of input datasets that have been used in the existing literature for monitoring water bodies.

2.2 *Based on Type of Application*

Approaches for monitoring the extent of water bodies have been applied in a variety of applications regions, such as lakes, ponds, reservoirs, glacial lakes, thaw lakes, rivers, river basins, wetlands, mangroves, shallow water, and tidal flats. Table 2 provides a summary of water monitoring approaches that have been used for different types of water bodies.

A number of existing approaches have been used for monitoring water bodies at local scales, i.e. monitoring a single water body in a local region. These have been applied in diverse regions of the world. Other approaches have focused on monitoring water bodies at a regional scale, i.e. monitoring a set of water bodies in a particular

Table 2 Table of references categorizing the existing water monitoring approaches based on the type of water body they attempt at monitoring

Water body type	References
Lakes and reservoirs	[18, 20, 48, 52, 58, 62, 66, 74, 80, 89, 90, 94, 97, 98]
Ponds and watersheds	[73, 83]
Glacial lakes	[1, 12, 37, 41, 49, 53, 61, 65]
Thaw lakes	[26, 36]
Rivers	[25, 42, 50, 59, 60, 76, 78, 79]
River basins	[23, 24, 82]
Wetlands	[6, 35, 38, 46, 54, 57, 84–86]
Mangroves, shallow water, and tidal flats	[17, 67, 93]

region of the Earth. Tables 3 and 4 provide a summary of the local and regional scale approaches that have been applied in diverse regions of the Earth. These approaches have been highlighted on the world map in Fig. 5.

Few attempts have aimed at a global scale monitoring of water bodies, e.g. the global mapping of the extent of lakes at a global scale for a single snapshot in time in the year 2000 [89], and the monitoring of water dynamics in 34 reservoirs across the Earth [28]. However, these approaches either have limited coverage in space [28] or in time [89], and hence, a global scale monitoring of the dynamics of inland water bodies across all regions of the world and over long time periods is still missing.

2.3 *Based on Type of Algorithm*

On the basis of the type of algorithm used for mapping land and water bodies using remote sensing datasets, the existing approaches for water extent monitoring can be categorized as follows:

2.3.1 Manual Annotation Based

Manual approaches for water monitoring involve the use of a human expert that annotates land and water classes using high-resolution optical remote sensing datasets and satellite images. As an example, the evolution of glacial lakes in the northern Patagonian icefields was monitored using manual mapping of the extent of water bodies over multiple time-frames [53]. Mapping the geographical characteristics of China's wetland using manual approaches was further carried out in [57]. Glacial lakes in the Himalayan basin were manually mapped in [41], for monitoring glacial lake outburst flood (GLOF) events. Although these approaches can potentially pro-

Table 3 Summary of approaches that have been applied in different regions of the world at a local scale, arranged in alphabetical order of the name of the study region

Region of study	References
1 : Alaknanda River, Himalayas	[41]
2 : Erie Lake, North America	[18]
3 : Chapala Lake, Mexico	[52]
4 : Dongkemadi Glacier, Tibet	[37]
5 : Gomso Bay, Korea	[67]
6 : Koa Catchment, India	[42]
7 : Loisach River, Germany	[76]
8 : Mackenzie River, Canada	[82]
9 : Murray Lake, Australia	[59]
10 : Nangaparbat Massif, Pakistan	[12]
11 : New Orleans, USA	[78, 79]
12 : Oberaletsch Glacier, Alps	[61]
13 : Peace-Athabasca Delta, Canada	[84–86]
14 : Powder River Basin, Wyoming	[73]
15 : Bayi Lake, China	[97]
16 : Pearl River Delta, China	[100]
17 : Poyang Lake, China	[38]
18 : Rio Negro River, Amazon	[24]
19 : Seyfe Lake, Turkey	[66]
20 : Terengganu, Malaysia	[22]
21 : Tewkesbury, UK	[70]
22 : Tiaoxi Watershed, China	[43]
23 : Vietnamese Mekong Delta	[68]

vide high estimation accuracy, they are time-consuming, labor-intensive, and lack in reproducibility of their results.

2.3.2 Unsupervised Learning

This includes automated algorithms that do not rely on the use of a reference or ground-truth information about the water or land classes, and are thus unsupervised in nature. Hence, they are not affected by the scarcity of labeled data, which is common in a number of real-world applications since obtaining plentiful and representative labeled samples is often time-consuming and expensive. However, unsupervised methods rely on certain assumptions about the input features and the land and water

Table 4 Summary of approaches that have been applied in different regions of the world at a regional scale, arranged in alphabetical order of the name of the study region

Region of Study	References
1 : Alaska	[26, 36]
2 : Amazon	[35, 54]
3 : Australia	[17, 46]
4 : Bangladesh	[40]
5 : China	[57, 80, 94]
6 : Himalayas	[49, 65]
7 : Kenya	[58]
8 : Mekong River, Asia	[23]
9 : Mississippi River, US	[50]
10 : Scandinavia	[48, 90]
11 : Panama	[93]
12 : Patagonia	[53]
13 : Southeast Asia	[99]
14 : Tibet	[62, 74, 98]
15 : Western USA	[83]

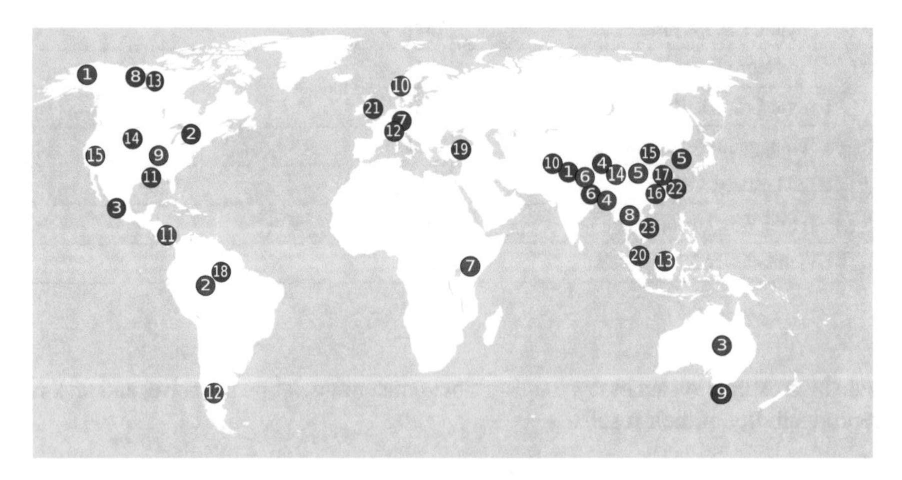

Fig. 5 A geo-referenced summary of existing local scale approaches (shown in *blue* and enlisted in Table 3) and regional scale approaches (shown in *red* and enlisted in Table 4) for monitoring the extent of water bodies

classes, such as the existence of an arbitrary threshold separating land from water signals, which are obtained using domain understanding at regional scales that may not generalize at global scales.

A large number of unsupervised approaches for monitoring water extent make use of thresholds on remote sensing signals for estimating the land and water classes, a technique termed as density slicing. Even though density slicing on a single remote sensing signal, e.g. a discriminatory optical band, is useful in distinguishing between land and water classes [25, 59, 67], it may not be sufficient for discriminating between all varieties of land and water classes [43]. Hence, a number of techniques for combining the information from multiple bands have been explored for improved detection of the extent of water bodies [54, 65, 68, 77]. This includes the use of non-linear transformations of multiple optical bands, termed as water indices, which are more discriminatory in distinguishing between land and water classes than the individual bands. Some of the indices that have been commonly used for monitoring water extent include the Normalized Difference Vegetation Index (NDVI) [88], the Normalized Difference Water Index (NDWI) [27], the Modified NDWI (MNDWI) [97], and the Tasseled Cap Wetness Index (TCW) [15]. Existing approaches have either used thresholds on a single water index [20, 23, 55], or combinations of multiple water indices and optical bands using intricately designed hand-coded rule systems [40, 43, 58, 61].

A major limitation of thresholding based methods is their dependence on the choice of an appropriate threshold that is commonly obtained using visual inspection on a small set of reference water bodies in the region of study. Since manual thresholding techniques and hand-coded rule systems rely on certain assumptions about the features which may not be generalizable at a global scale, their thresholds vary across different regions in space and lack robustness when applied to large regions. An approach for the automatic detection of a local threshold for every water body in the region of study was performed in [49] using an iterative thresholding method. Other automated approaches for the unsupervised detection of water extent include the use of clustering techniques, such as k-means [28, 99] or ISODATA algorithm [66, 73, 74]. Segmentation approaches that involve region-growing and agglomerative clustering operations have been explored for water extent mapping, using mean-shift segmentation [18], morphological segmentation [16, 64], and object-based unsupervised classification [35, 80]. In order to account for the heterogeneity in the land and water classes, target decomposition into multiple classes, such as clear, green, and turbid water has been used [80].

Advantages

- Unsupervised approaches are training-free approaches that do not rely on the availability of labeled data, which is often time-consuming, laborious, and expensive to obtain.
- Clustering and segmentation based approaches are adaptive in nature to the characteristics of the input dataset and hence offer better generalization abilities than supervised approaches at local to regional scales.

Challenges

- Since unsupervised techniques often rely on specific assumptions about the input dataset and the target application, they have limited applicability in scenarios where their underlying assumptions are violated. As an example, [28] recently proposed an unsupervised approach for characterizing water and land bodies as the two clusters obtained by clustering NDVI values using the k-means algorithm. This primarily assumes a bi-modal distribution of the data in the feature space, where each mode corresponds to either the land or the water class. However, in scenarios where the distribution of land and water classes are multi-modal in nature due to heterogeneity within the land and water bodies or if the land and water classes are obstructed due to clouds or aerosols, the underlying assumptions of such a clustering based technique is violated, leading to poor performance of the algorithm.
- Since unsupervised approaches do not explicitly model the differences in the characteristics of land and water bodies in different regions of the Earth, they suffer from poor performance in certain regions of the world that have distinct signatures of land and water classes. The presence of heterogeneity within the land and water classes further results in a high sensitivity of the model parameters, such as thresholds and clustering choices, used by unsupervised learning approaches in different regions and seasons of the world.
- Since thresholding based approaches often involve the manual selection of thresholds that vary from different regions and time-frames, their deployment at a global scale for large-scale water monitoring is time-consuming and expensive.

2.3.3 Supervised Learning

Supervised approaches for water extent monitoring involve the learning of a concept-based model of the relationship between remote sensing signals and a target attribute of interest, e.g. water and land classes or water surface fractions. This is achieved by the use of labeled training data for learning a model of relationship between the remote sensing signals and the target attribute, which is then applied on unseen test instances for estimating their target attribute. The performance of supervised approaches is thus dependent on the choice of the algorithm used for describing the relationship model as well as on the availability of adequate and representative training instances. Since, obtaining accurate ground-truth labeled data at a global scale and at frequent time intervals is time-consuming and expensive, the applicability of such approaches is generally limited to local to regional scales and at infrequent time-steps.

A range of supervised classification algorithms for mapping water bodies have been explored in the existing literature, including the maximum likelihood classifiers [1, 21], decision trees [26, 50, 51, 79, 90], rule-based algorithms [46], k-nearest neighbor classifiers using the Mahalanobis distance [84–86], support vector machines [37], artificial neural networks [12, 36, 100], self-organizing maps [2, 3], and fuzzy classification algorithms [72]. These have been shown to provide better performance in monitoring water bodies than unsupervised approaches, especially in regions where ground-truth is available.

Since supervised learning requires the use of representative training samples, their applicability has been limited to local and regional scales in space and short durations in time. A major challenge in applying supervised learning algorithms at a global scale is the presence of a rich variety in the land and water classes, which is difficult to capture in the training dataset and exploit in the model learning phase. Existing approaches for overcoming this challenge of heterogeneity among the land and water classes include target decomposition techniques, where the land and water classes are decomposed into multiple target land and water categories [37], thus converting a binary classification problem to a multi-class classification problem. Similarly, [90] accounted for the presence of heterogeneity within the land and water categories by considering 50 classes of water and land categories, which were further grouped into 4 classes, and eventually to two classes of water and land. Decision trees were used for multi-class classification and their results were evaluated in Sweden for a single snapshot in time in the year 2000, when accurate ground-truth information is available. They further applied their model for performing a global monitoring of water bodies for the year 2000, using LANDSAT optical imagery at 14.25 m resolution [89].

Advantages

- Since supervised approaches are able to utilize the discriminative information contained in the labeled training datasets, they generally offer better estimation accuracy as compared to unsupervised approaches that instead rely on certain assumptions about the input dataset.

Challenges

- Since a number of application domains suffer from limited availability of training data, the scope of application of supervised approaches is limited to regions which have representative samples of labeled data in the training dataset. This limits their applicability at a global scale and at frequent time-intervals.
- A number of supervised approaches involve the learning of a single model using all training instances, which is considered generalizable over all unseen test instances. However, due to the presence of heterogeneity within the land water classes at a global scale, the characteristics of the land and water bodies in the remote sensing signals vary differently for different regions on the Earth and for different time-steps. This requires learning different relationships between remote sensing signals and land and water classes for different regions in space and at different time-steps, which is challenging for traditional supervised learning approaches that do not explicitly model the heterogeneity in the data.

- Obtaining representative training samples for performing supervised learning is challenging due to the presence of heterogeneity within the land and water bodies. Existing techniques for sampling training instances do not take into account the heterogeneity within the two classes, which can result in an under-representative set of training instances that may lead to poor generalization performance.

2.3.4 Hybrid Learning

These approaches involve the use of both supervised and unsupervised techniques in conjunction for mapping the extent of water bodies. As an example, hybrid approaches for water fraction estimation make use of supervised approaches for estimating water and land labels, which are then used to compute water fractions using unsupervised approaches. In this context, Li et al. [51] used decision trees to estimate land and water classes, which was then used to estimate the water fraction using an unsupervised dynamic nearest neighbor searching (DNNS) algorithm to find the nearest pure land and water pixels along with the use of digital elevation data. Similarly, [78] used linear mixing models to estimate the water fraction at a pixel, using information about the land and water classes estimated using decision trees with feature transformations. The use of regression trees to estimate the water fraction at fine spatial resolutions has been explored in [79].

The use of supervised approaches to guide the behavior of unsupervised algorithms has been explored in [52], where polygons of water and land pixels generated using active learning schemes were used for obtaining labeled training data, which was then used as seeds in an unsupervised fuzzy clustering approach. A similar approach was used in [70], where labeled instances obtained using active learning schemes were used as seeds in an unsupervised clustering algorithm, that produced the target maps of land and water classes.

Supervised object-based methods have been explored in [26, 93], that use unsupervised segmentation approaches for constructing objects that are then used in a supervised learning framework. Instead of performing classification at a per-pixel level, these approaches use region-growing methods to segment the image into spatial objects, which are then classified into land or water classes on the basis of their aggregate optical remote sensing properties. This has the advantage of being robust to noise and outliers, as well as capturing the aggregate phenomena that is difficult to observe at fine spatial scales. Wang et al. [93] further explored an approach that combined pixel-based and object-based classification approaches.

3 Challenges and Opportunities in Monitoring Global Water Dynamics

We discuss the key challenges in the global monitoring of inland water dynamics using supervised learning algorithms and opportunities for future research in the following subsections. We consider supervised learning approaches, since they are able to provide better estimation performance than unsupervised approaches given the availability of representative training data.

3.1 Challenges

The major challenges in global monitoring of the extent of water bodies using supervised learning approaches can be summarized as follows:

- *Presence of noise*: Remote sensing datasets are often plagued with noise and outliers, due to cloud and aerosol contaminations. These can significantly impact the performance of a classifier and lead to inconsistencies in results. Shadows of mountains and clouds further show similar characteristics in remote sensing signals as water bodies, making it difficult to distinguish them from the water class using optical remote sensing datasets.
- *Presence of snow and ice*: Distinguishing water from snow, ice, and glaciers is additionally challenging due to their similar characteristics in optical remote sensing datasets. Furthermore, including training samples from snow and ice may lead to class confusion between the water class and the snow covered land class in the feature space, leading to poor performance of the classifier not only in snow covered regions of the Earth but across diverse regions in space at a global scale. Even though the use of radar data provides an opportunity for distinguishing water from snow and ice, radar datasets are expensive to obtain and are not available at a global scale and at regular time intervals.
- *Heterogeneity in space*: Due to the existence of a variety of land and water bodies across varying geographies, topographies and climatological conditions of the Earth, different land and water bodies show varying characteristics in remote sensing signals in diverse regions of the world. This results in a heterogeneity within the land and water classes at a global scale, which makes the learning of a classifier that distinguishes between all categories of land and water bodies challenging. Furthermore, the presence of heterogeneity demands the need for obtaining representative and diverse training samples from different regions of the world, in order to learn a classifier that is able to capture the heterogeneity in the data.
- *Heterogeneity in time*: A large number of land and water bodies further exhibit a high variability in their characteristics over time, either due to the presence of seasonal cycles or due to their inter-annual dynamics owing to changes in climate or due to human impacts. This results in a heterogeneity in the characteristics of land and water bodies over different time-steps, as the same land or water body may

show different characteristics in remote sensing signals at different time-steps. The presence of heterogeneity within the land and water classes over time restricts the usefulness of a static classifier that does not adapt its learning with the changing characteristics in the data over time.

- *Lack of representative training data*: Since obtaining training data is often time-consuming and expensive, global-scale ground truth information about the land and water bodies at frequent temporal scales is difficult to obtain. In the absence of representative training samples, it is challenging to generalize the performance of a classifier at a global scale over new varieties of test instances that had not been observed during the training phase. Furthermore, learning a classifier that can address the heterogeneity within the land and water classes requires the learning of models with high overall complexity, which further adds to the challenges of limited training data, as such models can be prone to over-fitting in the absence of representative training samples.

3.2 Opportunities

We present the opportunities in addressing some of the challenges in global monitoring of water bodies using supervised learning approaches, and demonstrate them using illustrative examples. In all our experiments presented in this section, we used the optical remote sensing dataset (MCD43A4) publically available at 500 m resolution for every 8 days, starting from Feb 18, 2000, via the MODIS repository. This dataset has seven reflectance bands covering visible, infrared, and thermal parts of the electromagnetic spectrum, which can be used as input features for discriminating between land and water bodies. However, it suffers from the presence of noise, outliers, and missing values, due to cloud or aerosol obstructions etc.

Ground truth information about the extent of lakes was obtained via the Shuttle Radar Topography Mission's (SRTM) Water Body Dataset (SWBD) dataset, which provides a mapping of all water bodies for a large fraction of the Earth (60 °S to 60 °N), but for a short duration of 11 days around Feb 18, 2000 (the closest date at MODIS scale). The SWBD dataset, publically available through the MODIS repository as the MOD44W product, thus provides a label of land or water for every MODIS pixel at 250 m for a single date, Feb 18, 2000. Even though the SWBD dataset suffers from inaccuracies in some regions of the world, it provides a reasonable proxy of the ground truth that can be used for training and evaluating classification models in absence of any other high quality global ground truth.

We consider a global set of 180 lakes collected from 33 different MODIS tile divisions across the globe (highlighted in red in Fig. 6) as our evaluation dataset. These lakes were selected as they represent diverse categories of land and water bodies at a global scale, and further have lower numbers of missing values and ground truth inconsistencies. For each lake, we created a buffer region of 20 pixels at 500 m resolution around the periphery of the water body, and used the buffer region as well as the interior of the water body to construct the evaluation dataset. After

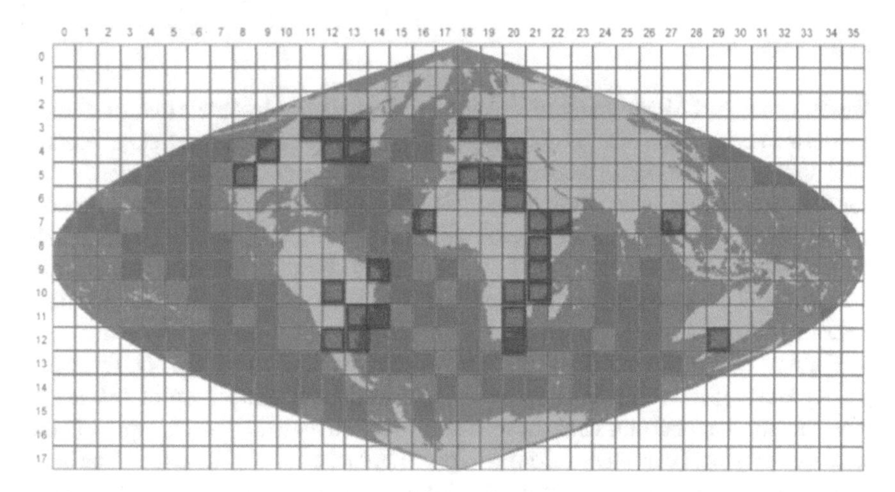

Fig. 6 The 33 MODIS tile divisions (highlighted as *red boxes*) that were used for constructing the evaluation dataset

removing instances at the immediate boundaries of the water bodies for which the ground truth might not be accurate, and ignoring instances with missing values, the evaluation set comprised of $\approx$2.6 million data instances, where every instance had an associated binary label of water (positive) or land (negative).

We considered Support Vector Machines with a linear kernel (SVM) as the choice of the classification algorithm in all the experiments presented in this section. The regularization parameter of SVM was chosen to be 0.5 in all experiments to ensure adequate regularization. We used the misclassification error rate as the evaluation metric for assessing the classification performance of an algorithm in distinguishing between water or land.

3.2.1 Using Local Models for Capturing the Heterogeneity in Space

The use of local models refers to the learning of a separate classification model distinguishing water and land pixels for every lake in the evaluation dataset, using labeled instances only within the lake for training. Due to the presence of heterogeneity within the land and water classes at a global scale, different lakes have different compositions of land and water bodies, which vary in their degree of separability in the feature space in different regions of the Earth. Hence, a classification model that is trained to distinguish between a particular pair of land and water bodies in a given lake may have difficulties in classifying the water and land bodies in a different lake, with widely varying characteristics of land and water classes in the feature space. The learning of local models is thus aimed at capturing the local patterns of differences in the characteristics of land and water classes, which are different for different lakes.

Demonstration

We demonstrate the importance of learning local classification models that are trained and tested on the same lake by contrasting it with the learning of other classification models that are trained on a lake and tested on some other lake. Specifically, we consider 50 % of the water pixels and an equal number of land pixels in every lake for training an SVM classifier for that lake, while the remainder of land and water pixels in that lake are considered for testing. The learned classifier in a lake, Lake i, can then be applied on the test instances in a lake, Lake j, to obtain the misclassification error rate, $E(i, j)$.

Figure 7 presents a matrix of the misclassifcation error rates, $E(i, j)$, for every pair of lakes, (Lake i, Lake j), in the evaluation dataset comprising of 180 lakes. Note that the diagonal entries in the matrix correspond to $E(i, i)$, which represent the error rate of a local classifier trained on a particular lake and tested on the same lake. Figure 7 shows that there is a grouping structure among the lakes which show lower error rates among the groups but high error rates across the groups. This is indicative of the heterogeneous categories of land and water bodies present at a global scale, which can be grouped into different categories of land and water bodies across the 180 lakes. It can be further observed from Fig. 7 that the error rates of local classifiers, $E(i, i)$, are always lower than the error rates of classifiers trained on a particular lake and tested on a different lake. Further, even though the classifiers for a number of lake pairs, (Lake i, Lake j), show similar error rates as the local classifiers, there exists multiple groups of lake pairs with significantly higher error rates than the local classifiers. This highlights the fact that for such pairs of lakes, (Lake i, Lake j), the training instances in Lake i are not representative of the test instances in Lake j, leading to poor generalization performance, $E(i, j)$.

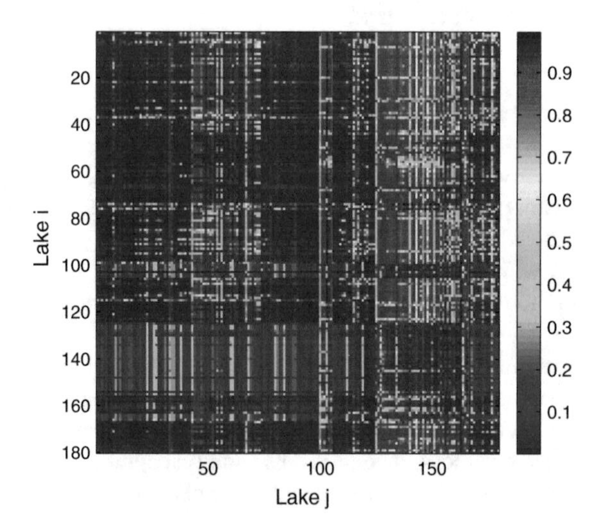

Fig. 7 Matrix of misclassification error rates where the entry at the ith row and jth column, $E(i, j)$, denotes the misclassification error rate of the classifier learned using training instances from Lake i, when evaluated on the test instances in Lake j

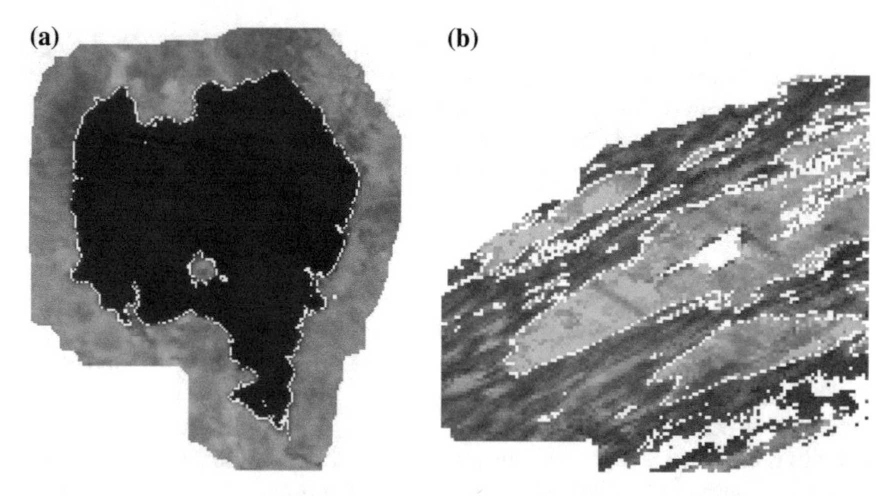

Fig. 8 False color composites (using the 7, 5 and 4th bands, as *red*, *green* and *blue* colors respectively) of Lake Tana and Lac La Ronge Lake. Error rate of the classifier learned in Lake Tana and tested in Lac La Ronge Lake is 0.67, and the error rate of the classifier learned in Lac La Ronge Lake and tested in Lake Tana is 0.48. **a** Lake Tana, Ethiopia. **b** Lac La Ronge Lake, Saskatchewan, Canada

We explore one such pair of lakes that are illustrative of the heterogeneity within the land and water bodies across the 180 lakes. Figure 8 shows an example of a pair of lakes, Lake Tana in Ethiopia (Fig. 8a) and Lac La Ronge Lake in Saskatchewan, Canada (Fig. 8b), which show an error rate of 0.67 when the learned classifier from Lake Tana is applied on Lac La Ronge Lake, and an error rate of 0.48 when the vice-versa is applied. The high magnitutdes of these error rates is indicative of the differences in land and water bodies across the two lakes that have been collected from diverse regions of the Earth. This motivates the need for learning local models that cater to the local characteristics of water and land categories in a given lake, and thus are able to capture the heterogeneity within the land and water classes in space.

3.2.2 Using Global Models for Capturing the Heterogeneity in Time

The heterogeneity in the characteristics of land and water bodies across time can either be attributed to seasonal cycles of the Earth or dynamic intra-annual changes occurring in the water body due to climate change or human-induced factors. This often leads to the emergence of different land and water categories at a future time-step that were not observed on a particular date in the year 2000 for which the SWBD ground truth data is available. Using a local model for a lake that only considers the training instances in the lake for a single date in the year 2000 can potentially lead to the incorrect classification of the emerging land and water categories at future time-steps, since the training data in the year 2000 is not representative of such classes. However, by obtaining adequate number of training samples from a diverse set of lakes at a global scale, as opposed to using local training samples from a single lake, we can expect to observe a broad variety of land and water categories that can potentially cover the new land and water categories that emerge at future time-steps. Such a model can be termed as a global classification model, which captures sufficient diversity of land and water bodies at a global scale in its training dataset.

Demonstration

Since the ground-truth information about the 180 lakes is available via the SWBD dataset for a single date in the year 2000, we do not have labeled data to validate the classification performance of a classifier at any of the 180 lakes going forward in time. However, we can still visually explore the differences in the results of comparative algorithms at a given lake at a particular time-step and compare them with false color composite images [63] of the lake using information in the optical datasets at the same time-step. We consider two classification algorithms: (a) a global model that uses training samples from all the 180 lakes in the evaluation dataset, and (b) a local model that uses training samples only from its respective local lake. We monitor the performance of the global and local SVM classifiers for the Mar Chiquita Lake, which is a salt lake in Argentina that has been slowly diminishing in area since 2003. Both the global and local classifiers are trained using equal number of $\approx$8000 positive (water) and negative (land) training samples.

Figure 9 shows the performance of the global and local classifiers at Mar Chiquita lake on September 21, 2000, when the shrinking of the lake extent had not begun and the characteristics of land and water bodies were in agreement with the training instances available on Feb 18, 2000. In fact, the differences between the classification output of the global classifier (Fig. 9b) and the classification output of the local classifier (Fig. 9c) is small and only occurs at the periphery of the lake as shown in Fig. 9d. However, Fig. 10 shows the performance of the global classifier (Fig. 10b) and the local classifier (Fig. 10c) for this lake on January 1, 2012, which can be seen to be significantly different in Fig. 10d. It can be seen from the false color composite image of the lake at this time-step (Fig. 10a) that the lake has undergone a major reduction in its surface area from 2000 to 2012. This has lead to the emergence of a new category of barren land around the shrinking periphery of the lake, which had not been observed by a local classifier at the same lake in the year 2000. Hence, the local classifier incorrectly classifies a large fraction of the new land category as water as shown as red in Fig. 10d. However, the results of the global classifier can be observed to be in better agreement with the false color composite. This can be attribute to the fact that the global classifier had observed training samples from a diverse set of lakes across the world, which was representative of the new category of land class that emerged in this lake in 2012. The global classifier was thus able to accurately map the extent of the lake in 2012 and thus identify the shrinking of this lake, as opposed to the local classifier that is unable to detect the shrinking of this lake.

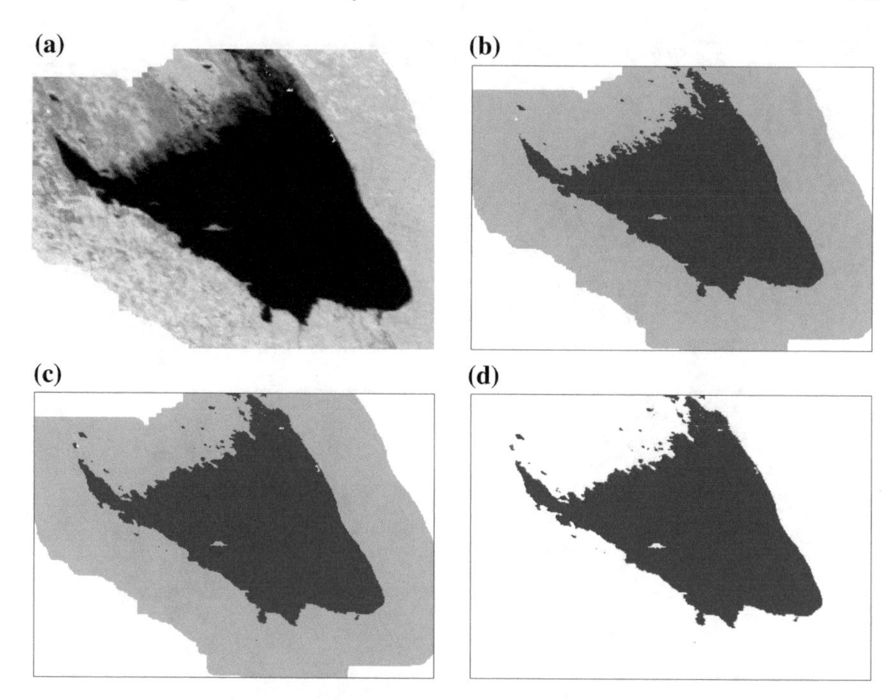

Fig. 9 Performance of global and local classifiers at Mar Chiquita Lake, Argentina on September 21, 2000. Color code for figures **b** and **c**: *green*—land, *blue*—water, and *white*—missing from evaluation. Color code for figure **d**: *blue*—pixels detected by both global and local classifiers as water, *red*—pixels detected only by the local classifier as water; and *white*—everything else. **a** False color composite. **b** Results of global. **c** Results of local. **d** Differences in results

4 Conclusion

In this chapter, we discussed the importance of monitoring the dynamics of inland water bodies at a global scale and its impact on human and ecosystem sustainability. We presented a survey of the existing efforts in monitoring the extent of water bodies, and categorized them on the basis of the type of input dataset used, the type of application considered, and the type of algorithm used. We highlighted the major challenges in monitoring the extent of inland water bodies at a global scale and further presented opportunities for using supervised learning approaches in overcoming some of these challenges, which can be helpful in framing future directions of research in this area.

A major challenge in the global monitoring of inland water dynamics is the presence of heterogeneity within the land and water classes in different regions across the world and at different time-steps. Since supervised learning approaches require the use of labeled training data, it is important to devise sampling strategies for obtaining training instances that are representative of the variety of water and land classes at a global scale. For example, active learning approaches that are cognizant of the heterogeneity within the classes while obtaining training instances can be

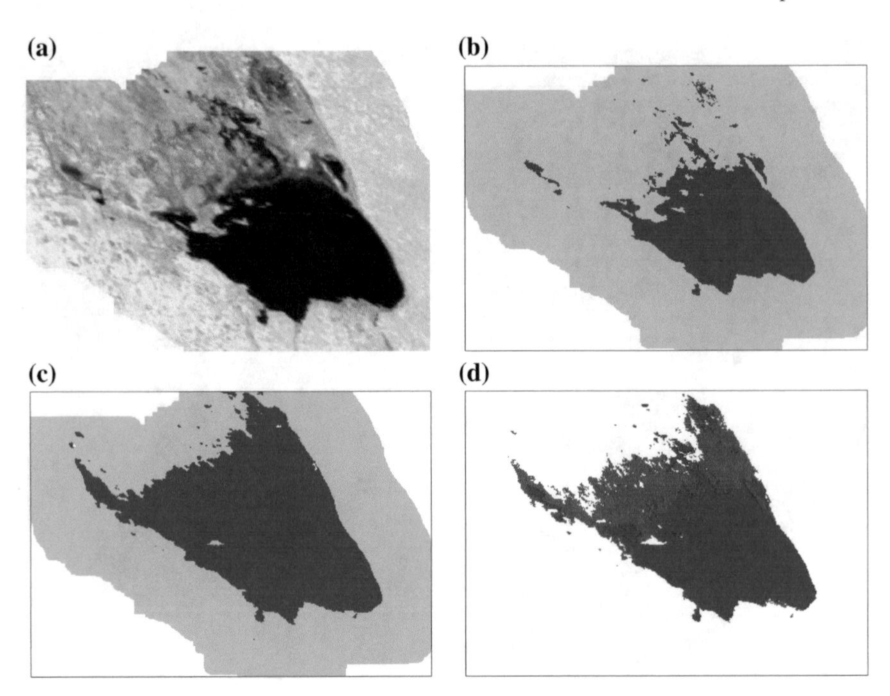

Fig. 10 Performance of global and local classifiers at Mar Chiquita Lake, Argentina on January 1, 2012. Color code for figures **b** and **c**: *green*—land, *blue*—water, and *white*—missing from evaluation. Color code for figure **d**: *blue*—pixels detected by both global and local classifiers as water, *red*—pixels detected only by the local classifier as water; and *white*—everything else. **a** False color composite. **b** Results of global. **c** Results of local. **d** Differences in results

explored. The importance of learning local and global models for monitoring global water dynamics had been presented in Sect. 3.2 using illustrative examples. Future research can focus on developing classification models that strike a balance between the local and global models, and appropriately leverage the advantages of both of these approaches by utilizing information about the context of classification. As an example, local models that can adapt themselves to the characteristics of the unlabeled test instances at a given lake and at a particular time-step could be explored using semi-supervised learning approaches.

Classical schemes for evaluating the performance of a classifier consider the overall classification performance over all test instances in the dataset, using evaluation metrics such as accuracy, precision, recall, or receiver-operating curves. For the problem of global monitoring of inland water bodies, this corresponds to evaluating the performance of the classifier over all locations and time-steps across all lakes in the world. Such an evaluation would tend to be biased towards few very large lakes with large interior portions of water that are easily distinguishable from land. However, the presence of heterogeneity in land and water classes is most prominent at the peripheries of small or moderately sized lakes, and such lakes tend to be more dynamic as compared to larger lakes, making their monitoring equally important from an operational perspective. Hence, it is important to devise evaluation procedures that perform a lake-wise analysis of the classification performance, as opposed to evaluating the overall classification performance.

Acknowledgments This work was supported by the National Science Foundation Awards 1029711 and 0905581, and the NASA Award NNX12AP37G.

References

1. Adam, S.: Glacier snow line mapping using ERS-1 SAR imagery. Remote Sens. Environ. **61**(1), 46–54 (1997)
2. Alvarez-Guerra, M., González-Piñuela, C., Andrés, A., Galán, B., Viguri, J.R.: Assessment of Self-Organizing Map artificial neural networks for the classification of sediment quality. Environ. Int. **34**(6), 782–90 (2008)
3. Astel, A., Tsakovski, S., Barbieri, P., Simeonov, V.: Comparison of self-organizing maps classification approach with cluster and principal components analysis for large environmental data sets. Water Res. **41**(19), 78–4566 (2007)
4. Balthrop, C., Hossain, F.: Short note: a review of state of the art on treaties in relation to management of transboundary flooding in international river basins and the global precipitation measurement mission. Water Policy **12**(5), 635–640 (2010)
5. Barnett, P.T., Adam, J.C., Lettenmaier, D.P.: Potential impacts of a warming climate on water availability in snow-dominated regions. Nature **438**(7066), 303–309 (2005)
6. Bartsch, A., Wagner, W., Scipal, K., Pathe, C., Sabel, D., Wolski, P.: Global monitoring of wetlands-the value of ENVISAT ASAR Global mode. J. Environ. Manage. **90**(7), 33–2226 (2009)
7. Battin, J.T., Luyssaert, S., Kaplan, L.A., Aufdenkampe, A.K., Richter, A., Tranvik, L.J.: The boundless carbon cycle. Nat. Geosci. **2**(9), 598–600 (2009)
8. Birkett, C.M.: Radar altimetry: a new concept in monitoring lake level changes. Am. Geophys. Union Trans. **75**(24), 273 (1994)
9. Birkett, C.M.: Surface water dynamics in the Amazon Basin: application of satellite radar altimetry. J. Geophys. Res. **107**(D20), 8059 (2002)
10. Birkett, C.M.: Contribution of the TOPEX NASA Radar Altimeter to the global monitoring of large rivers and wetlands. Water Resour. Res. **34**(5), 1223–1239 (1998)
11. Birkett, C.M., Mason, I.M.: A new global lakes database for a remote sensing program studying climatically sensitive large lakes. J. Great Lakes Res. **21**(3), 307–318 (1995)
12. Bishop, M.P., Shroder, J.F., Jr, Hickman, B.L.: SPOT panchromatic imagery and neural networks for information extraction in a complex mountain environment. Geocarto Int. (1999)
13. Burns, N.M., Rutherford, J.C., Clayton, J.S.: A monitoring and classification system for new Zealand Lakes and Reservoirs. Lake Reserv. Manag. **15**(4), 255–271 (1999)
14. Cole, J.J., Prairie, Y.T., Caraco, N.F., McDowell, W.H., Tranvik, L.J., Striegl, R.G., Duarte, C.M., Kortelainen, P., Downing, J.A., Middelburg, J.J., et al.: Plumbing the global carbon cycle: integrating inland waters into the terrestrial carbon budget. Ecosystems **10**(1), 172–185 (2007)
15. Crist, E.P., Cicone, R.C.: A physically-based transformation of thematic mapper data—the tm tasseled cap. IEEE Trans. Geosci. Remote Sens. **GE-22**(3), 256–263 (1984)
16. Daya Sagar, B.S., Gandhi, G., Prakasa Rag, B.S.: Applications of mathematical morphology in surface water body studies. Int. J. Remote Sens. **16**(8), 1495–1502 (1995)
17. Dekker, A.G., Phinn, S.R., Anstee, J.: Intercomparison of shallow water bathymetry, hydro-optics, and benthos mapping techniques in Australian and Caribbean coastal environments. Limnol. and Oceanogr. Methods (2011)

18. Di, K., Wang, J., Ma, R., Li, R.: Automatic shoreline extraction from high-resolution IKONOS satellite imagery. In: Proceeding of ASPRS 2003 Annual Conference (2003)
19. Doxaran, D., Froidefond, J.-M., Lavender, S., Castaing, P.: Spectral signature of highly turbid waters. Remote Sens. Environ. **81**(1), 149–161 (2002)
20. Duan, Z., Bastiaanssen, W.G.M.: Estimating water volume variations in lakes and reservoirs from four operational satellite altimetry databases and satellite imagery data. Remote Sens. Environ. **134**, 403–416 (2013)
21. Feyisa, G.L., Meilby, H., Fensholt, R., Proud, S.R.: Automated Water Extraction Index: a new technique for surface water mapping using Landsat imagery. Remote Sens. Environ. **140**, 23–35 (2014)
22. Foody, G.M., Muslim, A.M., Atkinson, P.M.: Super-resolution mapping of the waterline from remotely sensed data. Int. J. Remote Sens. **26**(24), 5381–5392 (2005)
23. Frappart, F., Minh, K.D., L'Hermitte, J., Cazenave, A., Ramillien, G., Toan, T.L., Mognard-Campbell, N.: Water volume change in the lower Mekong from satellite altimetry and imagery data. Geophys. J. Int. **167**(2), 570–584 (2006)
24. Frappart, F., Papa, F., Famiglietti, J.S., Prigent, C., Rossow, W.B., Seyler, F.: Interannual variations of river water storage from a multiple satellite approach: a case study for the Rio Negro River basin. J. Geophys. Res. **113**(D21), D21104 (2008)
25. Frazier, P.S., Page, K.J., et al.: Water body detection and delineation with landsat tm data. Photogram. Eng. Remote Sens. **66**(12), 1461–1468 (2000)
26. Frohn, R.C., Hinkel, K.M., Eisner, W.R.: Satellite remote sensing classification of thaw lakes and drained thaw lake basins on the North Slope of Alaska. Remote Sens. Environ. **97**(1), 116–126 (2005)
27. Gao, Bo-cai: NDWIA normalized difference water index for remote sensing of vegetation liquid water from space. Remote Sens. Environ. **58**(3), 257–266 (1996)
28. Gao, H., Birkett, C., Lettenmaier, D.P.: Global monitoring of large reservoir storage from satellite remote sensing. Water Resour. Res. **48**(9), n/a–n/a (2012)
29. Giordano, M.A., Wolf, A.T.: Sharing waters: post-rio international water management. In: Natural Resources Forum, vol. 27, pp. 163–171. Wiley Online Library (2003)
30. Gleason, C.J., Smith, L.C.: Toward global mapping of river discharge using satellite images and at-many-stations hydraulic geometry. Proc. Natl. Acad. Sci. **111**(13), 91–4788 (2014)
31. Gleick, H.P.: Global freshwater resources: soft-path solutions for the 21st century. Science **302**(5650), 8–1524 (2003)
32. Grabs, T., Seibert, J., Bishop, K., Laudon, H.: Modeling spatial patterns of saturated areas: a comparison of the topographic wetness index and a dynamic distributed model. J. Hydrol. **373**(1–2), 15–23 (2009)
33. Grossmann, M.: Cooperation on africa's international waterbodies: information needs and the role of information-sharing. Editors **173** (2006)
34. Harrison, J.A., Maranger, R.J., Alexander, R.B., Giblin, A.E., Jacinthe, P.-A., Mayorga, E., Seitzinger, S.P., Sobota, D.J., Wollheim, W.M.: The regional and global significance of nitrogen removal in lakes and reservoirs. Biogeochemistry **93**(1–2), 143–157 (2009)
35. Hess, L.: Dual-season mapping of wetland inundation and vegetation for the central Amazon basin. Remote Sens. Environ. **87**(4), 404–428 (2003)
36. Hinkel, K.M., Eisner, W.R., Bockheim, J.G., Nelson, F.E., Peterson, K.M.: Spatial extent, age, and carbon stocks in drained thaw lake basins on the barrow peninsula, alaska. Arct. Antarct. Alp. Res. **35**(3), 291–300 (2003)
37. Huang, L., Li, Z., Tian, B.-S., Chen, Q., Liu, J.-L., Zhang, R.: Classification and snow line detection for glacial areas using the polarimetric SAR image. Remote Sens. Environ. **115**(7), 1721–1732 (2011)
38. Hui, F., Bing, X., Huang, H., Qian, Y., Gong, P.: Modelling spatial-temporal change of poyang lake using multitemporal landsat imagery. Int. J. Remote Sens. **29**(20), 5767–5784 (2008)
39. Immerzeel, W.W., Droogers, P., De Jong, S.M., Bierkens, M.F.P.: Large-scale monitoring of snow cover and runoff simulation in himalayan river basins using remote sensing. Remote Sens. Environ. **113**(1), 40–49 (2009)

40. Islam, A.S., Bala, S.K., Haque, M.A.: Flood inundation map of Bangladesh using MODIS time-series images. J. Flood Risk Manag. **3**(3), 210–222 (2010)
41. Jain, S.K., Lohani, A.K., Singh, R.D., Chaudhary, A., Thakural, L.N.: Glacial lakes and glacial lake outburst flood in a Himalayan basin using remote sensing and GIS. Nat. Hazards **62**(3), 887–899 (2012)
42. Jain, S.K., Lohani, A.K., Singh, R.D., Chaudhary, A., Thakural, L.N.: Delineation of flood-prone areas using remote sensing techniques. Water Resour. Manage. **19**(4), 333–347 (2005)
43. Jiang, Z., Qi, J., Shiliang, S., Zhang, Z., Jiaping, W.: Water body delineation using index composition and HIS transformation. Int. J. Remote Sens. **33**(11), 3402–3421 (2012)
44. Kloiber, S.M., Brezonik, P.L., Bauer, M.E.: Application of Landsat imagery to regional-scale assessments of lake clarity. Water Res. **36**(17), 4330–4340 (2002)
45. Kloiber, S.M., Brezonik, P.L., Olmanson, L.G., Bauer, M.E.: A procedure for regional lake water clarity assessment using Landsat multispectral data. Remote Sens. Environ. **82**(1), 38–47 (2002)
46. Knight, A.W., Tindall, D.R., Wilson, B.A.: A multitemporal multiple density slice method for wetland mapping across the state of queensland, australia. Int. J. Remote Sens. **30**(13), 3365–3392 (2009)
47. Lacava, T., Cuomo, V., Di Leo, E.V., Pergola, N., Romano, F., Tramutoli, V.: Improving soil wetness variations monitoring from passive microwave satellite data: the case of April 2000 Hungary flood. Remote Sens. Environ. **96**(2), 135–148 (2005)
48. Larsen, S., Andersen, T., Hessen, D.: Climate change predicted to cause severe increase of organic carbon in lakes. Glob. Change Biol. **17**(2), 1186–1192 (2011)
49. Li, J., Sheng, Y.: An automated scheme for glacial lake dynamics mapping using Landsat imagery and digital elevation models: a case study in the Himalayas. Int. J. Remote Sens. **33**(16), 5194–5213 (2012)
50. Li, S., Sun, D., Goldberg, M., Stefanidis, A.: Derivation of 30-m-resolution water maps from TERRA/MODIS and SRTM. Remote Sens. Environ. **134**, 417–430 (2013)
51. Li, S., Sun, D., Yunyue, Y., Csiszar, I., Stefanidis, A., Goldberg, M.D.: A new Short-Wave infrared (SWIR) method for quantitative water fraction derivation and evaluation with EOS/MODIS and Landsat/TM data. IEEE Trans. Geosci. Remote Sens. **51**(3), 1852–1862 (2013)
52. Lira, J.: Segmentation and morphology of open water bodies from multispectral images. Int. J. Remote Sens. **27**(18), 4015–4038 (2006)
53. Loriaux, T., Casassa, G.: Evolution of glacial lakes from the Northern Patagonia Icefield and terrestrial water storage in a sea-level rise context. Glob. Planet. Change **102**, 33–40 (2013)
54. Martinez, J., Letoan, T.: Mapping of flood dynamics and spatial distribution of vegetation in the Amazon floodplain using multitemporal SAR data. Remote Sens. Environ. **108**(3), 209–223 (2007)
55. McFeeters, S.K.: The use of the Normalized Difference Water Index (NDWI) in the delineation of open water features. Int. J. Remote Sens. **17**(7), 1425–1432 (1996)
56. Micklin, P.P.: Desiccation of the aral sea: a water management disaster in the soviet union. Science **241**(4870), 1170–1176 (1988)
57. Niu, Z.G., Gong, P., Cheng, X., Guo, J.H., Wang, L., Huang, H.B., Shen, S.Q., Wu, Y.Z., Wang, X.F., Wang, X.W., Ying, Q., Liang, L., Zhang, L.N., Wang, L., Yao, Q., Yang, Z.Z., Guo, Z.Q.,Dai, Y.J.: Geographical characteristics China's of wetlands derived from remotely sensed data. Sci. China Ser. D: Earth Sci. **52**(6), 723–738 (2009)
58. Ouma, Y.O., Tateishi, R.: A water index for rapid mapping of shoreline changes of five East African Rift Valley lakes: an empirical analysis using Landsat TM and ETM+ data. Int. J. Remote Sens. **27**(15), 3153–3181 (2006)
59. Overton, I.C.: Modelling floodplain inundation on a regulated river: integrating GIS, remote sensing and hydrological models. River Res. Appl. **21**(9), 991–1001 (2005)
60. Palmer, M.A., Reidy Liermann, C.A., Nilsson, C., Flörke, M., Alcamo, J., Lake, P.S., Bond, N.: Climate change and the world's river basins: anticipating management options. Front. Ecol. Environ. **6**(2), 81–89 (2008)

61. Paul, F., Huggel, C., Kääb, A.: Combining satellite multispectral image data and a digital elevation model for mapping debris-covered glaciers. Remote Sens. Environ. **89**(4), 510–518 (2004)
62. Phan, V.H., Lindenbergh, R., Menenti, M.: ICESat derived elevation changes of Tibetan lakes between 2003 and 2009. Int. J. Appl. Earth Obs. Geoinformation **17**, 12–22 (2012)
63. Prost, G.L.: Remote Sensing for Geologists: A Guide to Image Interpretation. CRC Press (2002)
64. Pulvirenti, L., Chini, M., Pierdicca, N., Guerriero, L., Ferrazzoli, P.: Flood monitoring using multi-temporal COSMO-SkyMed data: Image segmentation and signature interpretation. Remote Sens. Environ. **115**(4), 990–1002 (2011)
65. Quincey, D.J., Richardson, S.D., Luckman, A., Lucas, R.M., Reynolds, J.M., Hambrey, M.J., Glasser, N.F.: Early recognition of glacial lake hazards in the Himalaya using remote sensing datasets. Global and Planet. Change **56**(1–2), 137–152 (2007)
66. Reis, S.: Temporal monitoring of water level changes in Seyfe Lake using remote sensing. Hydrol. Process. **22**(22), 4448–4454 (2008)
67. Ryu, J., Won, J., Min, K.: Waterline extraction from Landsat TM data in a tidal flat: A case study in Gomso Bay, Korea. Remote Sens. Environ. **83**(3), 442–456 (2002)
68. Sakamoto, T., Van Nguyen, N., Kotera, A., Ohno, H., Ishitsuka, N., Yokozawa, M.: Detecting temporal changes in the extent of annual flooding within the Cambodia and the Vietnamese Mekong Delta from MODIS time-series imagery. Remote Sens. Environ. **109**(3), 295–313 (2007)
69. Sass, G.Z., Creed, I.F., Bayley, S.E., Devito, K.J.: Understanding variation in trophic status of lakes on the Boreal Plain: a 20 year retrospective using Landsat TM imagery. Remote Sens. Environ. **109**(2), 127–141 (2007)
70. Schumann, G.J.-P., Neal, J.C., Mason, D.C., Bates, P.D.: The accuracy of sequential aerial photography and SAR data for observing urban flood dynamics, a case study of the UK summer 2007 floods. Remote Sens. Environ. **115**(10), 2536–2546 (2011)
71. Shibuo, Y., Jarsjö, J., Destouni, G.: Hydrological responses to climate change and irrigation in the aral sea drainage basin. Geophys. Res. Lett. **34**(21) (2007)
72. Simpson, J.J., Keller, R.H.: An improved fuzzy logic segmentation of sea ice, clouds, and ocean in remotely sensed arctic imagery. Remote Sens. Environ. **54**(3), 290–312 (1995)
73. Sivanpillai, R., Miller, S.N.: Improvements in mapping water bodies using ASTER data. Ecol. Inf. **5**(1), 73–78 (2010)
74. Song, C., Huang, B., Ke, L.: Modeling and analysis of lake water storage changes on the Tibetan Plateau using multi-mission satellite data. Remote Sens. Environ. **135**, 25–35 (2013)
75. Stave, K.A.: A system dynamics model to facilitate public understanding of water management options in Las Vegas, Nevada. J. Environ. Manage. **67**(4), 13–303 (2003)
76. Steinbacher, F., Pfennigbauer, M., Aufleger, M., Ullrich, A.: High resolution airborne shallow water mapping. ISPRS-Int. Arch. Photogrammetry Remote Sens. Spat. Inf. Sci. **1**, 55–60 (2012)
77. Subramaniam, S., Babu, A.V.S., Roy, P.S.: Automated water spread mapping using resourcesat-1 awifs data for water bodies information system. IEEE J. Sel. Top. Appl. Earth Obs. Remote Sens. **4**(1):205–215, 2011
78. Sun, D., Yunyue, Y., Goldberg, M.D.: Deriving water fraction and flood maps from MODIS images using a decision tree approach. IEEE J. Sel. Top. Appl. Earth Obs. Remote Sens. **4**(4), 814–825 (2011)
79. Sun, D., Yunyue, Y., Zhang, R., Li, S., Goldberg, M.D.: Towards operational automatic flood detection using EOS/MODIS data. Photogram. Eng. Remote Sens. **78**(6), 637–646 (2012)
80. Sun, F., Sun, W., Chen, J., Gong, P.: Comparison and improvement of methods for identifying waterbodies in remotely sensed imagery. Int. J. Remote Sens. **33**(21), 6854–6875 (2012)
81. Temimi, M., Leconte, R., Chaouch, N., Sukumal, P., Khanbilvardi, R., Brissette, F.: A combination of remote sensing data and topographic attributes for the spatial and temporal monitoring of soil wetness. J. Hydrol. **388**(1–2), 28–40 (2010)

82. Temimi, M., Leconte, R., Brissette, F., Chaouch, N.: Flood monitoring over the Mackenzie River Basin using passive microwave data. Remote Sens. Environ. **98**(2–3), 344–355 (2005)
83. Tidwell, V.C., Moreland, B.D., Zemlick, K.M., Roberts, B.L., Passell, H.D., Jensen, D., Forsgren, C., Sehlke, G., Cook, M.A., King, C.W., Larsen, S.: Mapping water availability, projected use and cost in the western United States. Environ. Res. Lett. **9**(6), 064009 (2014)
84. Töyrä, J., Pietroniro, A.: Towards operational monitoring of a northern wetland using geomatics-based techniques. Remote Sens. Environ. (2005)
85. Töyrä, J., Pietroniro, A., Martz, L.W.: Multisensor hydrologic assessment of a freshwater wetland. Remote Sens. Environ. **75**(2), 162–173 (2001)
86. Töyrä, J., Pietroniro, A., Martz, L.W., Prowse, T.D.: A multi-sensor approach to wetland flood monitoring. Hydrol. Process. **16**(8), 1569–1581 (2002)
87. Tranvik, L.J., Downing, J.A., Cotner, J.B., Loiselle, S.A., Striegl, R.G., Ballatore, T.J., Dillon, P., Finlay, K., Fortino, K., Knoll, L.B., et al.: Lakes and reservoirs as regulators of carbon cycling and climate. Limnol. Oceanogr. **54**(6), 2298–2314 (2009)
88. Tucker, C.J.: Red and photographic infrared linear combinations for monitoring vegetation. Remote Sens. Environ. **8**(2), 127–150 (1979)
89. Verpoorter, C., Kutser, T., Seekell, D.A., Tranvik, L.J.: A global inventory of lakes based on high-resolution satellite imagery. Geophys. Res. Lett. **41**(18), 6396–6402 (2014)
90. Verpoorter, C., Kutser, T., Tranvik, L.: Automated mapping of water bodies using Landsat multispectral data. Limnol. Oceanogr.: Methods **10**, 1037–1050 (2012)
91. Vörösmarty, C.J., Green, P., Salisbury, J., Lammers, R.B.: Global water resources: vulnerability from climate change and population growth. Science **289**(5477), 284–288 (2000)
92. Vörösmarty, C.J., McIntyre, P.B., Gessner, M.O., Dudgeon, D., Prusevich, A., Green, P., Glidden, S., Bunn, S.E., Sullivan, C.A., Liermann, C.R., et al.: Global threats to human water security and river biodiversity. Nature **467**(7315), 555–561 (2010)
93. Wang, L., Sousa, W.P., Gong, P.: Integration of object-based and pixel-based classification for mapping mangroves with IKONOS imagery. Int. J. Remote Sens. **25**(24), 5655–5668 (2004)
94. Wang, X., Gong, P., Zhao, Y., Yue, X., Cheng, X., Niu, Z., Luo, Z., Huang, H., Sun, F., Li, X.: Water-level changes in China's large lakes determined from ICESat/GLAS data. Remote Sens. Environ. **132**, 131–144 (2013)
95. Worm, B., Barbier, E.B., Nicola Beaumont, J., Duffy, E., Folke, C., Halpern, B.S., Jackson, J.B.C., Lotze, H.K., Micheli, F., Palumbi, S.R., et al.: Impacts of biodiversity loss on ocean ecosystem services. Science **314**(5800), 787–790 (2006)
96. Baiqing, X., Cao, J., Hansen, J., Yao, T., Joswia, D.R., Wang, N., Wu, G., Wang, M., Zhao, H., Yang, W., Liu, X., He, J.: Black soot and the survival of Tibetan glaciers. Proc. Natl. Acad. Sci. **106**(52), 8–22114 (2009)
97. Hanqiu, Xu: Modification of normalised difference water index (NDWI) to enhance open water features in remotely sensed imagery. Int. J. Remote Sens. **27**(14), 3025–3033 (2006)
98. Zhang, G., Xie, H., Kang, S., Yi, D., Ackley, S.F.: Monitoring lake level changes on the Tibetan Plateau using ICESat altimetry data (2003–2009). Remote Sens. Environ. **115**(7), 1733–1742 (2011)
99. Zhang, S., Gao, H., Naz, B.S.: Monitoring reservoir storage in south asia from multisatellite remote sensing. Water Resour. Res. (2014)
100. Zhu, X.: Remote sensing monitoring of coastline change in pearl river estuary. 22nd Asian Conference on Remote Sensing, vol. 5, p. 9 (2001)

Installing Electric Vehicle Charging Stations City-Scale: How Many and Where?

Marjan Momtazpour, Mohammad C. Bozchalui, Naren Ramakrishnan and Ratnesh Sharma

Abstract Electric Vehicles (EVs) are touted as the sustainable alternative to reduce our over-reliance on fossil fuels and stem our excessive carbon emissions. As the use of EVs becomes more widespread, planners in large metropolitan areas have begun thinking about the design and installation of charging stations city-wide. Unlike gas-based vehicles, EV charging requires a significant amount of time and must be done more periodically, after relatively shorter distances. We describe a KDD framework to plan the design and deployment of EV charging stations over a city. In particular, we study this problem from the economic viewpoint of the EV charging station owners. Our framework integrates user route trajectories, owner characteristics, electricity load patterns, and economic imperatives in a coordinated clustering framework to optimize the locations of stations and assignment of user trajectories to (nearby) stations. Using a dataset involving over a million individual movement patterns, we illustrate how our framework can answer many important questions about EV charging station deployment and profitability.

M. Momtazpour (✉) · N. Ramakrishnan
Department of Computer Science, Virginia Tech, Blacksburg, VA 24060, USA
e-mail: marjan@cs.vt.edu

N. Ramakrishnan
e-mail: naren@cs.vt.edu

M.C. Bozchalui · R. Sharma
NEC Laboratories America, Inc, Cupertino, CA 95014, USA
e-mail: mohammad@nec-labs.com

R. Sharma
e-mail: ratnesh@nec-labs.com

J. Lässig et al. (eds.), *Computational Sustainability*,
Studies in Computational Intelligence 645,
DOI 10.1007/978-3-319-31858-5_8

1 Introduction

In the last decade, electric vehicles (EVs) have been considered a promising solution for some environmental and economical issues. Fast decline of fossil fuels and global warming have increased the interest of policy makers in developed countries to use sustainable approaches to energy production, distribution, and consumption [1]; EVs have been touted for their potential to dramatically reduce fossil fuel consumption and CO_2 emissions [2].

To operationalize and encourage EV usage, charging stations should be installed in multiple areas of a city. In large metropolitan areas with a significant number of EVs, charging stations must be installed in carefully selected locations. As a matter of fact, charging an EV is different from refueling a traditional gas-based car: EV charging takes much longer and places a significant amount of load on the electric grid [2]. Furthermore, compared to traditional cars, EVs must be recharged after relatively shorter distances. Proper placement of charging stations can result in optimal distribution of electricity load, maximization of revenue of service providers, and lead to increased availability of charging stations, and reduced range anxiety.

While charging station placement is an important task for EV deployment in urban areas, there is a relatively small number of prior research in this area (e.g., see [3–5], and our own work [6]) and all aim to locate charging stations to maximize the meeting of demands. In a comprehensive planning effort, however, it is crucial to consider economic factors in design of charging infrastructure for EVs to ensure financial feasibility as well as long-term economic growth. Various business models can be considered for EV charging station infrastructure, and in fact, EV charging infrastructure installation will be driven by models that reflect the economic benefits on top of policy objectives.

In our previous work [6], we propose a solution for charging station placement problem without specific assignment of EVs to charging stations. In [6], we developed a coordinated clustering formulation to identify a set of locations that can be considered as the best candidates for charging stations. The locations were determined to be those that have a low electricity load, and where a significant number of EV owners spend a considerable duration of time. The drawback of our proposed method in [6] is that it did not consider a concrete economic model for charging station placement. In fact, in [6], charging stations are placed based on the stay points of EV owners and distribution of electricity consumption in the city. Therefore, that approach may result in placements which are economically sub-optimal. Furthermore, in [6], we did not consider the trajectory of EVs, which in turn results in unacceptable detours.

In this book chapter, we propose a new integrated framework where the centralized assignment of EVs is addressed simultaneously with the charging station placement problem. This integrated framework solves an optimization problem that simultaneously considers revenue of charging station owners and the trajectory of EV owners. In this work, an economic model is formulated that takes into account the costs and benefits of installing and operating charging stations from their owner's perspective. In this model, charging station owners provide infrastructure, and own and operate EV charging stations. No extra incentive is considered for the charging station owners and it is assumed that they will be charged the same rates as other mid-sized commercial customers for buying electricity from the utility company. Charging stations sell electricity to EV owners at a fixed, flat rate. Furthermore, we use trajectory mining to find routes that could host popular locations where EV owners might desire to recharge their cars. We applied trajectory clustering on this dataset which helps us to install charging stations proximal to high-traffic roads, in order to reduce possible detours to reach charging stations. The trajectory of each individual in a typical day is derived through the use of APIs such as Google Maps. The results of this step are integrated to our final optimization equation to situate charging stations near high-traffic roads in order to reduce possible detours to reach charging stations. Finally, the economical model, results of trajectory mining, and information about each individual driving path of EV owners are fed into an integrated optimization problem. This optimization problem attempts to maximize the revenue of charging station owners, minimize distances of charging stations to high-traffic routes, minimize distances of charging stations to stay points of EVs, and minimize number of failure to find an appropriate charging station for an EV. Furthermore, using KL-Divergence, the optimization problem tries to place charging stations in a way that results in a uniform distribution of charging assignments.

We outline a KDD framework, involving coordinated clustering, to design and deploy EV charging stations over a city. Our key contributions are:

1. An integration of diverse datasets, including synthetic populations (capturing over 1.5 million individuals), their profiles, and trajectories of driving, to inform the choice of locations that are most promising for EV charging station placement. We solve the 'How many?' and 'Where?' problem using a coordinated clustering framework that integrates multiple considerations. We focus on the modeling of downtown areas since previous studies have shown that public EV charging infrastructures should be focused on big urban centers [2]. We use trajectory mining to detect popular roads EV owners are likely to use when they need to recharge their vehicles, and integrate this information in charging station deployment. In particular, our framework situates charging stations near high-traffic roads in order to reduce possible detours to reach charging stations.
2. Unlike our prior work [6], we formulate the EV charging station placement problem in both economic and user terms: the financial benefits to an EV charging station owner and the convenience benefits to EV owners are integrated into our framework. Empirical results reveal key distinctions between taking economic factors into account versus otherwise.
3. We conduct extensive empirical investigations into the practical feasibility of EV charging station placement w.r.t. multiple considerations: e.g., how many users in a population are serviced, how effectively are stations utilized, differences among varying types of charging infrastructure, and the need for storage units in charging stations.

2 Related Work

Charging infrastructure design: Relevant prior work in this area include [3–7]. Frade et al. developed a maximum covering model to locate charging stations to maximize demand [3]. In [5], a two-step model is proposed to create demand clusters by hierarchical clustering, then a simple assignment strategy is used to assign charging stations to demand clusters. In [4], a game-theoretic approach is used to investigate interactions among availability of public charging and route choices of EVs. In our prior work [6], we developed a coordinated clustering formulation to identify a set of locations that can be considered as the best candidates for charging station placement. The locations were determined to be those that have a low existing load, and where a significant number of EV owners spend a considerable duration of time. In [7], behavioral models are developed to predict when and where vehicles are likely to be parked, and aims to reflect parking demands in the optimization assignment.

Interactions with the smart grid: In addition to the problem of charging station placement, EV penetration in urban areas has been explored with respect to interactions between grid infrastructure and urban populations. City behavior is simulated by agent-based systems in terms of agents with a view toward having decentralized systems and maximizing profits [1]. Swanson et al. in [8] investigated the use of linear discriminant analysis (LDA) in assessing the probable level of EV adoption. Energy storage systems, systems that are used when there is not enough power available from grid, are addressed in [9]. In [10], a solution is proposed to balance energy production against its consumption. In addition, authors in [11] try to design a general architecture in smart grid to have a significant gains in net cost/profit with particular emphasis on electric vehicles.

Mobility modeling: There are many studies that consider mobility of vehicles in urban areas and in most of the cases, GPS datasets have been used as a popular source for modeling and mining in urban computing contexts, e.g., [12–14]. Example applications include anomaly detection [12] and taxi recommender systems [14]. In taxi recommender systems in particular [14], the ultimate goal is to maximize taxi-driver profits and minimize passengers' waiting times. Mining mobility patterns of cars and people has been used to determine points of interest for tourists [15] and for routing and route recommendation [16]. In [13], Yuan et al. proposed a method to discover areas with different functionalities based on people movements. Finally, in [17], clusters of moving objects in a noisy stadium environment are detected using the DBSCAN algorithm [18].

To the best of our knowledge, the problem tackled in this paper is unique, and the methodology we propose integrates a variety of data sources with data mining/optimization techniques.

3 Methodology

The datasets utilized by our approach and the overall methodology are depicted in Fig. 1. As shown, one of the primary datasets we consider is a synthetic population dataset representing the city of Portland which contains details of 1,615,860 people and 243,423 locations out of which 1,779 are located in the downtown area. Detailed information about this dataset is available at [19]. Next, information about mobility of people is provided in terms of start and end points and time of travel. Using this information, we can determine the trajectory of every individual in a typical day through the use of APIs such as Google Maps. A total of 8,922,359 movements are available in this dataset. Finally, we have available electricity consumption data to determine the initial load of each building based on the number of residents of the building at a specified time (organized by NEC Labs, America).

The first step of our methodology is to discover location functionalities and to characterize electricity loads. As in our previous work [6], we utilize an information bottleneck type approach [20] to characterize locations and integrated the electricity load information to characterize usage patterns across locations. In this step, we cluster locations based on geographical proximity such that resulting clusters are highly informative of location functionalities. Then, we integrate information about electricity load profiles to characterize electricity usage patterns.

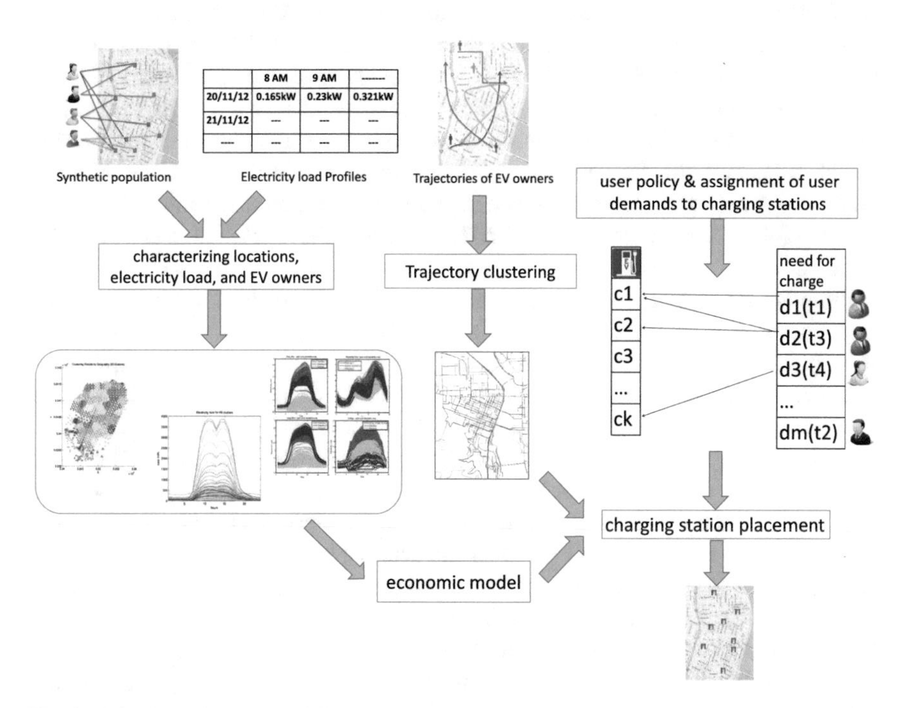

Fig. 1 Overview of our methodology

Second, we use trajectory clustering to find routes that could host popular locations where EV owners might desire to recharge their cars. To determine trajectories, we define a specific subset of people who are characterized using high-income attributes (as the likely owners of EVs). After locating the homes of these users, we can determine their trajectories and their start/stop locations. Based on this data, we can estimate their travel distances, and in turn estimate charging requirements of EVs, during a day. Since the maximum distance that a fully charged EV can travel is less than 100 km [2], it is highly likely that a significant number of them will need to be recharged en-route to their destinations. By clustering the trajectories, we can plan to install charging stations proximal to high-traffic roads, in order to reduce possible detours to reach charging stations.

Third, we develop an economic model that encapsulates costs of purchasing energy from the grid and other such expenses. Fourth, we identify demands based on our expectations about how users will behave. Finally, all this information provides the raw material for defining the charging station placement problem using clustering and optimization. Each of these stages are detailed next.

3.1 Characterizing Locations and EV Owners

The first step of the proposed data mining approach for EV infrastructure design is to characterize EV users and locations in the respective area. This step is similar to that in our previous work [6] and, hence, we provide an abridged summary of it.

Based on current trends, only a small percentage of people (6 % of people in the US) use EVs [21, 22]; in our study we explored a hypothetical scenario that considers a penetration for EVs in the Portland area to be 6.31 % of 329,218 people in our dataset. This assumption is realistic if various penetration scenarios in forecasted EV adoption between years 2012 and 2022 are to be believed [23], and can be easily modified.

From the synthetic population dataset, we can identify the locations a person visits, the duration of stay at each (stay points), and the purpose of the visit (e.g., work, leisure). We first begin by characterizing location with a view toward defining the specific purpose of the location. We focus on 1779 locations in the downtown Portland area whose attributes are given by a 9-length profile vector $P = [p_1, p_2, \ldots, p_9]$, where p_i is the number of travels incident on that location for the ith purpose. Specific purposes of each location (and cluster of nearby locations) can be used to determine electricity load distribution patterns. To uncover such patterns, we first cluster locations geographically and then characterize each of the discovered clusters using typical data available from public data sources such as the California End User Survey (CEUS). In addition to these kinds of patterns, we compute the electricity load leveraging the these patterns but w.r.t. our network model of the urban environment (by considering the average square footage occupied by one person in each specific location). Based on some exploratory data analysis, we selected a weekday (specifically, 18th March, 2011) and used the electricity load data of this day to map to the network model. More details are available at [6].

3.2 *Trajectory Mining*

The emergence of GPS-equipped devices has sprung a veritable cottage industry in the area of location and trajectory mining. One broad aim of trajectory mining (clustering) is to find similar routes in a dataset, but other applications have also been explored (e.g., see [24–26]). Most research in trajectory clustering is inspired by density-based clustering approaches such as DBSCAN and OPTICS. Leveraging such clustering methods, the authors in [27] propose a new framework (Traclus) for trajectory clustering which aims to discover common sub-trajectories. In applications where we have regions of special interest, finding common sub-trajectories is beneficial. In Traclus, each trajectory is partitioned into a set of line segments. Then, similar line segments are grouped together to form clusters of sub-trajectories [27]. This method has been proven to be effective in extracting similar portions of trajectories even when the whole trajectories are not similar. We employ this approach here to detect potential sub-routes where EV owners are more likely to travel and, thus, in need of charging.

3.3 *Economic Model for Profit Maximization*

The principal goal of this paper is to place charging stations in appropriate locations in order to maximize profits of charging station owners. Tran et al. [28] have studied cooperation of companies for profit maximization in dynamic systems. They have used regression and hierarchical agglomerative clustering to reveal optimal organizational substructures. Such approaches are not applicable here since they assume the locations (of markets) to be known and place restrictive assumptions on pricing schema.

The primary goal of charging station owners is to maximize revenue and profits by attracting enough customers during a day. The profit is defined as the difference between expense and income (revenue). Let us assume that R_i is the profit for charging station (location) i, which is the difference between the payments that the charging station owner will receive (S_i) and the costs spent on providing service to customers (C_i), as shown in Eq. 1. As Eq. 2 illustrates, C_i in turn consists of two elements: the static costs C_0, and the dynamic costs C_p.

$$R_i = S_i - C_i \tag{1}$$

$$C_i = C_p + C_0 \tag{2}$$

Static cost, C_0, is the initial cost for setting up a charging station, which includes the operational cost for installation and for storage units. Here, we calculate these costs for a single day, and thus assume an amortization function that estimates the installation cost for one day (e.g. if the installation cost will amortize in six years: $C_0 = \frac{InstallationCost}{6\times 365}$). Dynamic cost C_p, is the cost for the energy that the charging station will buy from the grid in order to service EV owners. The dynamic cost C_p, consists of two parameters: the cost of buying energy from the grid during the day (morning to evening), and the costs associated with recharging storage (during the night). These two parameters are denoted by C_b and C_r, respectively.

$$C_p = C_r + C_b \tag{3}$$

3.3.1 Calculating C_r

C_r is the payment that a charging station owner pays for charging the storage during the night (if needed). Typically, storage will be charged at night and used during the day and it should be sized to cover a day's net load.

$$C_r = P_{buy,night} \times StorageSize \tag{4}$$

where $P_{buy,night}$ is the price of off-peak hours that storage owner will pay to recharge the storage. $StorageSize$ is calculated through the following steps.

Suppose f is the load of building after considering EVs. Thus f is the Initial load of building $InitLoad_{i,t}$, and the load imposed by EVs. $Dload_{d,t}$ is the amount of electricity needed for user d at time t and n_i is the number of EVs receiving service by charging station i during a day.

$$f(t) = InitLoad_{i,t} + \sum_{d=1}^{n_i} (Dload_{d,t}) \tag{5}$$

In order to calculate the amount of storage for a particular charging station, we must calculate the number of EVs serviced by this charging station at each particular hour n_i. Here, we assume capacity of each building is constant and equals the maximum value of load of the building before introducing EVs:

$$capacity_i = \max_{0\leq t\leq 24} InitLoad_i \tag{6}$$

The size of required storage should be calculated from the area below the curve of new electricity load (that is f) (kW $\times$ h) and above the capacity (net peak load)(kW). X is the difference of load after EVs and capacity of building. Clearly, $StorageSize$ is a summation of X over time:

$$X(t) = \begin{cases} f(t) - capacity_i & \text{if } f(t) > capacity_i \\ 0 & \text{otherwise} \end{cases} \tag{7}$$

$$StorageSize = \int_0^{24} (X(t))\, dt \tag{8}$$

3.3.2 Calculating C_b

Calculation of C_b consists of three elements: Basic charges, Energy charges, and Demand charges [29].

$$C_b = C_{basic} + C_{energy} + C_{demand} \tag{9}$$

Basic charges is a constant charge (\$240 per month[1]) [29] and energy charges is a multiplication of energy purchased at time t (kWh) in TOU rate at time t (\$/kWh):

$$C_{energy} = \int_0^{24} (Y(t) - InitLoad_{i,t}) \times P_{buy,t}\, dt \tag{10}$$

where $P_{buy,t}$ is determined based on time of the day (TOU rate) and Y is the amount of load of a building when storage is placed:

$$Y(t) = \begin{cases} capacity_i & \text{if } f(t) > capacity_i \\ f(t) & \text{Otherwise} \end{cases} \tag{11}$$

Demand charges involves facility capacity charges and on-peak demand charges:

$$C_{demand} = C_{FC}/30 + C_{OnPeak\ demand} \tag{12}$$

Facility capacity charges (C_{FC}) for one month is calculated in Eq. 13 [29]:

$$C_{FC} = \begin{cases} capacity_i \times 2.41 & \text{if } capacity_i \leq 200 \\ 482 + (capacity_i - 200) \times 2.14 & \text{otherwise} \end{cases} \tag{13}$$

On-peak demand charges is the maximum on-peak demand of the charging station (in kW) times per kW monthly on-peak demand rates (\$/kW):

$$C_{OnPeak\,demand} = \max_{OnPeak\ t} Y(t) \times 2.67 \tag{14}$$

[1] In this paper all rates are in US dollar.

3.3.3 Calculating S_i

The income of charging station owner is calculated based on the summation of energies that he sells to EV owners over a day:

$$S_i = \sum_{d=1}^{n_i} \int_{t=0}^{24} (P_{sell,t} \times Dload_{d,t})\mathrm{d}t \tag{15}$$

where P_{sell} is price per kW.

The ultimate goal of charging station owner is to maximize his profit (maximize R_i).

3.4 *Modeling Users for Demand Assignment*

Before describing how we model users, it is necessary to review the types of charging stations since it is intricately connected to user behavior. The two basic types are level 2 chargers (240 V AC charging) and DC chargers (500 V). The former are more widespread (can even be installed in residential locations), whereas the latter are speedier to charge (and can be found in business and government buildings). We model users in the following manner: Let us assume that a user desires to travel from location A to location B and that he will stay for a certain time in each location. If during traveling from A to B, he runs out of charge, he will first seek an available charging station in the neighborhood of A. If he can find such a charging station, he will charge there, whether he stays at least 4 h (to charge with level 2) or less (DC). Otherwise, if he could not find any charging stations, or if charging stations are fully occupied at that time, we assume that he is aware of the availability of charging stations in neighborhood of B. This part is the same as before. If there is no charging station in A or in B, he has to charge his car by DC somewhere else along his route. The use of popular routes from trajectory clustering is helpful here where users know that there are charging stations along popular roads.

There are various strategies to the demand assignment problem. For example, [30] solved a task assignment problem with linear programming to maximize resource utilization in load balancing problem in multiple machines. Here, to assign users to charging stations, we use a typical first-in-first-out approach with the goal of uniform distribution of users over charging stations. For this purpose, we start from 1:00 AM to 12:00 AM and we assign each user to the least busy charging station which is located in its neighborhood. Algorithm 1 shows pseudo-code for assigning demands to charging stations for a particular hour.

Algorithm 1: Assignment of Users to Charging Stations

```
Input: Charging Stations, User Demands
Output: Assignment matrix
for each demand, d_i do
    for each charging station, CS_j do
        if distance(d_i, CS_j) ≤ r_0 then
            A_{i,j} = 1;
        end
    end
end
for each charging station, CS_j do
    Δ_j = AvailableSlots_{CS_j} − Σ_i(A_{i,j});
end
for k = 1 to K do
                    /* K is number of charging stations */;
    m = arg max_j(Δ_j);
    for each demand d_i do
        if A_{i,m} = 1 and AvailableSlots_{CS_m} > 0 then
            Assign d_i to charging station m;
            AvailableSlots_{CS_m} = AvailableSlots_{CS_m} − 1;
        end
    end
    Δ_m = −∞;
end
```

The ultimate goal from the user's point of view is to maximize the number of assigned demands as well as reducing costs associated with recharging EVs. Our user policy attempts to reduce the number of failures, i.e., the number of times that EV owners run out of charge and need to switch to traditional gas-based fuel. Also, this policy reduces the cost of charging since charging with level 2 has a higher priority compared to DC charging.

3.5 Charging Station Placement Using Clustering and Optimization

In addition to maximizing charging station owners' profits, we aim to minimize the number of failed (unassigned) demands. To this end, we aim to place charging stations next to major arterial roads and nearby stay points to provide better service for future EVs. Furthermore, we aim to have similar schedules for all charging stations to reduce very crowded or very under-utilized stations. Based on these goals, we can formulate an optimization function as a linear combination of several measures:

$$
\begin{aligned}
F(X) = & -\alpha \times \sum_{i=1}^{K} R_i + \beta \times N_{fail} \\
& + \gamma \times \sum_{t} D_{KL}(\zeta_t \,||\, U(\frac{1}{K}))/24 \\
& + \eta \times \frac{1}{K} \sum_{i=1}^{K} \sum_{p \in \phi} Distance(CS_i, p) \\
& + \theta \times \frac{1}{K} \sum_{i=1}^{K} \sum_{r \in \tau} Distance(CS_i, r).
\end{aligned} \quad (16)
$$

where D_{KL} Kullback Leibler distance and U is the uniform distribution. The goal is to uniformly distribute demands over charging stations. Here R_i is the profit for charging station i. N_{fail} is the total number of failed demands because either their distance from their nearest charging stations was more than r_0 or because the nearest charging stations were fully occupied. τ is a set of trajectory representatives and ϕ is set of stay points. *Distance* calculates the distance of charging stations to popular roads and stay points. α, β, γ, η, and θ are constant coefficients. $\zeta_t = [\zeta_t(1), \ldots, \zeta_t(K)]$ captures the distribution of demands over charging stations at time t. $\zeta_t(i)$ is computed as follows:

$$
\zeta_t(i) = \frac{W_t(i)}{\sum_i W_t(i)} \quad (17)
$$

where $W_t(i)$ is the number of assigned demands to charging station i at time t.

In our work, we further focus on a downtown modeling scenario and thus restrict charging station locations to be in such areas:

$$
\begin{aligned}
& \text{Minimize } F(X) \\
& s.t. X_i \in \text{Downtown}
\end{aligned} \quad (18)
$$

where $X = \{X_1, X_2, \ldots, X_K\}$ contains coordinates of K charging stations.

This set of charging stations contains prototypes of K clusters such that each charging station will cover a certain area and also, distance between charging stations will be maximized. Furthermore, in each area, a charging station will be responsible for future demands in that vicinity.

To optimize the objective function, we first find initial prototypes (representing charging stations) using the k-means algorithm (with geographic coordination of locations as features). Next, we use a bound-constrained optimization (simulated annealing with a maximum iteration of 500) to identify the best prototypes that minimize the objective function and also satisfy the inequality constraints, i.e. points must fall into the downtown region. Simulated annealing is used here because the search space (set of building locations) is discrete. At each iteration of simulated annealing, assignment of users to current prototypes is done with respect to the specified parameters. Calculation of profit and other parameters is done at this step. After convergence, we calculate the profit, storage size, utilization, and assigned ratio of trajectories for this final solution.

4 Experiments

Our evaluation is focused on answering the following questions:

1. Which routes are popular among EV owners?
2. How many public charging stations are necessary to serve EV owner needs?
3. What are the load profiles of the designed charging stations?

Table 1 shows parameter settings for our experiments. The price of selling energy to customer is 49 cents per kilowatt hour [31]. Also, $P_{buy,night}$ and $P_{buy,t}$ are calculated based on [31]. Also, installation cost of storage is set to 100 dollars per kWh and the installation cost of chargers in charging station is 4000 dollars [32]. For the time period of amortization, we assume six years based on [2]. We assume that available slots in each charging station is at most 10 EVs at each hour. We also assume that people can charge their EVs if they can find a charging station 800 m (i.e. walking distance) away from their current location ($r_0 = 800$ m). Furthermore, in our experiments, EV owners are assumed to have chargers in their houses and, hence, are presumed to use public charging stations during the day (and recharge again during the night [2]). On-peak hours are determined based on the nature of our dataset, that is from 6 AM to 10 AM and from 5 PM to 8 PM.

Table 1 Parameter settings used in our experiments

Parameter	Value
Time of charging (level 2 (220 V))	4 h
Time of charging (DC)	1 h
C_0 for storage	100 $/kWh
C_0 for each charger	4000 $
C_{basic}	$8 per month
Life time of utility	6 years
Number of charger in charging station	10
Electricity load (level 2)	3.3 kW
Electricity load (DC)	50 kW
P_{sell}	0.49 $/kWh
$P_{buy,night}$	(5.420 + 0.277) cents
$P_{buy,t}$	6.454 × on-peak + 5.697 × off-peak cents
r_0	800 m
$\alpha, \beta, \gamma, \theta, \eta$	1

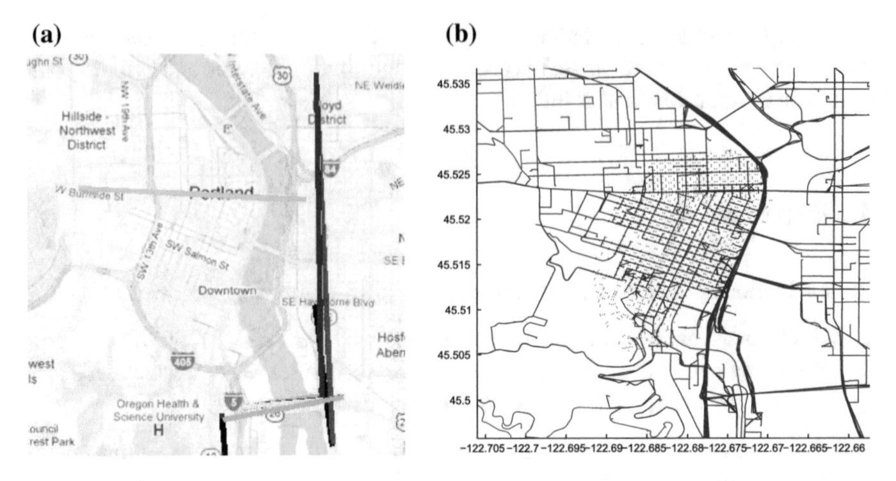

Fig. 2 **a** Trajectory representatives. **b** Schematic view of trajectories that assigned (*blue*) versus others (*red*)

4.1 *Trajectory Clustering*

We prune our dataset by removing users that do not enter or cross the downtown area of Portland. Next, we calculate those major routes that users take when they need to recharge their vehicles. Based on our previous work [6], a high proportion of users charge once during their daily travel and hence the probable routes for each user (when they needs to recharge their car) falls between two sequential stay points of the user. Our processing leads to 1259 trajectories. After extracting the actual trajectories from Google Maps, we use the Traclus algorithm [27] (with epsilon equal to 0.01 and minLns equal to 3) to cluster these trajectories. The result consists of 16 clusters. As Fig. 2a illustrates, representative trajectories mostly fall within the boundaries of downtown.

4.2 *Ideal Number of Charging Stations*

In order to compare the performance of charging stations suggested by our proposed method, we compare it with k-means clustering where only geographic coordinates of locations are considered. Figure 3 illustrates how the total profit of all charging stations changed by increasing number of charging stations. It appears that by deploying a certain number of charging stations, the total profit in our proposed method is much higher than the location-based algorithm ($5000 per day). Also, profit will begin to remain stable when the number of charging stations increases up to a certain threshold (25). It should be noted that, since the location-based k-means algorithm works solely on geographic coordinates, it will not consider the initial load values of buildings. This may cause randomness to the results.

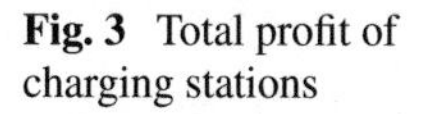

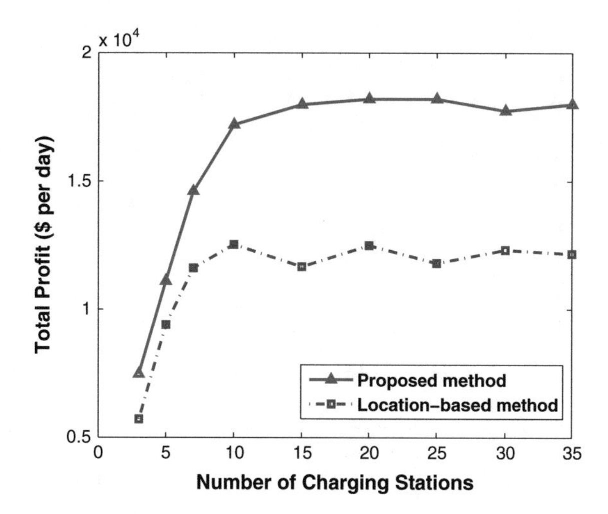

Fig. 3 Total profit of charging stations

The size of storage and utilization of charging stations (chargers and storages) are important issues in charging station and storage deployment. Utilization can be determined as $U_i = \frac{t_d}{24[h/d]}$ where $\frac{t_d}{[h/d]}$ is the daily time in use of the facility [2]. As Fig. 4a illustrates, time-based utilization of chargers is often less than 50 % but we assume satisfaction as long as the station profits exceed a certain threshold.

As the number of charging stations increases, we expect the total number of required storage units also increase. This expectation is shown in Fig. 5. Clearly, the proposed method works better than location-based k-means since the total profit and total utilization are higher. While the total storage is higher in our proposed method, time-based utilization of storages is higher than that of location-based (Fig. 4b). The time-based utilization measure depicts the percentage of the time storage units are used in charging stations.

The number of assigned users will not increase as the number of charging stations goes beyond a certain value (20). This is demonstrated in Fig. 4c. In our method, the ratio of assigned users is often more than 90 %. This ratio will vary if we change the radius of users' attention (r_0). As Fig. 4d illustrates, by increasing the allowed distance between the nearest charging station and users (r_0), the ratio of assigned users will increase. Here, number of charging stations is set to 15. This ratio in the proposed method is higher than location-based k-means because vicinity of charging stations to the common trajectories were considered in the optimization function.

To explore the profit of charging stations individually, we assess the number of charging stations with non-positive profit in each setting. As Fig. 6 illustrates, the number of such charging stations in location-based k-means is greater than in our proposed method.

Based on our results, the optimum number of charging stations which yields the highest profit and utilization is 15. For higher penetration rate, this method can be re-run to find a suitable number of charging stations. Here, we continue our experiments with 15 charging stations.

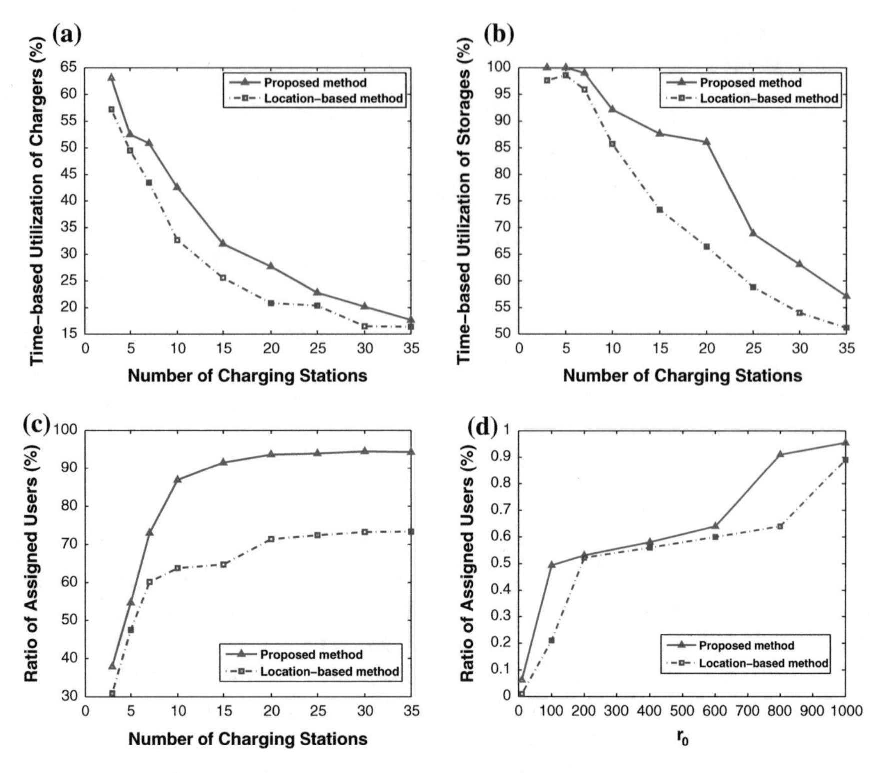

Fig. 4 Comparison of proposed method and location-based method: **a** Average time-based utilization of chargers. **b** Average time-based utilization of storages. **c** Total ratio of assigned users and **d** Total ratio of assigned users based on distance to charging stations

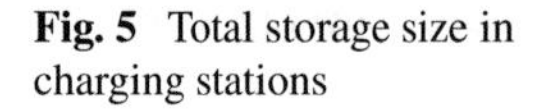

Fig. 5 Total storage size in charging stations

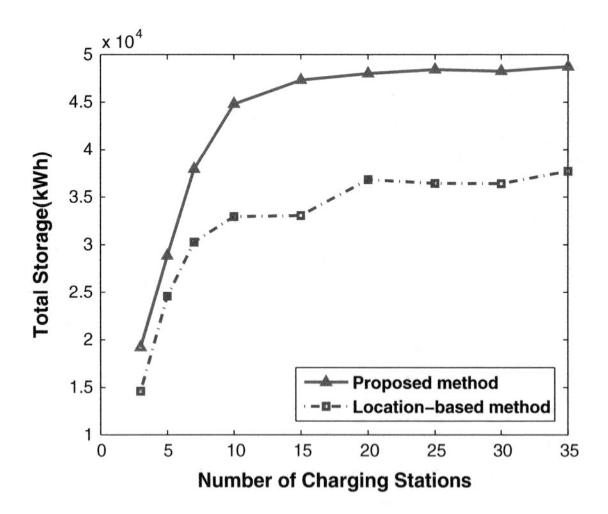

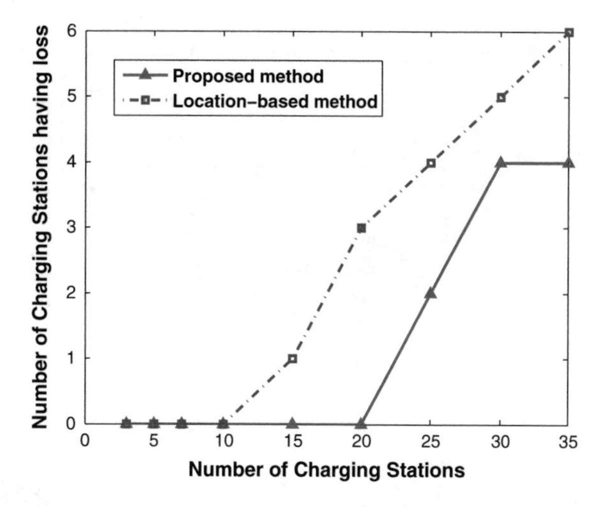

Fig. 6 Number of charging stations with loss

4.3 *Profile of Individual Charging Station*

After determining the charging stations, we can cluster other locations by considering charging stations to be the prototypes of location clusters. This strategy will be beneficial to understand which regions are covered by which charging station. Clusters of locations are shown in Fig. 7. One interesting result here is that charging stations 2 and 15 are deployed in the same locations, pointing to the potential of this location.

The daily profit, storage size, and utilization of chargers and storages in charging stations are shown in Fig. 8. A notable result here is that charging station 8 is not efficient as others since it has a low profit ($100) due to low storage and low utilization.

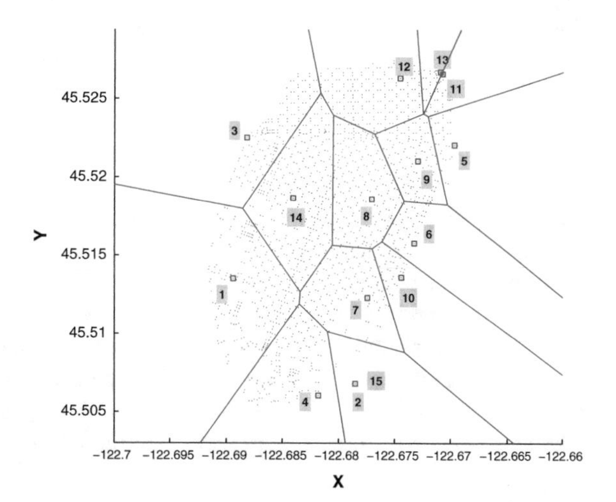

Fig. 7 Voronoi diagram of charging stations and their associated coverage area

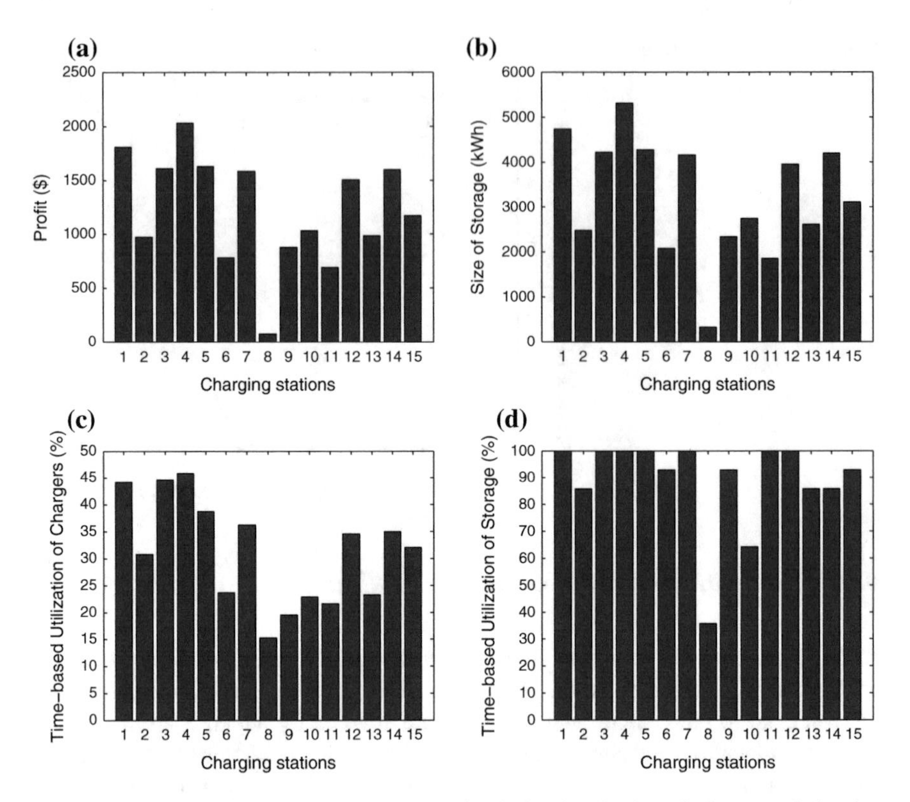

Fig. 8 Performance of charging stations: **a** Profit of charging Stations. **b** Storage of charging stations. **c** Time-based utilization of chargers and **d** Time-based utilization of storages

Most of the EVs need to be charged at peak hours (9 AM–8 PM). The number of EVs at each time slot in each charging station is shown in Fig. 9a. Since most downtown activities occur during afternoon and nights, most of the demands are concentrated between 12 PM to 8 PM. Also, the number of charging stations for each type of charging is shown in Fig. 9b, c, for level 2 and DC, respectively. Since those locations that people stay at least 4 h are outside of downtown, the demand for level 2 is lower than that for DC. Based on these results, we can determine the required number of chargers in each station. In our experiments, we assume that each charging station is able to have at most 10 chargers. As Fig. 9 illustrates, in charging station 1, we can organize it to have 2 chargers for level 2 and 9 chargers for DC. Conversely, for charging station 8, we do not need any level 2 chargers and only require 3 chargers for level 3 (DC).

Profiles of charging stations can be clustered with respect to their loads at different times. To this effect, we used the K-SC clustering approach originally proposed for time series data [33]. Here, the value of electricity load before adding EV, after adding EV, and after storage deployment during 24 h were considered as a sequence

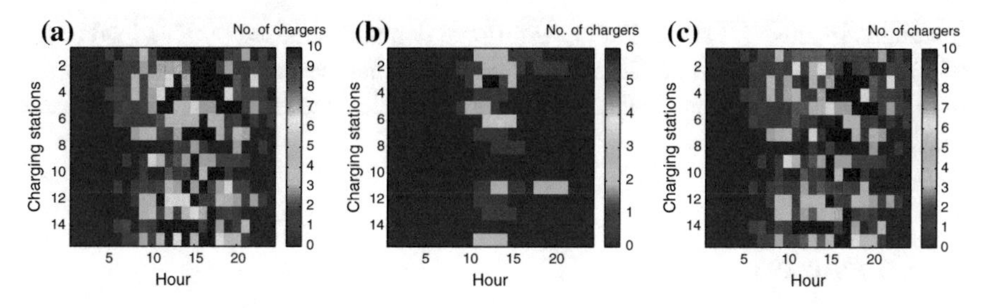

Fig. 9 **a** Number of EVs getting charged at each time slot. **b** Number of EVs getting charged at each time slot by level 2. **c** Number of EVs getting charged at each time slot by level 3 (DC)

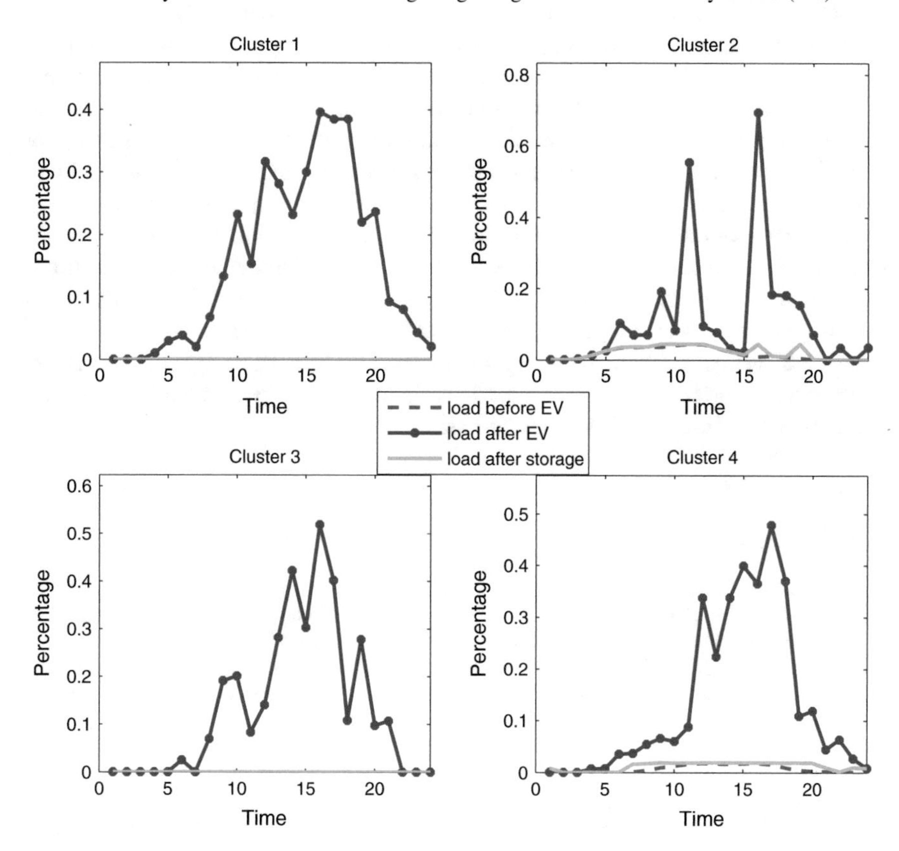

Fig. 10 Clustering of load profiles of charging stations

of 24×3 elements. Profiles of prototypes of four clusters are shown in Fig. 10. This figure is important in understanding the behavior of charging stations in order to make a decision between using a mobile storage versus a stationary one. Locations in cluster 1 and 3 are places where no one enters them (such as a parking lot).

Locations in cluster 2 show that the additional demand imposed by EVs lead to use of storage in 8–12 h and 16–20 h. Based on this profile, we can place mobile storage in locations where storage is needed during a specific time rather than an entire day (e.g., cluster 4).

5 Conclusion

Effective usage of the next generation of smart grids requires a comprehensive understanding of the interactions between networks of urban environments and electric systems. In this paper, we proposed a framework to design charging and storage infrastructure for electric vehicles in an urban environment. There is an inherent trade-off between user expectations and the expectations of charging stations owners, which is captured in our framework and aids in the selection of the number of charging stations along with their placement. More constraints such as availability of parking space, effects of charging stations on electrical substations, different pricing schema in charging stations are being considered for integration into our framework. Results of this research illustrate the efficiency of our approach in terms of profit maximization and energy usage. While we studied the effect of different parameters on the performance of charging station placement, there are other factors that can be considered in this problem. In this regard, the impact of different EV penetration rates and the use of probabilistic framework in assignment strategy of drivers to charging stations can be considered as future works.

References

1. Ramchurn, S.D., Vytelingum, P., Rogers, A., Jennings, N.R.: Putting the 'smarts' into the smart grid: a grand challenge for artificial intelligence. Commun. ACM **55**(4), 86–97 (2012)
2. Wirges, J., Linder, S., Kessler, A.: Modelling the development of a regional charging infrastructure for electric vehicles in time and space. Eur. J. Transp. Infrastruct. Res. **12**(4), 391–416 (2012)
3. Frade, I., Ribeiro, A., Goncalves, G., Antunes, A.: An optimization model for locating electric vehicle charging stations in central urban areas. In: Proceedings of the TRB 90th Annual Meeting, Transportation Research Board (2010)
4. He, F., Wu, D., Yin, Y., Guan, Y.: Optimal deployment of public charging stations for plug-in hybrid electric vehicles. Transp. Res. Part B Methodol. **47**, 87–101 (2013)
5. Ip, A., Fong, S., Liu, E.: Optimization for allocating BEV recharging stations in urban areas by using hierarchical clustering. In: Proceedings of the 6th International Conference on Advanced Information Management and Service, pp. 460–465 (2010)
6. Momtazpour, M., Butler, P., Hossain, M.S., Bozchalui, M.C., Ramakrishnan, N., Sharma, R.: Coordinated clustering algorithms to support charging infrastructure design for electric vehicles. In: Proceedings of the ACM SIGKDD International Workshop on Urban Computing, UrbComp '12, pp. 126–133 (2012)

7. Chen, T.D., Kockelman, K.M., Khan, M.: The electric vehicle charging station location problem: a parking-based assignment method for seattle. In: Proceedings of the Transportation Research Record (2013)
8. Swanson, J., Aslin, R., Yuccel, Z.: Electric vehicle penetration study using linear discriminant analysis. In: Proceedings of the Center for Research in Regulated Industries (CRRI), pp. 1–9 (2011)
9. Hoffman, M., Sadovsky, A., Kintner-Meyer, M., DeSteese, J.: Analysis tools for sizing and placement of energy storage in grid applications. Technical report PNNL-19703. Pacific Northwest National Laboratory Richland, Washington (2010)
10. Makarov, Y., Pengwei, D., Kintner-Meyer, M., Chunlian, J., Illian, H.: Sizing energy storage to accommodate high penetration of variable energy resources. IEEE Trans Sustain. Energ. **3**(1), 34–40 (2012)
11. Bayram, I., Michailidis, G., Devetsikiotis, M., Bhattacharya, S., Chakrabortty, A., Granelli, F.: Local energy storage sizing in plug-in hybrid electric vehicle charging stations under blocking probability constraints. In: Proceedings of the IEEE International Conference on Smart Grid Communications, pp. 78–83 (2011)
12. Liu, W., Zheng, Y., Chawla, S., Yuan, J., Xie, X.: Discovering spatio-temporal causal interactions in traffic data streams. In: KDD '11, August 2011
13. Yuan, J., Zheng, Y., Xie, X.: Discovering region of different functions in a city using human mobility and POI. In: KDD '12 (2012)
14. Yuan, J., Zheng, Y., Zhang, C., Xie, W., Xie, X., Huang, Y.: T-drive: driving directions based on taxi trajectories. In: Proceedings of the ACM SIGSPATIAL GIS 2010 (2010)
15. Zheng, Y., Zhang, L., Xie, X., Ma, W.: Mining correlation between locations using human location history. In: Proceedings of the ACM SIGSPATIAL GIS 2009 (2009)
16. Takahashi, R., Osogami, T., Morimura, T.: Large-scale nonparametric estimation of vehicle travel time distributions. In: SDM '12, pp. 12–23, April 2012
17. Rosswog, J., Ghose, K.: Detecting and tracking coordinated groups in dense, systematically moving, crowds. In: SDM '12, pp. 1–11 (2012)
18. Ester, M., Kriegel, H., Sander, J., Xu, X.: A density-based algorithm for discovering clusters in large spatial databases with noise. In: KDD '96, pp. 226–231 (1996)
19. Synthetic Data Products for Societal Infrastructures and Proto-Populations: data set 1.0. Technical report NDSSL-TR-06-006. Network Dynamics and Simulation Science Laboratory, Virginia Tech (2006)
20. Tishby, N., Pereira, F.C., Bialek, W.: The information bottleneck method. In: Proceedings of the 37th Annual Allerton Conference on Communication, Control and Computing, pp. 368–377 (1999)
21. Munro, N.: Obama hikes subsidy to wealthy electric car buyers. http://dailycaller.com/2012/02/13/obama-hikes-subsidy-to-wealthy-electric-car-buyers/ Accessed 16 May 2012
22. Simply Hired, Inc. Portland jobs. http://www.simplyhired.com/a/local-jobs/city/l-Portland,+OR Accessed 16 May 2012
23. KEMA, Inc. Distributed energy storage: serving national interests, advancing wide-scale DES in the United States. Technical report 20130065. National Alliance for Advanced Technology Batteries (2012)
24. Andrienko, G., Andrienko, N., Rinzivillo, S., Nanni, M., Pedreschi, D., Giannotti, F.: Interactive visual clustering of large collections of trajectories. In: Proceedings of the IEEE Symposium on Visual Analytics Science and Technology, pp. 3–10 (2009)
25. Monreale, A., Pinelli, F., Trasarti, R., Giannotti, F.: WhereNext : a location predictor on trajectory pattern mining. In KDD, pp. 637–645 (2009)
26. Zheng, Y., Zhang, L., Xie, X., Ma, W.: Mining interesting locations and travel sequences from gps trajectories. In: WWW **49** (2009)
27. Lee, J., Han, J., Kyu-Young, W.: Trajectory clustering : a partition-and-group framework. In: Proceedings of the 2007 ACM SIGMOD International Conference on Management of data, pp. 593–604. Beijing (2007)

28. Tran, N., Giraud-Carrier, C., Seppi, K., Warnick, S.: Cooperation-based clustering for profit-maximizing organizational design. In: Proceedings of the 2006 IEEE International Joint Conference on Neural Network Proceedings pp. 1813–1817. IEEE (2006)
29. Portland General Electric Company. Schedule 85 large nonresidential standard service (201–1,000 kW). http://www.portlandgeneral.com/our_company/corporate_info/regulatory_documents/pdfs/schedules/sched_085.pdf. Accessed 5 Feb 2013
30. Kumar, V., Saxena, P.C., Katti, C.P.: A clustering approach for task assignment problem. Int. J. Comput. Appl. **47**(7), 46–49 (2012)
31. Holly Yan. Public charging stations fuel desire for electric cars. http://www.cnn.com/2012/10/24/us/public-car-chargers/index.html. Accessed 12 Feb 2013
32. Kley, F., Dallinger, D., Wietschel, M.: Assessment of future EV charging infrastructure. In: Proceedings of the International Advanced Mobility Forum (2010)
33. Yang, J., Leskovec, J.: Patterns of temporal variation in online media. In: ACM WSDM'11, pp. 177–186 (2011)

Computationally Efficient Design Optimization of Compact Microwave and Antenna Structures

Slawomir Koziel, Piotr Kurgan and Adrian Bekasiewicz

Abstract Miniaturization is one of the important concerns of contemporary wireless communication systems, especially regarding their passive microwave components, such as filters, couplers, power dividers, etc., as well as antennas. It is also very challenging, because adequate performance evaluation of such components requires full-wave electromagnetic (EM) simulation, which is computationally expensive. Although high-fidelity EM analysis is not a problem for design verification, it becomes a serious bottleneck when it comes to automated design optimization. Conventional optimization algorithms (both gradient-based and derivative-free ones such as genetic algorithms) normally require large number of simulations of the structure under design, which may be prohibitive. Considerable design speedup can be achieved by means of surrogate-based optimization (SBO) where a direct handling of the expensive high-fidelity model is replaced by iterative construction and re-optimization of its faster representation, a surrogate model. In this chapter, we review some of the recent advances and applications of SBO techniques for the design of compact microwave and antenna structures. Most of these methods are tailored for a design problem at hand, and attempt to utilize its particular aspects such as a possibility of decomposing the structure. Each of the methods exploits an underlying low-fidelity model, which might be an equivalent circuit, coarse-discretization EM

S. Koziel (✉) · A. Bekasiewicz
Engineering Optimization & Modeling Center,
School of Science and Engineering, Reykjavik University,
Menntavegur 1, 101 Reykjavik, Iceland
e-mail: koziel@ru.is

A. Bekasiewicz
e-mail: bekasiewicz@ru.is

P. Kurgan
Faculty of Electronics, Telecommunications and Informatics,
Gdansk University of Technology, 11/12 Narutowicza Street,
80-233 Gdansk, Poland
e-mail: piotr.kurgan@eti.pg.gda.pl

J. Lässig et al. (eds.), *Computational Sustainability*,
Studies in Computational Intelligence 645,
DOI 10.1007/978-3-319-31858-5_9

simulation data, and approximation model, or a combination of the above. The common feature of the presented techniques is that a final design can be obtained at the cost of a few evaluations of the high-fidelity EM-simulated model of the optimized structure.

Keywords Microwave engineering · Simulation-driven design · Surrogate modeling · Surrogate-based optimization · Compact structures · Expensive optimization problems

1 Introduction

Small size is one of the most important requirements imposed upon modern wireless communication system blocks, with particular emphasis on microwave passive components [1, 2], including, among others, filters [3–6], couplers [7–10], power dividers [11–14], as well as antennas [15–18]. Satisfying strict electrical performance specifications and achieving compact size are normally conflicting objectives [19–22]. These difficulties can be alleviated to some extent, e.g., through replacing conventional transmission lines by their more compact counterparts such as slow-wave resonant structures (SWRSs) [23–26], or, in case of antennas, by introducing certain topological modifications (e.g., stubs and slits in ground planes of ultra-wideband antennas [27–29]). At the same time, the use of traditional design techniques based on equivalent circuit models does not lead to reliable results due to considerable electromagnetic (EM) couplings between circuit components within highly compressed layouts that cannot be accurately accounted for at the network level. Reliable evaluation of the structure performance is only possible by means of CPU-intensive and time-consuming full-wave EM simulations.

Unfortunately, EM analysis may be computationally expensive, even when using vast computing resources: simulation time with fine discretization of the structure may be from 15–30 min for simple passive microwave circuits and small antennas [24, 25, 28], to a few hours for more complex structures (e.g., miniaturized Butler matrix [30]). This creates a serious bottleneck for automated, EM-simulation-driven design optimization of compact circuits: conventional optimization methods (such as gradient-based routines [31, 32], or derivative-free methods, e.g., pattern search [33] or population-based metaheuristics [34–36]) normally require large number of objective function evaluations, each of which is already expensive. The use of adjoint sensitivities [37, 38] allows—to some extent—for reducing design optimization cost, however, this technology is not yet widespread in microwave and antenna community, especially in terms of its availability through commercial simulation software packages (with some exceptions, e.g., [39, 40]). Heuristic simulation-based design approaches, typically exploiting parameter sweeps guided by engineering experience, tend not to work for compact structures because the latter are characterized by many designable parameters as well as optimum parameter setups are often counter-intuitive.

Reducing the computational efforts of microwave/antenna design processes, especially in the context of compact structures, is of fundamental importance for design automation and lowering the overall design cost (both in terms of computational resources and time), which, consequently, has implications for economy, environmental protection, as well as the quality of life. Thus, it is important from the point of view of computational sustainability.

Probably the most promising approach in terms of computationally efficient design is surrogate-based optimization (SBO) [41, 42]. In SBO, direct optimization of the expensive high-fidelity simulation model is replaced by iterative updating and re-optimization of its computationally cheap representation, the surrogate. As a result of shifting the optimization burden into the surrogate, the overall design cost can be greatly reduced. The high-fidelity model is referenced rarely, to verify the prediction produced by the surrogate and to improve the latter. Various SBO methods differ mostly in the way the surrogate is created. There is a large class of function approximation modeling techniques, where the surrogate is created by approximating sampled high-fidelity model data. The most popular methods in this group include polynomial approximation [43], radial basis function interpolation [44], kriging [43, 45], support vector regression [46], and artificial neural networks [47, 48]. Approximation models are very fast but a large number of training samples are necessary to ensure reasonable accuracy. Also, majority of approximation techniques suffer from the so-called *curse of dimensionality* (i.e., an exponential growth of required number of training samples with the dimensionality of the design space, [36]). Depending on the model purpose, this initial computational overhead may (e.g., multiple-use library models) or may not (e.g., one-time optimization) be justified.

An alternative approach to creating surrogate models is by correcting an underlying low-fidelity (or coarse) model. The latter is a simplified representation of the structure (system) under design. It can be obtained, among others, by using a different level of physical description of the system (e.g., equivalent circuit versus full-wave electromagnetic simulation in case of microwave or antenna structures), or through the same type of simulation as utilized by the high-fidelity model but with coarser structure discretization, relaxed convergence criteria, etc. As opposed to approximation models, low-fidelity models are stand-alone models embedding some knowledge about the system of interest [49]. Therefore, physics-based surrogates normally exhibit much better generalization capability [25, 50]. Consequently, considerably smaller amount of training data is required to ensure sufficient accuracy of the model. Some more or less known SBO techniques exploiting physics-based surrogates include approximation model management optimization (AMMO) framework [51], space mapping (SM) [52, 53], manifold mapping [54, 55], and simulation-based tuning [56, 57].

In this chapter, we review some of the recent advances and applications of SBO techniques for the design of compact microwave and antenna structures. While some of the discussed methods are rather standard (e.g., space mapping with additive response correction utilized for ultra-wideband antenna optimization), others are more tailored for a design problem at hand, and attempt to utilize its particular aspects such as a possibility of decomposing the structure of interest. Each of the methods exploits an underlying low-fidelity model, which might be an equivalent circuit, coarse-discretization EM simulation data, and approximation model, or a combination of the above. The common feature of the presented techniques is that a final design can be obtained at the cost of a few evaluations of the high-fidelity EM-simulated model of the optimized structure.

The chapter is organized as follows. In Sect. 2, we briefly highlight the design challenges for miniaturized microwave and antenna components and emphasize the necessity of using—in the design process—high-fidelity electromagnetic simulations, as well as the need for computationally efficient techniques aimed at reducing computational cost of the process. In Sect. 3, we formulate the microwave design optimization problem, discuss electromagnetic simulation models, and introduce the concept of surrogate-based optimization (SBO). Sections 4, 5, and 6 showcase three selected case studies concerning the design of compact branch-line couplers, compact radio frequency (RF) components, and miniaturized ultra-wideband antennas. While each of these sections describes optimization techniques exploiting the SBO paradigm, methodological details are developed to handle problem-specific design challenges. We also present numerical results, as well as comparisons with benchmark techniques. An emphasis is on computational savings that can be obtained by using a suitable combination of surrogate-modeling and variable-fidelity simulations (both at the equivalent circuit and full-wave electromagnetic level). Section 7 concludes the chapter.

2 Challenges of Compact Microwave Structure Design

Reliable development of compact microwave circuits is the subject of intense research in the field of microwave and antenna engineering [58–60]. Miniaturized microwave and antenna components can be extensively utilized as fundamental building blocks of modern wireless communication systems that are continuously challenged with ever more severe specifications: higher performance, smaller size, lighter weight, and lower cost. The fundamental problem of small-size microwave and antenna structure development process is to find a design that satisfies a given specification, within a tight computational budget, and providing highly accurate results. However, the accomplishment of this goal poses extremely difficult obstacles from the methodological point of view.

A typical microwave circuit with a compact footprint is constructed from non-uniform transmission lines that mimic the electrical performance of their conventional counterparts in a limited frequency range, but at the same time offer reduced physical dimensions, which can lead to a smaller size of the circuit [59–81]. Most commonly, T-networks [59, 61–67] or π-networks [59, 65, 68–73] are chosen for the purpose of non-uniform transmission line realization. Such simple circuits can be analyzed with ease by means of transmission line theory, which leads to relatively accurate results, assuming the lack of cross-coupling effects and a negligible influence of other high-frequency phenomena (e.g., anisotropy of the substrate, current crowding, etc.) on the performance of a microwave component. This supposition, however, is valid only for conventional circuits [80]. It should be underlined that—in case of highly compact microwave and antenna structures with complex and closely fit building blocks within a layout—the use of simplified theory-based models is limited to providing initial design solutions that indispensably require further EM fine-tuning [7, 61–66, 71, 72, 76, 77, 79]. In order to produce accurate results, it is preferable to apply a high-fidelity EM analysis from the early stages of the design process of miniaturized microwave components [81, 82]. Similarly, the use of theoretical models in contemporary antenna engineering is an outdated practice that cannot provide information on complex EM phenomena taking place in the antenna structure under consideration. This is particularly true in case of unconventional antennas with complex and compact footprints. For the above-mentioned reasons, EM-simulation-driven design is nowadays a necessity in computer-aided design of microwave and antenna devices [79, 83].

On the other hand, exploitation of EM simulation tools, either throughout the entire design process or in the design closure eventually leads to high demands on enormous computational resources. The main issue here lies in the numerical cost associated with the high-fidelity EM analysis of the entire compact component. In case of conventional EM-driven design approaches, based either on laborious parameter sweeps [9] or direct single-objective optimization [79, 83] (gradient-based or derivative-free), this becomes impractical or even prohibitive when handling computational demands of miniaturized passives. The problem becomes even more profound when the design process is realized in more general setting, i.e., when it entails adjustment of designable parameters of the structure to satisfy multiple, often conflicting objectives such as size, bandwidth, phase response, etc. [20], in case of microwave components, and size, return loss, gain, etc., in case of antenna structures [84]. This, however, illustrates a multi-objective optimization problem, which is far more challenging than the single-objective one. For typical examples of compact microwave or antenna structures—that are characterized by a number of designable parameters—the design process aimed at finding a set of trade-offs between conflicting objectives cannot be accomplished using traditional multi-objective algorithms such as population-based metaheuristics [34, 35], because these require a massive number of objective function evaluations (thousands or tens of thousands), which in case of high-fidelity EM models is prohibitive within a reasonable time-frame [36, 84].

3 Optimization Problem Formulation. Electromagnetic Models. Surrogate-Based Optimization

In this section, we formulate the microwave/antenna optimization problem, discuss various types of models utilized in microwave engineering, as well as recall the basics of surrogate-based optimization (SBO) [32]. The specific SBO techniques developed and utilized for expedited design of compact microwave and antenna structures are discussed in Sects. 4 through 6.

3.1 Microwave Optimization Problem

The microwave design optimization problem can be formulated as

$$x^* = \arg\min_{x} U\left(R_f(x)\right) \tag{1}$$

where $R_f(x) \in R^m$, denotes the response vector of a high-fidelity (or fine) model of the device or system of interest. In microwave engineering, the response vector may contain, for example, the values of so-called scattering parameters [1] evaluated over certain frequency band. The objective function $U : R^m \to R$ is formulated so that a better design corresponds to a smaller value of U. Often, the design specifications are formulated in a minimax sense [85], e.g., in the form of minimum/maximum levels of performance parameters for certain frequency bands.

In many situations, it is necessary to consider several design objectives at a time (e.g., gain, bandwidth, layout area). The objectives usually conflict with each other and the design process aims at finding a satisfactory trade-off between them. However, genuine multi-objective optimization exceeds the scope of this chapter. Thus, although multiple objectives are actually present in all the design cases discussed in Sects. 4–6, they are handled by a priori preference articulation, i.e., selecting the primary objective and controlling the others either through appropriately defined constraints or penalty functions [13].

3.2 Simulation Models in Microwave Engineering

The most important simulation models utilized in microwave/antenna engineering include equivalent circuit (or network) models and electromagnetic (EM) simulation ones.

For decades, circuit models have been important tools microwave structure design. They provide a simplified structure representation by means of analytical equations that origin from the transmission line theory. The behavior of a structure is described using its complex impedance, capacitance and/or inductance [1, 86]. Construction of a circuit model representation of a structure is based on interconnection of basic building blocks (i.e., transmission lines, coupled lines, bends, tees, etc.) in such a way that the behavior of a model mimics the behavior of a real structure [87, 88], i.e., coupled lines represents the coupling within the circuit and tee represents the interconnection between transmission lines within a circuit, etc. Such a representation is very useful for variety of structures characterized by modular construction (e.g. filters, couplers, matching transformers, etc.). Unfortunately, limited diversity of building blocks prohibits the design of unusual geometries. Moreover, circuit models are inaccurate that forces the utilization of electromagnetic simulations for final tuning of the structure. Another problem is that they lack of capability to estimate radiation field, which turns them useless for the design of antenna structures.

Antenna structures can be modeled using quasi-static representation [89]. Utilization of such models allows for i.e., estimation of antenna operating frequency, radiation pattern, and/or impedance matching. Empirical models are particularly useful if antenna design requires tremendous number of model evaluations (e.g., in multi-objective optimization [21, 84]). Despite fast evaluation, quasi-static representation suffers from considerable inaccuracy. Moreover, it is available only for some conventional designs, which turns empirical antenna representation not very popular nowadays.

The most generic and accurate way of representing microwave/antenna structures is full-wave electromagnetic simulation. The electromagnetic solvers utilize advanced meshing techniques aimed at discretization of the design into a set of subproblems, which are evaluated by solving Maxwell's equations [89, 90]. Despite its accuracy, such a representation suffers from considerable computational cost. Reduction of the simulation time can be obtained by introducing certain simplifications into the EM model, e.g., sparse mesh, modeling of metallization as infinitely thin sheet, neglecting losses of dielectric substrate, or utilization of perfect electric conductor instead of finite-conductivity metals. Usually the simplified model is 10 up to 50 times faster than the high-fidelity one. Coarse-mesh EM models are useful auxiliary tools utilized by many surrogate-based optimization methods [15, 91].

A conceptual illustration of a high- and low-fidelity EM model as well as a circuit model of an exemplary structure is provided in Fig. 1.

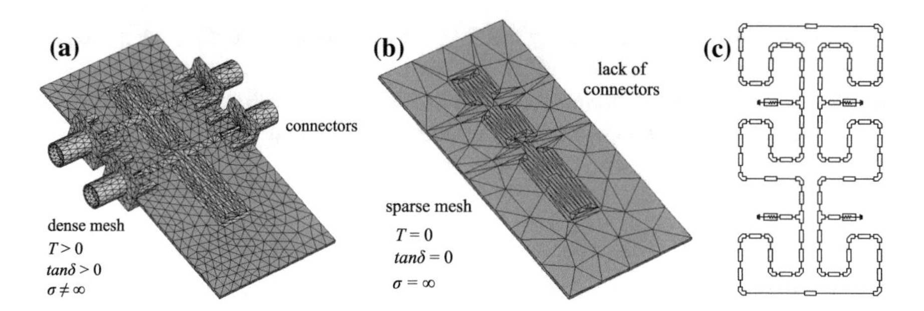

Fig. 1 Various representations of the same rat-race coupler structure. **a** High-fidelity EM model representation; **b** simplified low-fidelity EM model (simplifications include: lack of connectors, coarse mesh, neglected metallization thickness and dielectric losses, as well as perfect conductivity of metallization); **c** circuit representation of the coupler—the fastest, yet the most inaccurate

3.3 Surrogate-Based Optimization

Our major concern is to reduce the cost of solving the optimization problem (1). As mentioned before, in many cases, direct handling of the high-fidelity model $\boldsymbol{R}_f$ is not even possible either due to the high individual cost of evaluating $\boldsymbol{R}_f$ or very large number of model evaluations when solving (1). Therefore, we are interested in surrogate-based optimization (SBO) [32] methods, where the sequence $\boldsymbol{x}^{(i)}$, $i = 0, 1, \ldots$, of approximate solutions to (1) is found by means of an iterative procedure [92]

$$\boldsymbol{x}^{(i+1)} = \arg\min_{\boldsymbol{x}} U\left(\boldsymbol{R}_s^{(i)}(\boldsymbol{x})\right) \tag{2}$$

Here, $\boldsymbol{x}^{(i+1)}$ is the optimal design of the surrogate model $\boldsymbol{R}_s^{(i)}$, $i = 0, 1, \ldots \boldsymbol{R}_s^{(i)}$ is assumed to be a computationally cheap and sufficiently reliable representation of the high-fidelity model $\boldsymbol{R}_f$, particularly in the vicinity of the current design $\boldsymbol{x}^{(i)}$. Under these assumptions, the algorithm (2) is likely to produce a sequence of designs that quickly approach $\boldsymbol{x}^*$.

Because $\boldsymbol{R}_f$ is evaluated rarely (usually once per iteration), the surrogate model is supposedly fast, and the number of iterations for a well-performing algorithm is substantially smaller than for most conventional optimization methods, the process (2) may lead to substantial reduction of the computational cost of solving (1). Moreover, if the surrogate model satisfies zero- and first-order consistency conditions with the high-fidelity model [51], i.e., $\boldsymbol{R}_s^{(i)}(\boldsymbol{x}^{(i)}) = \boldsymbol{R}_f(\boldsymbol{x}^{(i)})$ and $\boldsymbol{J}_{Rs}^{(i)}(\boldsymbol{x}^{(i)}) = \boldsymbol{J}_{Rf}(\boldsymbol{x}^{(i)})$ with $\boldsymbol{J}$ being a Jacobian of the respective model (verification of the latter requires $\boldsymbol{R}_f$ sensitivity data), and the algorithm (2) is embedded in the trust region framework [93], then it is provably convergent to a local optimum of original problem (1). Convergence can also be guaranteed if the algorithm (2) is enhanced by properly selected local search methods [31].

Various SBO techniques mostly differ by the way of constructing the surrogate model. The specific techniques utilized to handle the design cases presented in this chapter are elaborated on in Sects. 4–6. Interested reader is referred to the literature (e.g., [19, 22, 85]) to find out more about other possible options.

4 Case Study I: Expedited Design of Compact Branch-Line Couplers

In this section, we present a design methodology based on [94], dedicated to efficient solving of complex and numerically demanding design problems of popular microwave components, i.e., branch-line couplers with compact footprints. This technique enables a cost-efficient and accelerated design optimization of the microwave component of interest together with high accuracy of the results. The desired performance of the circuit under design as well as its miniaturized layout are achieved by adjusting designable parameters of the non-uniform building blocks of the structure.

The proposed method can be divided into two separate stages: (i) concurrent EM optimization of non-uniform constitutive elements of the branch-line coupler (so-called composite cells), where cross-coupling effects between the adjacent cells are absent, and (ii) fine-tuning of the entire branch-line coupler exploiting space-mapping-corrected surrogate model, constituted by cascaded local response surface approximation models of the separate cells. The first stage is realized at a low computational cost (just a few evaluations of the entire branch-line coupler), because the mutual EM coupling between the cells is not taken into account. The second stage, in turn, accounts for any cross-coupling phenomena and other interactions between building blocks of the structure under design. We showcase the efficiency of the proposed technique by a numerical case and its experimental validation.

4.1 Design Problem and Coupler Structure

Microstrip branch-line couplers (BLCs) are vital components widely used in many practical microwave and radio frequency circuits, such as balanced mixers [95], Butler matrixes [42], and others. An ideal BLC offers perfect transmission characteristics at a the operating frequency: an equal power splitting between the output ports with 90° phase shift, perfect matching with system impedance Z_0, and perfect isolation. In the vicinity of the operating frequency, usually within the 10 % bandwidth, the performance of the circuit is still acceptable, however not ideal. A conventional BLC is composed of four quarter-wavelength uniform transmission lines, two of which are defined by Z_0 characteristic impedance, while the rest are determined by $0.707 \cdot Z_0$ characteristic impedance. In this case study, the task is to design a 3-dB BLC for operating frequency f_0 = 1 GHz using Taconic RF-35 ($\varepsilon_r = 3.5$, $h = 0.508$ mm, tan$\delta = 0.0018$) as a dielectric substrate. The intended bandwidth is 0.96–1.04 GHz with return loss and isolation $|S_{11}|$, $|S_{41}| \leq -20$ dB. For comparison purposes, a conventional BLC has been designed on the basis of [82]. Its exterior dimensions are: 45.6 mm × 48.1 mm. We use BLC parameterization shown in Fig. 2. It can be observed that the complementary building blocks almost completely fill the interior of the BLC. Dimensions of horizontal and vertical cell (denoted Cell$_1$ and Cell$_2$, respectively) are given by vectors $\boldsymbol{x}_1$ and $\boldsymbol{x}_2$. All parameters of Cell$_1$, $x_1, x_2, \ldots, x_7$, are independent, whereas Cell$_2$ is described by both independent, $x_8, x_9, \ldots, x_{12}$, and dependent, y_1, y_2, y_3, parameters. The latter depend on specific parameters of Cell$_1$ and the predefined distance d between the cells. Parameterization of Cell$_1$ and Cell$_2$ is given by: $f_1(\boldsymbol{x}) = f_1([x_1\, x_2 \ldots x_{12}]^T) = [x_1\, x_2 \ldots x_7]^T$ and $f_2(\boldsymbol{x}) = f_2([x_1\, x_2 \ldots x_{12}]^T) = [x_8\, x_9 \ldots x_{12}\, y_1\, y_2\, y_3]^T$, where $y_1 = 0.5 \cdot (x_7 + d/2 - x_8 - 2 \cdot x_{10} - 2.5 \cdot x_{12})$, $y_2 = x_1 + x_3 + x_4 + x_6 - d - x_{12}$ and $y_3 = x_8 + x_{12} - d$. Geometrical dependence between the non-uniform building blocks enables the preservation of consistency of the dimensions and a high miniaturization of the entire structure. In this example, $d = 0.2$ mm.

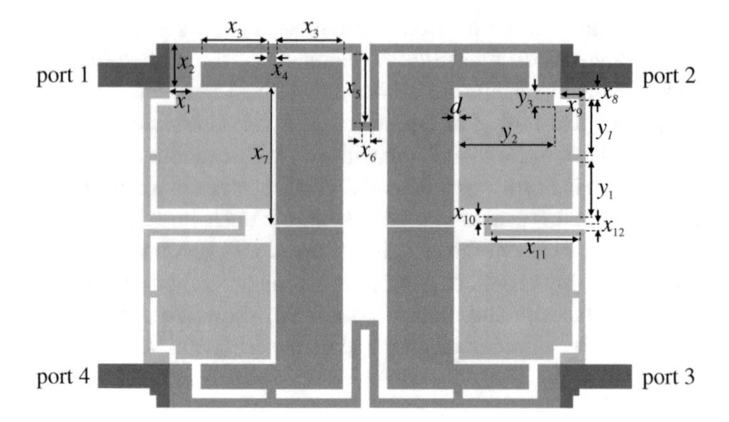

Fig. 2 Parameterized BLC composed of Cell_1 (*dark gray color*, port impedance $Z_0 = 35.35\ \Omega$) and Cell_2 (*light gray color*, port impedance $Z_0 = 50\ \Omega$)

4.2 Design Methodology

In this section, we present a specific surrogate-based optimization scheme dedicated for solving computationally complex design problems of compact branch-line couplers. In particular, we provide a general description of the proposed approach, together with a mathematical formulation of its subsequent steps, i.e., concurrent cell optimization, response surface approximation model construction, and surrogate-based design refinement.

4.2.1 General Design Scheme

An overview of the proposed design method is shown in Fig. 3a. The numerical efficiency of this technique stems from concurrent optimization detached non-uniform building blocks of a compact BLC, and subsequent surrogate-based fine-tuning of the entire component. The first stage is aimed at arriving at the optimized design of composite cells treated separately. It is noteworthy that the non-uniform building blocks of the compact coupler are geometrically dependent (to ensure the design consistency and size reduction of the BLC), but electromagnetically isolated (to lower the evaluation cost of their high-fidelity EM model). This has proven to be exceptionally beneficial from the numerical standpoint as it enabled the embedment—within a single optimization procedure—of two high-fidelity, yet reasonably cheap EM models representing BLC building blocks of interest instead of one CPU-intensive EM model of the entire BLC under development. The use of ultimately accurate simulation tool available for an engineer ensures us that the solution obtained in this manner lies in the neighborhood of the BLC optimized design.

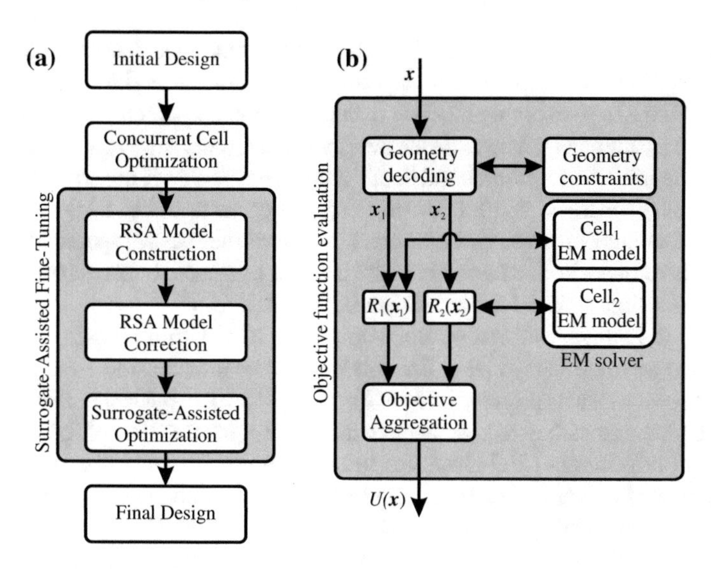

Fig. 3 **a** Proposed design procedure of highly miniaturized BLCs; **b** Objective function evaluation for concurrent cell optimization

The following fine-tuning of the BLC is accomplished as a surrogate-based optimization process with the underlying low-fidelity model being a cascade of local response surface approximations (RSAs) of the optimized non-uniform building blocks of the BLC. Formally, this is an iterative process, however, as given by the case study (cf. Sect. 4.3), one iteration is completely sufficient.

The main benefit of the proposed method lies in the low computational cost required to obtain the optimized design, which is realized within a single and fully automated process. The most expensive model in the design problem under consideration—namely, high-fidelity EM model of the entire compact BLC—is evaluated only twice, i.e., at the initial design produced by the first stage, and for the verification of the final design.

4.2.2 Concurrent Cell Optimization

The simulation-driven design methodology discussed in this section offers a fully automated development of compact BLCs. However, it requires a manual setup of non-uniform building blocks of a miniaturized BLC. Wide collections of such composite cells can be found in the literature, e.g., in [82]. A typical BLC is composed of quarter-wavelength uniform transmission lines of Z_0 and $0.707 \cdot Z_0$ characteristic impedances to achieve an equal power split between the output ports, where Z_0 is the system impedance [80].

Thus, in order to construct a small-size BLC, two complementary constitutive cells are needed. These are separated by a predefined distance d and intended to fill the interior of the BLC in a highly efficient manner. Geometry parameters of the cells are encoded in a parameter vector $\boldsymbol{x}$. Some designable parameters are independent; others are dependent to ensure geometrical fit of the cells. Formally, we describe this relation as $\boldsymbol{x}_1 = f_1(\boldsymbol{x})$ and $\boldsymbol{x}_2 = f_2(\boldsymbol{x})$. In practice, f_1 is a projection (i.e., $\boldsymbol{x}_1$ is composed of selected elements of the vector $\boldsymbol{x}$), whereas f_2 determines clearly specified geometry constraints to make both composite cells fit into a compact coupler layout. Specific realizations of these functions are shown in Sect. 4.1.

During the concurrent cell optimization stage of the entire design process, we exploit two separate high-fidelity EM models to take into consideration all internal electromagnetic effects present inside of a non-uniform cell, but neglecting any cross-coupling phenomena between adjacent cells. Such a formulation of the design task allows for performing a CPU-cheap concurrent optimization of both composite cells. The aggregated objective function $U(\boldsymbol{x})$ for the cells is evaluated using given design specifications and cell response vectors $R_i(\boldsymbol{x}_i)$, $i = 1, 2$ (cf. Fig. 3b).

Design specifications pertaining to the cells under concurrent optimization are imposed by the transmission line theory [80]. We aim at obtaining a required phase shift ϕ_c, $\arg(S_{21})$ at the operating frequency f_0 as well as to minimize the return loss $|S_{11}|$ at f_0 and around it. The task is formulated as follows:

$$\boldsymbol{x}_c^* = \arg\min_{\boldsymbol{x}} U(\boldsymbol{x}) \tag{3}$$

where

$$U(\boldsymbol{x}) = \max_{f_0 - df \le f \le f_0 + df} \{|S_{11.1}(\boldsymbol{x}_1; f)|, |S_{11.2}(\boldsymbol{x}_2; f)|\} + \\ + \beta \sum_{i=1}^{2} \left[\arg(S_{21.i}(\boldsymbol{x}_i; f_0)) - \phi_c\right]^2 \tag{4}$$

Here, $\boldsymbol{x}$ is a vector of all designable parameters, $\boldsymbol{x}_i$, $i = 1, 2$, is a geometry parameter vector of the ith cells, β is a penalty factor (here, we use $\beta = 10^4$), whereas $S_{11.i}(\boldsymbol{x}_i; f)$ and $S_{21.i}(\boldsymbol{x}_i; f)$ denote explicit dependence of S-parameters on frequency for the ith cell. Therefore, the process (3) aims at minimizing $|S_{11}|$ in the vicinity of f_0 (in practice, at three frequency points: $f_0 - df$, f_0, and $f_0 + df$), while forcing $\arg(S_{21})$ to obtain the value of ϕ_c. The value of β is set to ensure that even small violation of the phase requirement results in a meaningful contribution of the penalty function. Here, for $\beta = 10^4$, deviation of $\arg(S_{21.i})$ from ϕ_c by 0.01° results in the penalty function value of 1, which is a few percent of the primary cost function value; for deviation by 0.1°, the penalty function becomes a dominant component of U. Thus, formulation (4) allows for obtaining the required phase shift with good accuracy. The use of three frequency points is motivated by the necessity of ensuring an adequate reflection response not only at the operating frequency but also in some vicinity of it. Moreover, minimization of (4) leads to obtaining more or less symmetric response around f_0 with $|S_{11}|$ being equal (or almost equal) at both $f_0 - df$ and $f_0 + df$. The problem (3) is solved using a pattern search algorithm (cf. [33]).

4.2.3 Response Surface Approximation (RSA) Models

Local RSA models of the non-uniform building blocks of a compact BLC, exploited during the fine-tuning process (cf. Sect. 4.2.4), are developed in the neighborhood of their optimized designs, $\boldsymbol{x}_{c.i}{}^*$, $i = 1, 2$, defined as $[\boldsymbol{x}_{c.i}{}^* - d\boldsymbol{x}_i, x_{c.i}{}^* + d\boldsymbol{x}_i]$. Each model uses $2n_i + 1(n_i = \dim(\boldsymbol{x}_i))$ EM simulations for each composite cell at $\boldsymbol{x}_{c.i}^{(0)} = x_{c.i}{}^*$ and at the perturbed designs $\boldsymbol{x}_{c.i}^{(k)} = [\boldsymbol{x}_{c.i.1}{}^* \ldots x_{c.i.\lceil k/2\rceil}{}^* + (-1)^k dx_{i.\lceil k/2\rceil} \ldots x_{c.i.ni}{}^*]^T$, $k = 1, \ldots, 2n_i$, where $x_{c.i.k}{}^*$ and $dx_{i.k}$ are kth elements of the vectors $\boldsymbol{x}_{c.i}{}^*$ and $d\boldsymbol{x}_i$, respectively. The RSA model $\boldsymbol{R}_{c.i}(\boldsymbol{x})$ of the ith cell is a simple second-order polynomial without mixed terms

$$\boldsymbol{R}_{c.i}(\boldsymbol{x}) = c_{0.i} + \sum\nolimits_{k=1}^{n_1} c_{k.i} x_{i.k} + \sum\nolimits_{k=1}^{n_1} c_{(n+k).i} x_{i.k}^2 \tag{5}$$

with the parameters determined as least-square solution to the linear regression problem $\boldsymbol{R}_{c.i}(\boldsymbol{x}_{c.i}^{(k)}) = \boldsymbol{R}_{f.i}(\boldsymbol{x}_{c.i}^{(k)})$, $k = 0, 1, \ldots, 2n_i$, where $\boldsymbol{R}_{f.i}$ denotes the EM model of the ith non-uniform building block of a compact BLC. The RSA model $\boldsymbol{R}_c$ of the entire coupler is subsequently constructed by cascading $\boldsymbol{R}_{c.i}$ using ABCD matrix representation [1]. The particular choice of the RSA model comes from the fact that S-parameters of the individual cells are not highly nonlinear (the modeling is carried out for complex responses), and the model needs to be valid only in the vicinities of the optimized cell geometries.

4.2.4 Surrogate-Assisted Design Refinement

In order to account for cross-coupling effects between the adjacent BLC building blocks, as well as other phenomena (e.g., T-junction phase shifts), it is required to perform a final fine-tuning of the entire BLC. The tuning procedure is realized as a surrogate-based optimization process

$$\boldsymbol{x}^{(i+1)} = \arg\min_{\boldsymbol{x}} H\left(\boldsymbol{R}_s^{(i)}(\boldsymbol{x})\right) \tag{6}$$

Vectors $\boldsymbol{x}^{(i)}$, $i = 0, 1, \ldots$, approximate the solution of the direct design problem $\boldsymbol{x}^* = \mathrm{argmin}\{\boldsymbol{x} : H(\boldsymbol{R}_f(\boldsymbol{x}))\}$ (H encodes design specifications for the coupler), whereas $\boldsymbol{R}_s^{(i)}$ is the surrogate model at iteration i. $\boldsymbol{R}_s^{(i)}$ is constructed from the RSA model $\boldsymbol{R}_c$ using input space mapping [42]

$$\boldsymbol{R}_s^{(i)}(\boldsymbol{x}) = \boldsymbol{R}_c(\boldsymbol{x} + \boldsymbol{q}^{(i)}) \tag{7}$$

where $\boldsymbol{q}^{(i)}$ is the input SM shift vector acquired through the usual parameter extraction procedure $\boldsymbol{q}^{(i)} = \mathrm{argmin}\{\boldsymbol{q} : ||\boldsymbol{R}_f(\mathbf{x}^{(i)}) - \boldsymbol{R}_c(\boldsymbol{x}^{(i)} + \boldsymbol{q})||\}$, aiming at minimization of misalignment between the $\boldsymbol{R}_c$ and the EM model $\boldsymbol{R}_f$ of the structure under consideration. Upon completion of parameter extraction, the surrogate model becomes a very good representation of the high-fidelity model in the vicinity of the current design so that it can be successfully used to find the optimum design of the latter. Note that the high-fidelity model $\boldsymbol{R}_f$ is not evaluated until the fine-tuning stage. In practice, a single iteration (6) is sufficient. The overall cost of the coupler design process is therefore very low and usually corresponds to a few simulations of the entire structure of interest, including cell optimization and the cost associated with the development of the RSA models.

4.3 Results

Here, we use Sonnet ***em*** [96] to conduct all high-fidelity EM simulations, where the grid size is set to 0.025 mm × 0.0025 mm to provide sufficient accuracy of composite cell design solutions. EM simulations are performed on PC with 8-core Intel Xeon 2.5 GHz processor and 6 GB RAM. With this setup, single frequency simulation of an individual non-uniform building block of a compact BLC is, on average, 15 and 18 s. Simulation time of the entire BLC is approximately 90 min.

First, we perform concurrent optimization of the cells of Fig. 2. Design requirements, imposed on each constitutive cell, are theory-based [80]. Therefore we aim at finding appropriate cell designs that approximate electrical parameters of theoretical BLC building blocks. More specifically, $|S_{11}| \leq -20$ dB at 0.96, 1, and 1.04 GHz, when loaded with 35.35 Ω and 50 Ω resistances, respectively, and the phase shift $\phi_c = -90°$ at f_0. Both cells of Fig. 2 have been subjected to concurrent optimization, yielding $\boldsymbol{x}_c^* = [0.45\ 1.9\ 3.325\ 0.225\ 1.55\ 0.125\ 8.05\ 0.1\ 1.15\ 0.1\ 0.41]^T$. Corresponding constrained optima for individual cells are $f_1(\boldsymbol{x}_c^*) = [0.45\ 1.9\ 3.325\ 0.225\ 1.55\ 0.125\ 8.05]^T$ mm, and $f_2(\boldsymbol{x}_c^*) = [0.1\ 1.15\ 0.1\ 0.41\ 2.675\ 2.925\ 0.9]^T$ mm. The EM evaluation of the entire coupler structure at the design produced in this step illustrates a degraded performance due to cross-coupling effects that occur between adjacent cells. This issue is addressed by the fine-tuning procedure. For that purpose, we develop local RSA models of the respective optimized cells and use them to construct the coarse model of the entire coupler (by cascading ABCD matrices of the corresponding building blocks). Next, we execute the surrogate-assisted design refinement algorithm.

The differences between pre- and post-tuning coupler design solution are depicted in Fig. 4a, b. One can notice that the final post-tuning BLC, given by $f_1(\boldsymbol{x}^*) = [0.475\ 1.9\ 3.25\ 0.225\ 1.5\ 0.125\ 8.05]^T$ mm, and $f_2(\boldsymbol{x}^*) = [0.125\ 1.15\ 0.125\ 0.375\ 0.95\ 2.7\ 2.925\ 0.875]^T$ mm, shows perfect transmission characteristics in contradiction to the pre-tuning BLC design. The final design solution has been manufactured and measured.

Measurement results presented in Fig. 4c are in agreement with the simulated BLC performance. Minor discrepancy between simulation and measurement results is most likely due to the smooth metal surface and dielectric anisotropy included in EM simulation [97], as well as fabrication inaccuracy.

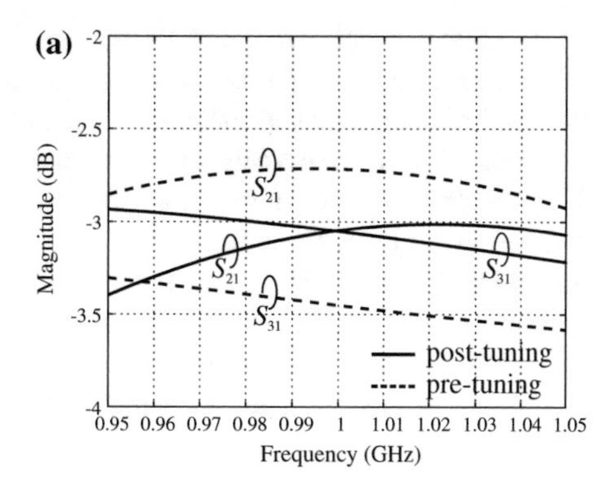

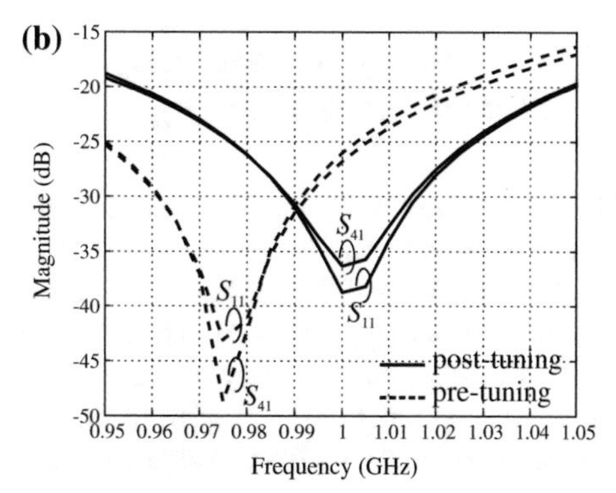

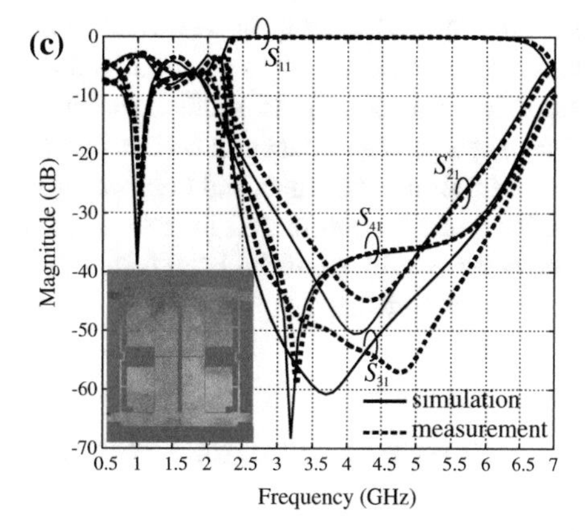

Fig. 4 **a–b** pre-tuning (*dashed*) versus post-tuning (*solid*) BLC S-parameters; **c** post-tuning BLC broadband performance—simulation versus measurement

It should be emphasized that the final BLC has reached a significant 83.7 % scale of miniaturization in comparison to a conventional BLC, together with ideal characteristics confirmed by measurement data—all at a low computational cost corresponding to about 5.6 full-wave analyses of the final compact BLC high-fidelity EM model. The design cost breakdown is as follows: concurrent cell optimization (150 cell evaluations at three frequencies each, ~225 min in total), simulation data for RSA model construction (21 cell evaluations at 10 frequencies each, ~105 min in total), and two simulations of the entire coupler (~180 min). The overall design cost of the proposed method is ~510 min. One should bear in mind that a direct optimization would require several hundred such evaluations, which is virtually infeasible.

5 Case Study II: Fast Design of Compact RF Circuits Using Nested Space Mapping

Design and optimization of microwave/RF circuits for space limited applications is a challenging task. Miniaturization of a structure is usually obtained by replacing its fundamental building blocks with composite elements that should be adjusted to obtain compact geometry [6, 7]. Unfortunately, implementation of such elements within a circuit significantly increases its complexity, and therefore a considerable amount of simulations is needed to find the optimum design. Consequently, direct utilization of high-fidelity EM simulations in the design process is usually prohibitive. This issue can be partially resolved by using equivalent circuit representations of the structure which are, however, of limited accuracy [2, 10]. In this section, we discuss a so-called nested space mapping (NSM) methodology that aims at using suitably corrected circuit models as well as structure decomposition for fast EM-driven design of compact microwave structures. NSM provides two levels of structure representation, i.e., inner- and outer-layer surrogate models. The former is applied at the level of each sub-circuit in order to provide their good generalization capability. The latter is utilized at the level of entire complex design to account for couplings between sub-circuits.

5.1 *Design Problem and Matching Transformer Structure*

Consider a impedance transformer realized in microstrip technology aimed to match 50-Ohm source to 130-Ohm load. The structure is supposed to ensure reflection $|S_{11}| \leq -15$ dB within 3.1 GHz to 10.6 GHz frequency. A circuit is designated to operate on a Taconic RF-35 dielectric substrate ($\varepsilon_r = 3.5$, $\tan\delta = 0.0018$, $h = 0.762$).

A conventional design that is sufficient to satisfy the aforementioned design specifications is composed of four 90° transmission line sections of various impedances [1]. The geometry of a structure and its circuit representation with highlighted sections

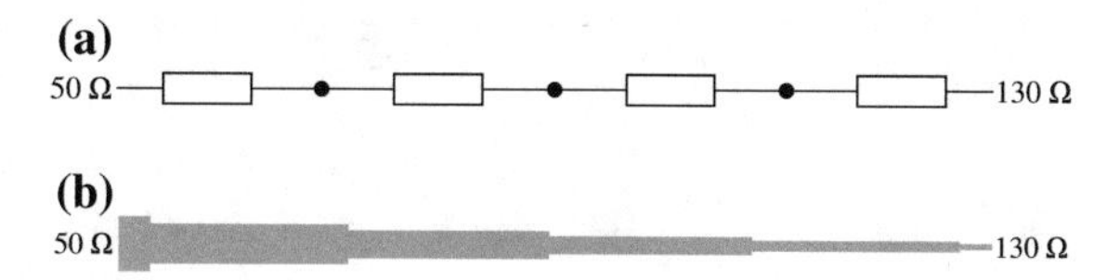

Fig. 5 Conventional 50-Ohm to 130-Ohm matching transformer: **a** circuit representation; **b** geometry of a structure

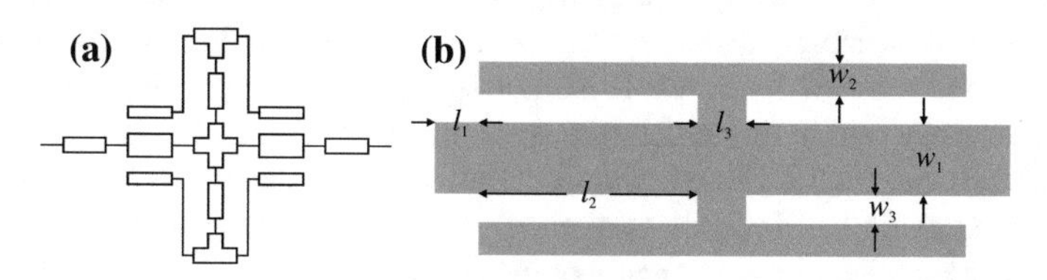

Fig. 6 A double-T composite structure: **a** circuit representation; **b** geometry with highlighted dimensions

are illustrated in Fig. 5. The simplicity of the design allows for its decomposition into four transmission line sections that could be substituted with composite elements of increased functionality [2]. A single, versatile component in the form of double-T composite may be utilized to construct miniaturized matching transformer [98]. The structure of interest is described by four independent design parameters: $\boldsymbol{y} = [l_1\, l_2\, w_1\, w_2]^T$ (variables $l_3 = 0.2$ and $w_3 = 0.2$ are fixed). Moreover, solution space for the component in the form of the following lower/upper $\boldsymbol{l}/\boldsymbol{u}$ bounds: $\boldsymbol{l} = [0.1\, 1\, 0.1\, 0.1]^T$ and $\boldsymbol{u} = [1\, 5\, 1\, 1]^T$ is defined to account for technology limitations (i.e., minimum feasible width of the composite element lines and the gaps between them equal to 0.1 mm). A high-fidelity model of the double-T composite element (~200,000 mesh cells and average evaluation time of 60 s) is implemented in CST Microwave Studio [39], whereas its circuit representation is constructed in Agilent ADS simulator [99]. Figure 6 illustrates a circuit and EM model of a double-T composite structure.

5.2 *Optimization Methodology*

Let $\boldsymbol{y}$ stand for the geometry parameters of a composite element, whereas $\boldsymbol{R}_{f.cell}(\boldsymbol{y})$ and $\boldsymbol{R}_{c.cell}(\boldsymbol{y})$ denote the responses of EM and circuit models, respectively. The nested space mapping method [25, 98] constructs a surrogate model of a miniaturized structure starting from the level of each composite element (a so-called local level space mapping). Consider $\boldsymbol{R}_{s.g.cell}(\boldsymbol{y}, \boldsymbol{p})$ as a generic surrogate model, constructed using $\boldsymbol{R}_{c.cell}$ and suitable space mapping transformations. Vector $\boldsymbol{p}$ denotes extractable space mapping parameters of the surrogate. The $\boldsymbol{R}_{s.cell}$ model is obtained using the following transformation

$$\boldsymbol{R}_{s.cell}(\boldsymbol{y}) = \boldsymbol{R}_{s.g.cell}(\boldsymbol{y}, \boldsymbol{p}^*) \tag{8}$$

where

$$\boldsymbol{p}^* = \arg\min_{\boldsymbol{p}} \sum_{k=1}^{N_{cell}} ||\boldsymbol{R}_{s.g.cell}(\boldsymbol{y}^{(k)}, \boldsymbol{p}) - \boldsymbol{R}_f(\boldsymbol{y}^{(k)})|| \tag{9}$$

Here, $\boldsymbol{y}^{(k)}$, $k = 1, \ldots, N_{cell}$, are the training designs obtained using a star-distribution scheme. The base set is composed of $N_{cell} = 2n + 1$, where n is a number of independent design variables. A surrogate model $\boldsymbol{R}_{s.g.cell}$ is usually constructed using a combination of input SM, implicit SM and frequency scaling [24] in such a way that $\boldsymbol{R}_{s.cell}$ is accurate within entire solution space of the model.

Let $\boldsymbol{R}_f(\boldsymbol{x})$ and $\boldsymbol{R}_c(\boldsymbol{x})$ denote the EM and circuit models of the entire composite structure with $\boldsymbol{x}$ being a vector of geometry parameters. Additionally, let $\boldsymbol{R}_{s.g}(\boldsymbol{x}, \boldsymbol{P})$ stand for a surrogate model of the entire compact structure, constructed of surrogate models of composite elements, i.e., $\boldsymbol{R}_{s.g}(\boldsymbol{x}, \boldsymbol{P}) = \boldsymbol{R}_{s.g}([\boldsymbol{y}_1; \ldots; \boldsymbol{y}_p], \boldsymbol{P}) = F(\boldsymbol{R}_{s.g.cell}(\boldsymbol{y}_1, \boldsymbol{p}^*), \ldots, \boldsymbol{R}_{s.g.cell}(\boldsymbol{y}_p, \boldsymbol{p}^*), \boldsymbol{P})$. Function F realizes a cascade connection of individual composite element responses [2, 98], whereas the vector $\boldsymbol{x}$ is a concatenation of component parameter vectors $\boldsymbol{y}_k$ (where $k = 1, \ldots, p$). The outer layer surrogate model parameter vector $\boldsymbol{P}$ is usually defined as a perturbation with respect to selected space mapping parameters $\boldsymbol{p}^*$ of each composite element.

An outer space mapping correction is applied at the level of entire compact structure $\boldsymbol{R}_{s.g}(\boldsymbol{x}, \boldsymbol{P})$, so that the final surrogate $\boldsymbol{R}_s^{(i)}$ utilized in the ith iteration of the surrogate-based optimization scheme (2) is as follows

$$\boldsymbol{R}_s^{(i)}(\boldsymbol{x}) = \boldsymbol{R}_{s.g}(\boldsymbol{x}^{(i)}, \boldsymbol{P}^{(i)}) \tag{10}$$

where

$$\boldsymbol{P}^{(i)} = \arg\min_{\boldsymbol{P}} ||\boldsymbol{R}_{s.g}(\boldsymbol{x}^{(i)}, \boldsymbol{P}) - \boldsymbol{R}_f(\boldsymbol{x}^{(i)})|| \tag{11}$$

Generally, vector $\boldsymbol{P}$ utilized in NSM approach is composed of considerably smaller number of space mapping parameters than in competitive techniques, which is due to good alignment of $\boldsymbol{R}_{s.cell}$ and $\boldsymbol{R}_{f.cell}$ provided by the inner SM layer. For more detailed description of nested space mapping methodology see e.g. [25, 98].

5.3 Results

For demonstration purposes, a compact 50-Ohm to 130-Ohm impedance transformer composed of a cascade connected double-T composite elements [98] is considered for optimization with respect to the design specifications of Sect. 5.1. A surrogate model $\boldsymbol{R}_{s.cell}$ is constructed using a total of 16 space mapping parameters including: 8 input space mapping, 6 implicit space mapping, and 2 frequency scaling ones [25, 41, 92]. Subsequently, a 9-point parameter extraction based on star-distribution

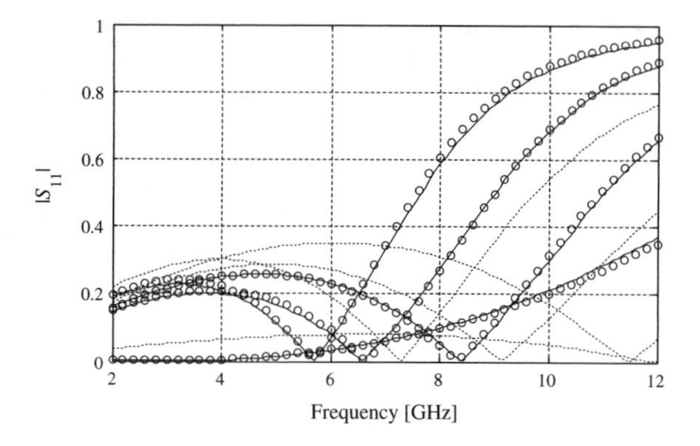

Fig. 7 Responses of double-T composite element at the selected test designs: coarse model (· · ··), fine model (—), NSM surrogate after multipoint parameter extraction (o o o). The plots indicate very good approximation capability of the surrogate

scheme is performed. Figure 7 illustrates a comparison of the model responses before and after multi-point parameter extraction step.

An outer layer surrogate model $\boldsymbol{R}_{s.g}$ of the compact matching transformer is composed of interconnected surrogate models of composite elements $\boldsymbol{R}_{s.cell}$. Its corresponding $\boldsymbol{R}_f$ counterpart is prepared in CST Microwave Studio (~1,060,000 mesh cells and average simulation time of 10 min). The initial set of design parameters is: $\boldsymbol{x}$ = [0.55 3.75 0.65 0.35 0.55 3.75 0.65 0.35 0.55 3.75 0.65 0.35 0.55 3.75 0.65 0.35]^T. Next, the NSM technique of Sect. 5.2 is utilized to obtain the final design. Optimized compact matching transformer is represented by the following vector of design parameters: $\boldsymbol{x}$ = [1.0 3.52 0.85 0.2 0.8 4.1 0.58 0.1 0.8 3.09 0.1 0.25 1 2.32 0.13 0.1]^T. Schematic representation of $\boldsymbol{R}_s$ model of compact structure in the form of cascade connection of $\boldsymbol{R}_{s.cell}$ models and geometry of a circuit are illustrated in Fig. 8.

The optimized design that satisfies all assumed requirements, i.e., provides 50-Ohm source to 130-Ohm load matching and $|S_{11}| \leq -15$ dB within defined oper-

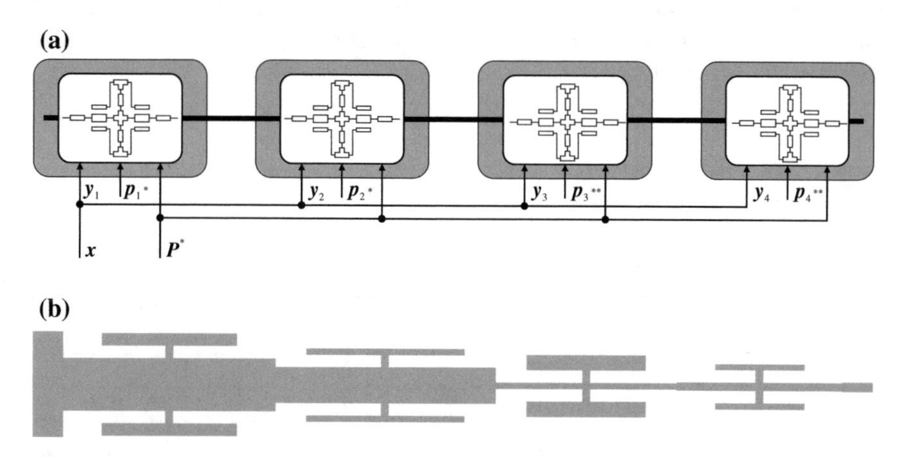

Fig. 8 A compact matching transformer: **a** schematic diagram of double-T composite element interconnections; **b** geometry of an optimized structure

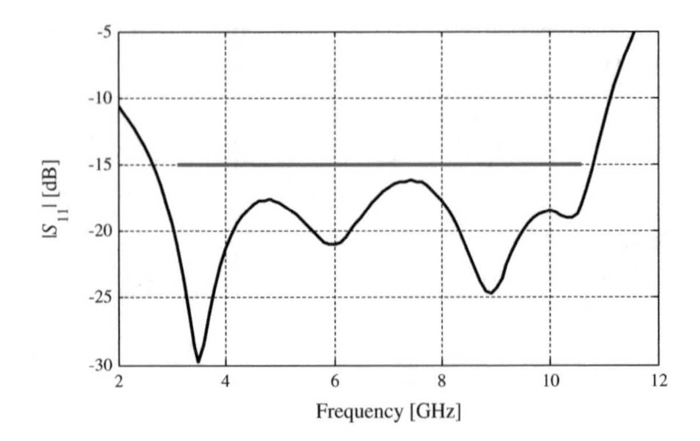

Fig. 9 Reflection response of the optimized compact matching transformer

ational bandwidth is obtained after only 3 iterations of nested space mapping algorithm. One should note that the optimized structure exhibits 15 % broader bandwidth than the assumed (lower and upper operating frequency is 2.7 GHz and 10.8 GHz, respectively). Moreover, reflection level for 3.1 GHz to 10.6 GHz bandwidth is below –16.2 dB, which is almost 15 % lower than the assumed value. Reflection characteristic of the optimized design is shown in Fig. 9.

The total cost of compact matching transformer design and optimization is about 40 min and it includes: 9 simulations of $\boldsymbol{R}_{f.cell}$ model during multi-point parameter extraction, 3 evaluations of the $\boldsymbol{R}_f$ model, and simulations of surrogate models that corresponds to a total of 0.2 $\boldsymbol{R}_f$. It should be emphasized that NSM technique outclasses other competitive methods, i.e., implicit space mapping [100], and sequential space mapping [2]. The results indicate that the computational cost of the method is almost 70 % smaller in comparison to implicit space mapping and sequential space mapping. For the sake of comparison, a direct optimization driven by pattern search algorithm [33] has been also conducted; however algorithm failed to find a design satisfying given specifications and it was terminated after 500 iterations. Table 1 gathers detailed data concerning the computational cost of techniques utilized for transformer optimization.

Table 1 Four section unconventional MT: design and optimization cost

Model type	Optimization algorithm			
	NSM	ISM	SSM	Direct search
$\boldsymbol{R}_{s.cell}$	$0.1 \times \boldsymbol{R}_f$	N/A	N/A	N/A
$\boldsymbol{R}_{f.cell}$	$0.6 \times \boldsymbol{R}_f$	N/A	N/A	N/A
$\boldsymbol{R}_s$	$0.1 \times \boldsymbol{R}_f$	$5.1 \times \boldsymbol{R}_f$	$1.7 \times \boldsymbol{R}_f$	N/A
$\boldsymbol{R}_f$	3	7	10[a]	500[b]
Total cost	$3.8 \times \boldsymbol{R}_f$	$12.1 \times \boldsymbol{R}_f$	$11.7 \times \boldsymbol{R}_f$	500[b] $\times \boldsymbol{R}_f$
Total cost [min]	38	121	N/A	5000

[a]The algorithm started diverging and was terminated after 10 iterations
[b]The algorithm failed to find a geometry satisfying performance specifications

6 Case Study III: Simulation-Driven Design of Small UWB Antennas

Antennas are crucial components in mobile communication systems. They play a role of an interface between the wireless medium and the transmission lines. Ultra-wideband (UWB) antennas are of particular interest, because they can provide high data transmission rates, and their broadband properties could be useful for reduction of transceiver section complexity [28]. On the other hand, such antennas are characterized by considerable footprint, which is a serious drawback: rapid development of handheld devices imposes strict requirements upon miniaturization of contemporary antenna structures. Miniaturization of ultra-wideband antennas is troublesome, since their performance—especially for lower frequency range—depends on the wideband impedance and current path within the ground plane [28]. While the former may be accounted for by appropriate construction of the feeding line, the latter requires complex modifications within the ground plane. Unintuitive and asymmetrical geometries of compact UWB antennas significantly influence the evaluation cost of their EM models, which varies from half an hour up to several hours. These problems make the design and optimization of compact ultra-wideband antennas a challenging task. In particular, their simulation-driven design directly based on high-fidelity EM models is impractical [84]. On the other hand, lack of theoretical insight for such structures makes EM simulators the only reliable tools for their performance evaluation. In this section, a design of compact UWB antenna with modified ground plane and feed line is discussed. Computational efficiency of the optimization process is ensured by the utilization of high- and low-fidelity EM models. Discrepancy between model responses is addressed using adaptively-adjusted design specification (AADS) technique [85].

6.1 *Antenna Structure and Design Problem*

Consider a conventional planar monopole antenna [101] consisting a circular shape radiator and a feeding structure in the form of 50-Ohm microstrip line. The structure is characterized by a considerable size of $45 \times 50\ \text{mm}^2$, which is necessary to achieve wideband impedance matching and sufficient current path within ground plane. The structure can be miniaturized by introduction of a tapered feed line and modification of the ground plane by means of a L-shaped ground plane stub [102, 103]. Geometry of a conventional monopole and its compact counterpart with modified geometry is illustrated in Fig. 10.

The antenna is described by seven independent design variables: $\boldsymbol{x} = [g_0\ \ g_1\ \ l_1\ \ rg_2\ \ c_0\ \ w_1]^T$, whereas $w_0 = 2.35$, $l_0 = 8$ remain fixed; $c_1 = c_0/2$ (all dimensions in mm). It is designated to work on Taconic TLP-5 dielectric substrate ($\varepsilon_r = 2.2$, $tan\delta = 0.0009$, $h = 0.762$ mm). In order to utilize AADS methodology for the design and optimization, two models of an antenna have been prepared

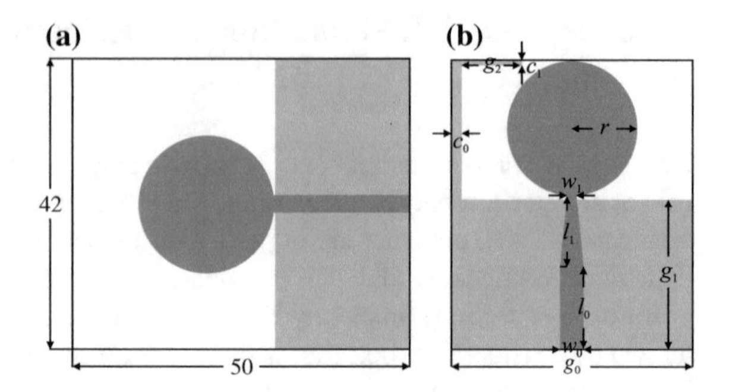

Fig. 10 Monopole UWB antenna: **a** initial design [101]; **b** a compact structure with modified geometry [28]

and simulated in CST Microwave Studio transient solver. The high-fidelity model $\boldsymbol{R}_f$ is represented by ~2,577,000 mesh cells and average evaluation time of 30 min, whereas its low-fidelity counterpart $\boldsymbol{R}_c$ is composed of ~75,000 mesh cells and its average simulation time is 47 s.

Two design cases are considered: (i) reduction of antenna footprint defined as a rectangle $R = A \times B$, where $A = g_0$ and $B = l_0 + l_1 + 2r$, and (ii) minimization of antenna reflection. One should note that operational properties of an antenna optimized within the former criteria, i.e., $|S_{11}| \leq -10$ dB are enforced by sufficiently defined penalty factor. Both objectives are considered within 3.1 GHz to 10.6 GHz frequency band.

6.2 Optimization Methodology

In order to provide reliable prediction of the antenna response, a misalignment between $\boldsymbol{R}_f$ and $\boldsymbol{R}_c$ responses is accounted for by means of adaptively adjusted design specifications (AADS) technique [85]. Majority of available surrogate-based optimization techniques (e.g., [22, 56, 100]) perform enhancement and corrections of the $\boldsymbol{R}_c$ model in order to minimize its misalignment with respect to $\boldsymbol{R}_f$. The AADS methodology utilizes knowledge about discrepancy between responses of $\boldsymbol{R}_c$ and $\boldsymbol{R}_f$ in order to modify design specifications so that they account for the response differences. AADS works very well for a problems that are defined in a minimax sense, e.g., $|S_{11}| \leq -10$ dB over a defined operational frequency, which is the case for the design of UWB antenna structures. A conceptual explanation of the method with highlight on the determination of characteristic points is provided in Fig. 11, whereas the algorithm flow is presented below:

1. Modify the original design specifications to account for the difference between the responses of $\boldsymbol{R}_f$ and $\boldsymbol{R}_c$ at their characteristic points.
2. Obtain a new design by optimizing the low-fidelity model $\boldsymbol{R}_c$ with respect to the modified specifications.

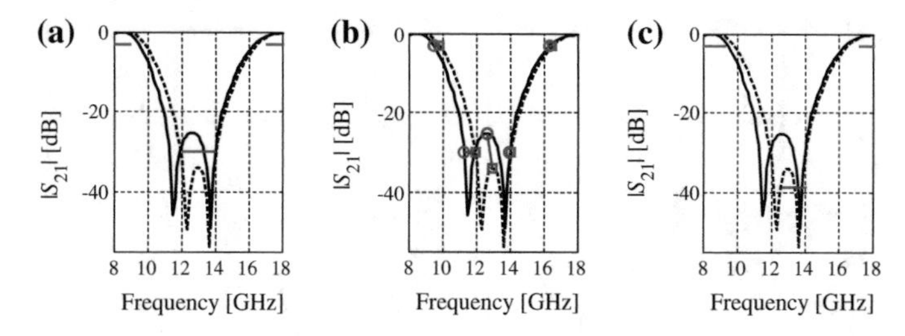

Fig. 11 Design optimization through AADS [85] on the basis of a bandstop filter $\boldsymbol{R}_f$ (—) and $\boldsymbol{R}_c$ (- - -): **a** initial responses and the original design specifications, **b** characteristic points of the responses corresponding to the specification levels (here, –3 and –30 dB) and to the local maxima, **c** responses at the initial design as well as the modified design specifications. The modification accounts for the discrepancy between the models: optimizing $\boldsymbol{R}_c$ w.r.t. the modified specs corresponds (approximately) to optimizing $\boldsymbol{R}_f$ w.r.t. the original specs

In the first step of algorithm, the specifications are altered in such a way that specifications for $\boldsymbol{R}_c$ model corresponds to the desired frequency properties of the high fidelity one. In the second stage, the $\boldsymbol{R}_c$ model is optimized with respect to the redefined design specifications. The optimal design is an approximated solution to the original problem defined for $\boldsymbol{R}_f$ model. Although the most considerable design improvement is normally observed after the first iteration, the algorithm steps may be repeated if further refinement of design specifications is required. In practice, the algorithm is terminated once the current iteration does not bring further improvement of the high-fidelity model design. One should emphasize that determination of appropriate characteristic points is crucial for the operation of the technique. They should account for local extrema of both model responses, at which specifications may not be satisfied. Moreover, due to differences between $\boldsymbol{R}_f$ and $\boldsymbol{R}_c$ models, redefinition of design specifications may be necessary at each algorithm iteration.

6.3 Results

The initial design of a compact monopole antenna of Sect. 6.1 is $\boldsymbol{x}^0 = [24\ 14.5\ 7\ 6.5\ 6\ 1\ 0.9]^T$ and corresponding footprint of a structure is 672 mm^2, whereas its maximal reflection within frequency band of interest is –10.8 dB. The design parameters of a structure optimized with respect to minimization of reflection are $\boldsymbol{x}^{(i)} = [26.58\ 13.83\ 6.19\ 6.24\ 7.79\ 0.33 0.7]^T$. The design is obtained after 3 iterations of the AADS algorithm. The second design—optimized towards minimization of footprint—has been obtained after only 2 iterations of the algorithm and the corresponding dimensions are $\boldsymbol{x}^{(ii)} = [19.78\ 13.63\ 5.81\ 5.84\ 7.89\ 0.33\ 0.72]^T$. The footprint of a structure is only 504 mm^2. It should be emphasized that the design optimized with respect to reflection is characterized by a maximal in-band $|S_{11}| = -13.5$ dB which is 20 % lower in comparison with the reference structure. Additionally, the

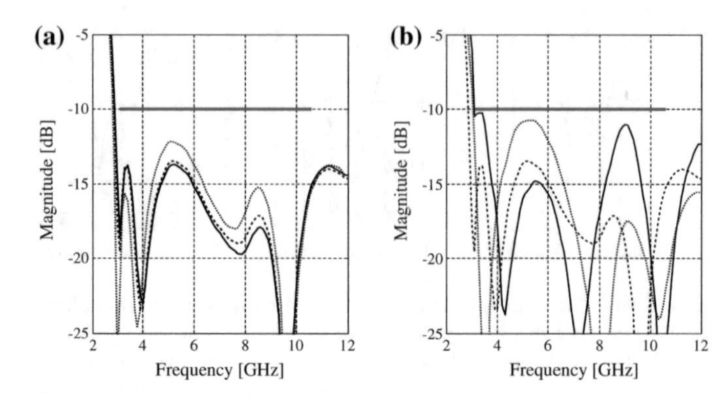

Fig. 12 Reflection response of optimized compact UWB monopole antennas. Initial design (· · · ·), first iteration (– – –), final result (——): **a** design (i); **b** design (ii)

Table 2 Design of a compact UWB antenna: optimization results

Antenna design	AADS		Direct-search (pattern search)	
	Design (i)	Design (ii)	Design (i)	Design (ii)
$\boldsymbol{R}_c$	300	200	N/A	N/A
$\boldsymbol{R}_f$	4	3	97	107
Total cost	$11.8 \times \boldsymbol{R}_f$	$8.2 \times \boldsymbol{R}_f$	$97 \times \boldsymbol{R}_f$	$107 \times \boldsymbol{R}_f$
Total cost [h]	5.9	4.1	48.5	53.5

antenna optimized with respect to minimization of the lateral area is 25 % smaller than the reference one. It should be also highlighted that the variation of size between both optimized structures is 29 %, whereas their maximal in-band reflection varies by 26 %. Frequency responses of both optimized antenna designs are shown in Fig. 12.

The optimization cost for the first case corresponds to 11.8 $\boldsymbol{R}_f$ (~5.9 h) and it includes: a total of 300 $\boldsymbol{R}_c$ evaluations for the design optimization and 4 $\boldsymbol{R}_f$ simulations for the response verification. The cost of obtaining the second design corresponds to about 8.2 $\boldsymbol{R}_f$ model simulations (~4.1 h): 200 $\boldsymbol{R}_c$ simulations for the design optimization and 3 $\boldsymbol{R}_f$ model evaluations. For the sake of comparison, both designs have been optimized using direct-search approach driven by pattern search algorithm [33]. The cost of design optimization towards minimum reflection and footprint miniaturization is 97 $\boldsymbol{R}_f$ and 107 $\boldsymbol{R}_f$, respectively. Design costs for these two cases are over 8 and over 13 times lower compared to direct search. A detailed comparison of design and optimization cost of both designs is gathered in Table 2.

7 Conclusion

This chapter highlighted several techniques for rapid EM-simulation-driven design of miniaturized microwave and antenna structures. Techniques such as nested space mapping, design tuning exploiting structure decomposition and local approximation models, or adaptively adjusted design specifications, can be utilized to obtain the optimized geometries of compact circuits in reasonable timeframe. The key is a proper combination of fast low-fidelity models, their suitable correction, as well as appropriate correction-prediction schemes linking the process of surrogate model identification and optimization. In all of the discussed schemes, the original, high-fidelity model is referred to rarely (for design verification and providing data for further surrogate model enhancement). In case of compact microwave structures, it is also usually possible to exploit structure decomposition, which further speeds up the design process. In any case, it seems that tailoring the optimization method for a given design problems gives better results than taking off-the-shelf algorithm. One of the open problems in the field discussed in this chapter include design automation, such as automated selection of the low-fidelity model, as well as controlling the convergence of the surrogate-based optimization process. This and other issues will be the subject of the future research.

References

1. Pozar, D.M.: Microwave Engineering, 4th edn. Wiley, Hoboken (2012)
2. Bekasiewicz, A., Kurgan, P., Kitlinski, M.: A new approach to a fast and accurate design of microwave circuits with complex topologies. IET Microw. Antennas Propag. **6**, 1616–1622 (2012)
3. Smierzchalski, M., Kurgan, P., Kitlinski, M.: Improved selectivity compact band-stop filter with Gosper fractal-shaped defected ground structures. Microw. Opt. Technol. Lett. **52**, 227–232 (2010)
4. Kurgan, P., Kitlinski, M.: Novel microstrip low-pass filters with fractal defected ground structures. Microw. Opt. Technol. Lett. **51**, 2473–2477 (2009)
5. Aznar, F., Velez, A., Duran-Sindreu, M., Bonache, J., Martin, F.: Elliptic-function CPW low-pass filters implemented by means of open complementary split ring resonators (OCSRRs). IEEE Microw. Wirel. Compon. Lett. **19**, 689–691 (2009)
6. Kurgan, P., Bekasiewicz, A., Pietras, M., Kitlinski, M.: Novel topology of compact coplanar waveguide resonant cell low-pass filter. Microw. Opt. Technol. Lett. **54**, 732–735 (2012)
7. Bekasiewicz, A., Kurgan, P.: A compact microstrip rat-race coupler constituted by nonuniform transmission lines. Microw. Opt. Technol. Lett. **56**, 970–974 (2014)
8. Opozda, S., Kurgan, P., Kitlinski, M.: A compact seven-section rat-race hybrid coupler incorporating PBG cells. Microw. Opt. Technol. Lett. **51**, 2910–2913 (2009)
9. Kurgan, P., Kitlinski, M.: Novel doubly perforated broadband microstrip branch-line couplers. Microw. Opt. Technol. Lett. **51**, 2149–2152 (2009)
10. Kurgan, P., Bekasiewicz, A.: A robust design of a numerically demanding compact rat-race coupler. Microw. Opt. Technol. Lett. **56**, 1259–1263 (2014)
11. Wu, Y., Liu, Y., Xue, Q., Li, S., Yu, C.: Analytical design method of multiway dual-band planar power dividers with arbitrary power division. IEEE Trans. Microw. Theory Tech. **58**, 3832–3841 (2010)

12. Chiu, L., Xue, Q.: A parallel-strip ring power divider with high isolation and arbitrary power-dividing ratio. IEEE Trans. Microw. Theory Tech. **55**, 2419–2426 (2007)
13. Bekasiewicz, A., Koziel, S., Ogurtsov, S., Zieniutycz, W.: Design of microstrip antenna subarrays: a simulation-driven surrogate-based approach. In: International Conference Microwaves, Radar and Wireless Communications (2014)
14. Wang, X., Wu, K.-L., Yin, W.-Y.: A compact gysel power divider with unequal power-dividing ratio using one resistor. IEEE Trans. Microw. Theory Tech. **62**, 1480–1486 (2014)
15. Koziel, S., Ogurtsow, S., Zieniutycz, W., Bekasiewicz, A.: Design of a planar UWB dipole antenna with an integrated balun using surrogate-based optimization. IEEE Antennas Wirel. Propag. Lett. (2014)
16. Shao, J., Fang, G., Ji, Y., Tan, K., Yin, H.: A novel compact tapered-slot antenna for GPR applications. IEEE Antennas Wirel. Propag. Lett. **12**, 972–975 (2013)
17. Koziel, S., Bekasiewicz, A.: Novel structure and EM-Driven design of small UWB monople antenna. In: International Symposium on Antenna Technology and Applied Electromagnetics (2014)
18. Quan, X., Li, R., Cui, Y., Tentzeris, M.M.: Analysis and design of a compact dual-band directional antenna. IEEE Antennas Wirel. Propag. Lett. **11**, 547–550 (2012)
19. Koziel, S., Ogurtsov, S.: Multi-objective design of antennas using variable-fidelity simulations and surrogate models. IEEE Trans. Antennas Propag. **61**, 5931–5939 (2013)
20. Yeung, S.H., Man, K.F.: Multiobjective Optimization. IEEE Microw. Mag. **12**, 120–133 (2011)
21. Kuwahara, Y.: Multiobjective optimization design of Yagi-Uda antenna. IEEE Trans. Antennas Propag. **53**, 1984–1992 (2005)
22. Koziel, S., Bekasiewicz, A., Couckuyt, I., Dhaene, T.: Efficient multi-objective simulation-driven antenna design using Co-Kriging. IEEE Trans. Antennas Propag. **62**, 5900–5905 (2014)
23. Kurgan, P., Filipcewicz, J., Kitlinski, M.: Design considerations for compact microstrip resonant cells dedicated to efficient branch-line miniaturization. Microw. Opt. Technol. Lett. **54**, 1949–1954 (2012)
24. Bekasiewicz, A., Koziel, S., Pankiewicz, B.: Accelerated simulation-driven design optimization of compact couplers by means of two-level space mapping. IET Microw. Antennas Propag. (2014)
25. Koziel, S., Bekasiewicz, A., Kurgan, P.: Nested space mapping technique for design and optimization of complex microwave structures with enhanced functionality. In: Koziel, S., Leifsson, L., Yang, X.S. (eds.) Solving Computationally Expensive Engineering Problems: Methods and Applications, pp. 53–86. Springer, Switzerland (2014)
26. Kurgan, P., Bekasiewicz, A.: Atomistic surrogate-based optimization for simulation-driven design of computationally expensive microwave circuits with compact footprints. In: Koziel, S., Leifsson, L., Yang, X.S. (eds.) Solving Computationally Expensive Engineering Problems: Methods and Applications, pp. 195–218. Springer, Switzerland (2014)
27. Li, J.-F., Chu, Q.-X., Li, Z.-H., Xia, X.-X.: Compact dual band-notched UWB MIMO antenna with high isolation. IEEE Trans. Antennas Propag. **61**, 4759–4766 (2013)
28. Koziel, S., Bekasiewicz, A.: Small antenna design using surrogate-based optimization. In: IEEE International Symposium on Antennas Propagation (2014)
29. Liu, Y.-F., Wang, P., Qin, H.: Compact ACS-fed UWB monopole antenna with extra Bluetooth band. Electron. Lett. **50**, 1263–1264 (2014)
30. Koziel, S., Kurgan, P.: Low-cost optimization of compact branch-line couplers and its application to miniaturized Butler matrix design. In: European Microwave Conference (2014)
31. Nocedal, J., Wright, S.: Numerical Optimization, 2nd edn. Springer, New York (2006)
32. Koziel, S., Yang X.S. (eds.): Computational optimization, methods and algorithms. Studies in Computational Intelligence, vol. 356, Springer, New York (2011)
33. Kolda, T.G., Lewis, R.M., Torczon, V.: Optimization by direct search: new perspectives on some classical and modern methods. SIAM Rev. **45**, 385–482 (2003)
34. Deb, K.: Multi-Objective Optimization Using Evolutionary Algorithms. Wiley, Chichester (2001)
35. Talbi, E.-G.: Metaheuristics—From Design to Implementation. Wiley, Chichester (2009)

36. Koziel, S., Bekasiewicz, A., Zieniutycz, W.: Expedite EM-driven multi-objective antenna design in highly-dimensional parameter spaces. IEEE Antennas Wirel. Propag. Lett. **13**, 631–634 (2014)
37. Bekasiewicz, A., Koziel, S.: Efficient multi-fidelity design optimization of microwave filters using adjoint sensitivity. Int. J. RF Microw. Comput. Aided Eng. (2014)
38. Koziel, S., Ogurtsov, S., Cheng, Q.S., Bandler, J.W.: Rapid EM-based microwave design optimization exploiting shape-preserving response prediction and adjoint sensitivities. IET Microw. Antennas Propag. (2014)
39. CST Microwave Studio: CST AG, Bad Nauheimer Str. 19, D-64289 Darmstadt, Germany (2011)
40. Ansys HFSS: ver. 14.0, ANSYS, Inc., Southpointe 275 Technology Dr., Canonsburg, PA (2012)
41. Bandler, J.W., Cheng, Q.S., Dakroury, S.A., Mohamed, A.S., Bakr, M.H., Madsen, K., Sondergaard, J.: Space mapping: the state of the art. IEEE Trans. Microw. Theory Tech. **52**, 337–361 (2004)
42. Koziel, S., Bandler, J.W., Madsen, K.: Towards a rigorous formulation of the space mapping technique for engineering design. In: Proceedings International Symposium Circuits and Systems (2005)
43. Queipo, N.V., Haftka, R.T., Shyy, W., Goel, T., Vaidynathan, R., Tucker, P.K.: Surrogatebased analysis and optimization. Prog. Aerosp. Sci. **41**, 1–28 (2005)
44. El Zooghby, A.H., Christodoulou, C.G., Georgiopoulos, M.: A neural network-based smart antenna for multiple source tracking. IEEE Trans. Antennas Propag. **48**, 768–776 (2000)
45. Siah, E.S., Sasena, M., Volakis, J.L., Papalambros, P.Y., Wiese, R.W.: Fast parameter optimization of large-scale electromagnetic objects using DIRECT with Kriging metamodeling. IEEE Trans. Microw. Theory Tech. **52**, 276–285 (2004)
46. Xia, L., Meng, J., Xu, R., Yan, B., Guo, Y.: Modeling of 3-D vertical interconnect using support vector machine regression. IEEE Microw. Wirel. Compon. Lett. **16**, 639–641 (2006)
47. Kabir, H., Wang, Y., Yu, M., Zhang, Q.J.: Neural network inverse modeling and applications to microwave filter design. IEEE Trans. Microw. Theory Tech. **56**, 867–879 (2008)
48. Tighilt, Y., Bouttout, F., Khellaf, A.: Modeling and design of printed antennas using neural networks. Int. J. RF Microw. Comput. Aided Eng. **21**, 228–233 (2011)
49. Koziel, S., Bandler, J.W.: Space mapping with multiple coarse models for optimization of microwave components. IEEE Microw. Wirel. Compon. Lett. **18**, 1–3 (2008)
50. Koziel, S., Bandler, J.W., Cheng, Q.S.: Robust trust-region space-mapping algorithms for microwave design optimization. IEEE Trans. Microw. Theory Tech. **58**, 2166–2174 (2010)
51. Alexandrov, N.M., Lewis, R.M.: An overview of first-order model management for engineering optimization. Optim. Eng. **2**, 413–430 (2001)
52. Amineh, R.K., Koziel, S., Nikolova, N.K., Bandler, J.W., Reilly, J.P.: A space mapping methodology for defect characterization. In: International Review of Progress in Applied Computational Electromagnetics (2008)
53. Koziel, S., Bandler, J.W.: SMF: a user-friendly software engine for space-mapping-based engineering design optimization. In: International Symposium Signals Systems Electronics (2007)
54. Koziel, S., Leifsson, L., Ogurtsov, S.: Reliable EM-driven microwave design optimization using manifold mapping and adjoint sensitivity. Microw. Opt. Technol. Lett. **55**, 809–813 (2013)
55. Echeverria, D., Lahaye, D., Encica, L., Lomonova, E.A., Hemker, P.W., Vandenput, A.J.A.: Manifold-mapping optimization applied to linear actuator design. IEEE Trans. Mangetics **42**, 1183–1186 (2006)
56. Cheng, Q.S., Rautio, J.C., Bandler, J.W., Koziel, S.: Progress in simulator-based tuning–the art of tuning space mapping. IEEE Microw. Mag. **11**, 96–110 (2010)
57. Rautio, J.C.: Perfectly calibrated internal ports in EM analysis of planar circuits. In: International Microwave Symposium Digest (2008)

58. Gilmore, R., Besser, L.: Practical RF Circuit Design for Modern Wireless Systems. Artech House, Norwood (2003)
59. Xu, H.-X., Wang, G.-M., Lu, K.: Microstrip rat-race couplers. IEEE Microw. Mag. **12**, 117–129 (2011)
60. Ahn, H.-R., Bumman, K.: Toward integrated circuit size reduction. IEEE Microw. Mag. **9**, 65–75 (2008)
61. Liao, S.-S., Sun, P.-T., Chin, N.-C., Peng, J.-T.: A novel compact-size branch-line coupler. IEEE Microw. Wirel. Compon. Lett. **15**, 588–590 (2005)
62. Liao, S.-S., Peng, J.-T.: Compact planar microstrip branch-line couplers using the quasi-lumped elements approach with nonsymmetrical and symmetrical T-shaped structure. IEEE Trans. Microw. Theory Tech. **54**, 3508–3514 (2006)
63. Tang, C.-W., Chen, M.-G.: Synthesizing microstrip branch-line couplers with predetermined compact size and bandwidth. IEEE Trans. Microw. Theory Tech. **55**, 1926–1934 (2007)
64. Jung, S.-C., Negra, R., Ghannouchi, F.M.: A design methodology for miniaturized 3-dB branch-line hybrid couplers using distributed capacitors printed in the inner area. IEEE Trans. Microw. Theory Tech. **56**, 2950–2953 (2008)
65. Ahn, H.-R.: Modified asymmetric impedance transformers (MCCTs and MCVTs) and their application to impedance-transforming three-port 3-dB power dividers. IEEE Trans. Microw. Theory Tech. **59**, 3312–3321 (2011)
66. Tseng, C.-H., Chang, C.-L.: A rigorous design methodology for compact planar branch-line and rat-race couplers with asymmetrical T-structures. IEEE Trans. Microw. Theory Tech. **60**, 2085–2092 (2012)
67. Ahn, H.-R., Nam, S.: Compact microstrip 3-dB coupled-line ring and branch-line hybrids with new symmetric equivalent circuits. IEEE Trans. Microw. Theory Tech. **61**, 1067–1078 (2013)
68. Eccleston, K.W., Ong, S.H.M.: Compact planar microstripline branch-line and rat-race couplers. IEEE Trans. Microw. Theory Tech. **51**, 2119–2125 (2003)
69. Chuang, M.-L.: Miniaturized ring coupler of arbitrary reduced size. IEEE Microw. Wirel. Compon. Lett. **15**, 16–18 (2005)
70. Chun, Y.-H., Hong, J.-S.: Compact wide-band branch-line hybrids. IEEE Trans. Microw. Theory Tech. **54**, 704–709 (2006)
71. Kuo, J.-T., Wu, J.-S., Chiou, Y.-C.: Miniaturized rat race coupler with suppression of spurious passband. IEEE Microw. Wirel. Compon. Lett. **17**, 46–48 (2007)
72. Mondal, P., Chakrabarty, A.: Design of miniaturised branch-line and rat-race hybrid couplers with harmonics suppression. IET Microw. Antennas Propag. **3**, 109–116 (2009)
73. Ahn, H.-R., Kim, B.: Small wideband coupled-line ring hybrids with no restriction on coupling power. IEEE Trans. Microw. Theory Tech. **57**, 1806–1817 (2009)
74. Sun, K.-O., Ho, S.-J., Yen, C.-C., van der Weide, D.: A compact branch-line coupler using discontinuous microstrip lines. IEEE Microw. Wirel. Compon. Lett. **15**, 501–503 (2005)
75. Lee, H.-S., Choi, K., Hwang, H.-Y.: A harmonic and size reduced ring hybrid using coupled lines. IEEE Microw. Wirel. Compon. Lett. **17**, 259–261 (2005)
76. Tseng, C.-H., Chen, H.-J.: Compact rat-race coupler using shunt-stub-based artificial transmission lines. IEEE Microw. Wirel. Compon. Lett. **18**, 734–736 (2008)
77. Wang, C.-W., Ma, T.-G., Yang, C.-F.: A new planar artificial transmission line and its applications to a miniaturized butler matrix. IEEE Trans. Microw. Theory Tech. **55**, 2792–2801 (2007)
78. Ahn, H.-R., Nam, S.: Wideband microstrip coupled-line ring hybrids for high power-division ratios. IEEE Trans. Microw. Theory Tech. **61**, 1768–1780 (2013)
79. Tsai, K.-Y., Yang, H.-S., Chen, J.-H., Chen, Y.-J.: A miniaturized 3 dB branch-line hybrid coupler with harmonics suppression. IEEE Microw. Wirel. Compon. Lett. **21**, 537–539 (2011)
80. Collin, R.E.: Foundations for Microwave Engineering. Wiley, New York (2001)
81. Gu, J., Sun, X.: Miniaturization and harmonic suppression rat-race coupler using C-SCMRC resonators with distributive equivalent circuit. IEEE Microw. Wirel. Compon. Lett. **15**, 880–882 (2005)

82. Kurgan, P., Filipcewicz, J., Kitlinski, M.: Development of a compact microstrip resonant cell aimed at efficient microwave component size reduction. IET Microw. Antennas Propag. **6**, 1291–1298 (2012)
83. Kurgan, P., Kitlinski, M.: Doubly miniaturized rat-race hybrid coupler. Mirow. Opt. Technol. Lett. **53**, 1242–1244 (2011)
84. Bekasiewicz, A., Koziel, S., Zieniutycz, W.: Design space reduction for expedited multi-objective design optimization of antennas in highly-dimensional spaces. In: Koziel, S., Leifsson, L., Yang, X.S. (eds.) Solving Computationally Expensive Engineering Problems: Methods and Applications, pp. 113–147. Springer, Switzerland (2014)
85. Koziel, S., Ogurtsov, S.: Rapid optimization of omnidirectional antennas using adaptively adjusted design specifications and kriging surrogates. IET Microw. Antennas Propag. **7**, 1194–1200 (2013)
86. Hong, J.-S., Lancaster, M.J.: Microstrip filters for RF/microwave applications. Wiley, New York (2001)
87. Koziel, S., Bekasiewicz, A.: Simulation-driven design of planar filters using response surface approximations and space mapping. In: European Microwave Conference (2014)
88. Lai, M.-I., Jeng, S.-K.: A microstrip three-port and four-channel multiplexer for WLAN and UWB coexistence. IEEE Trans. Microw. Theory Tech. **53**, 3244–3250 (2005)
89. Balanis, C.A.: Antenna Theory: Analysis and Design, 2nd edn. Wiley, New York (1997)
90. Milligan, T.A.: Modern Antenna Design, 2nd edn. Wiley, New York (2005)
91. Koziel, S., Ogurtsov, S.: Computational-budget-driven automated microwave design optimization using variable-fidelity electromagnetic simulations. Int. J. RF Microw. Comput. Aided Eng. **23**, 349–356 (2013)
92. Koziel, S., Cheng, Q.S., Bandler, J.W.: Space mapping. IEEE Microw. Mag. **9**, 105–122 (2008)
93. Conn, A.R., Gould, N.I.M., Toint, P.L.: Trust Region Methods. MPS-SIAM Series on Optimization (2000)
94. Koziel, S., Kurgan, P.: Rapid design of miniaturized branch-line couplers through concurrent cell optimization and surrogate-assisted fine-tuning. IET Microw. Antennas Propag. **9**, 957–963 (2015)
95. Mongia, R., Bahl, I., Bhartia, P.: RF and Microwave Coupler-line Circuits. Artech House, Norwood (1999)
96. Sonnet: version 14.54. Sonnet Software, North Syracuse, NY, Unites States (2013)
97. Rautio, J.C., Rautio, B.J., Arvas, S., Horn, A.F., Reynolds, J.W.: The effect of dielectric anisotropy and metal surface roughness. In: Proceedings Asia-Pacific Microwave Conference (2010)
98. Koziel, S., Bekasiewicz, A., Kurgan, P.: Rapid EM-driven design of compact RF circuits by means of nested space mapping. IEEE Microw. Wirel. Compon. Lett. **24**, 364–366 (2014)
99. Agilent, A.D.S.: Version 2011 Agilent Technologies, 1400 Fountaingrove Parkway, Santa Rosa, CA 95403-1799 (2011)
100. Bandler, J.W., Cheng, Q.S., Nikolova, N.K., Ismail, M.A.: Implicit space mapping optimization exploiting preassigned parameters. IEEE Trans. Microw. Theory Tech. **52**, 378–385 (2004)
101. Liang, J., Chiau, C.C., Chen, X., Parini, C.G.: Printed circular disc monopole antenna for ultra-wideband applications. Electron. Letters. **40**, 1246–1247 (2004)
102. Li, T., Zhai, H., Li, G., Li, L., Liang, C.: Compact UWB band-notched antenna design using interdigital capacitance loading loop resonator. IEEE Antennas Wirel. Propag. Lett. **11**, 724–727 (2012)
103. Li, L., Cheung, S.W., Yuk, T.I.: Compact MIMO antenna for portable devices in UWB applications. IEEE Trans. Antennas Propag. **61**, 4257–4264 (2013)

Sustainable Industrial Processes by Embedded Real-Time Quality Prediction

Marco Stolpe, Hendrik Blom and Katharina Morik

Abstract Sustainability of industrial production focuses on minimizing gas house emissions and the consumption of materials and energy. The iron and steel production offers an enormous potential for resource savings through production enhancements. This chapter describes how embedding data analysis (data mining, machine learning) enhances steel production such that resources are saved. The steps of embedded data analysis are comprehensively presented giving an overview of related work. The challenges of (steel) production for data analysis are investigated. A framework for processing data streams is used for real-time processing. We have developed new algorithms that learn from aggregated data and from vertically distributed data. Two real-world case studies are described: the prediction of the Basic Oxygen Furnace endpoint and the quality prediction in a hot rolling mill process. Both case studies are not academic prototypes, but truly real-world applications.

1 Introduction

The *United Nations Conference on Sustainable Development* took place in Rio de Janeiro, Brazil, 20 years after the 1992 Earth Summit in Rio. Sustainability is defined as guaranteeing a decent standard of living for everyone today without compromising the needs of future generations. Climate change, disasters and conflicts, ecosystem management, environmental governance, chemicals and waste, environment under review, and resource efficiency are the *United Nations' Environment Programmes*.

M. Stolpe (✉) · H. Blom · K. Morik
TU Dortmund University, Otto-Hahn-Strasse 12, 44227 Dortmund, Germany
e-mail: Marco.Stolpe@tu-dortmund.de

H. Blom
e-mail: Hendrik.Blom@tu-dortmund.de

K. Morik
e-mail: Katharina.Morik@tu-dortmund.de

J. Lässig et al. (eds.), *Computational Sustainability*,
Studies in Computational Intelligence 645,
DOI 10.1007/978-3-319-31858-5_10

To each of these topics, there exists a large variety of scientific studies and to many of them, data analysis contributes insights. Investigating climate change clearly needs to be based on data, their analysis and simulation, but also agriculture, biodiversity, resilience of water, soil, and air are to be based on reliable data. A low-carbon, resource-efficient economy was discussed as one important approach to "the future we want" as is the name of the *United Nations Sustainable Development Knowledge Platform*. In this chapter, we contribute to resource efficiency, namely in production processes. We focus on the industrial sector, because it uses more energy than any other end-use sector, with iron and steel production being the second largest consumer after the chemical industry [4]. There, even achieving only a small factor of enhancement in the production process leads to enormous resource savings.

In the remainder of the introduction, we present definitions of sustainable industrial processes and resilient systems (Sect. 1.1). Based on a discussion of related work, we then argue in favor of a full integration of real-time data analysis into existing technical systems (Sect. 1.2). At the end of this section, we give an outlook of the chapter.

1.1 Sustainability of Industrial Processes

Sustainability is often regarded an environmental issue focussing on emission collections, cleaning and waste management. For instance, the CO_2 emissions from steelmaking are investigated [74]. Recycling in the iron and steel making industry has early been handled [81]. The greenhouse gas emissions are considered an important indicator of sustainability (cf. Martins and colleagues [58]). The second indicator is the energy intensity. This has also been carefully studied for steelmaking, e.g. in [21], where a comprehensive reference process, from the coke oven and sinter plant over blast furnace, basic oxygen furnace and continuous casting to hotstrip mill, is analyzed with respect to energy consumption and supply by the reference power plant and the use of waste heat. Martins et al. state material efficiency as the third and the existence of an environmental management system as the fourth indicator of a sustainable industrial process. Today's steel companies operate certified environment management including recycling.[1]

The CO_2 emission and the energy and material consumption are to be minimized while at the same time the product quality and the economic value are to be maximized. The importance of the objectives are often weighted. If the ecological factor is greater than the economic factor, the solutions of the optimization save more resources. If the economy outweighs the ecology, solutions of the optimization create more value. In an evolutionary approach, the orthogonal objectives, resource use and created value, are optimized delivering a Pareto front of *all* possible weights for the objectives [23]. Depending on the needs of the society, there might be particular

[1]ISO 14001 for environment and ISO 50001 for energy management certify today's standards.

bounds of the two objectives. For instance, there might be a strict bound on the greenhouse gas emission, the energy, and the material consumption, which are not to be exceeded regardless of the economic value produced. On the other hand, there might be a certain number of products so necessary, that they have to be produced regardless the resource efficiency. Such bounds limit the overall Pareto front further. This optimization problem expresses the issue of sustainable industry, in general. In this chapter, we are not investigating this macro level of steel production in general, but focus on how to reduce the resource consumption in a steel factory through embedded data analysis.

Making processes traceable has been demanded for enhancing their sustainability [10]. The first and important step is to measure the processes [70]. A complete and sound collection of data allows to generate some overview of the process performance. Aggregating performance indicators allows the engineers to analyze the process [39]. However, working on the factory processes themselves requires an analysis on the basis of more detailed data. With respect to control, Alghazzawi and colleagues have put forward to monitor a process based on its data [3]. The stored measurements are a necessary prerequisite for evaluating processes on the basis of sound data. The importance of grounding steel processes on data has been put forward, by, e.g. Kano et al. [38]. They argue that quality improvement in steel industry becomes possible because of the analysis of data. Recording the data of the process allows to compare the engineering knowledge about the process with its actual operation. This comparison is one of the reasons for cyber-physical systems, which distributedly sense and interact with their environment, thus linking the *cyber* and *physical worlds*. Cyber-physical systems not only sense some factors of the production process (data acquisition), but may also perform local computations for data enhancement (cleansing, feature extraction) and finally interact with the controlling personnel or even control it partially themselves (model employment). Cyber-physical systems acquire and process local data at distributed sources. The value of the distributed data is created by their analysis. Hence, in this chapter, we deal with machine learning methods for the analysis of production data. Moreover, we not only deliver analysis results to the engineers, but directly influence the production process.

Sustainability is supported by quality prediction in two ways. First, quality prediction allows to stop processing as soon as the desired quality can no longer be achieved so that no work is vainly invested into material of insufficient quality. This saves energy and materials. Second, quality prediction allows to stop processing at the right moment. Stopping *before* having produced the optimal quality may require additional work that is more costly than to continue the process until the optimum. Stopping *after* the optimum may decrease the quality again, thus wasting resources. In this chapter, we present a case study for each types of quality prediction supporting sustainability.

1.2 Related Work on Sustainable Steel Factories

Steelmaking is based on metallurgical knowledge about the raw materials, knowledge about the processes and furnaces, and knowledge about process control and process variants. Thermodynamic and kinetic models of converter steelmaking are used for simulations that explain, for instance, the composition of the slag at the end of process [54]. An overview shows diverse blast furnace models ranging from first principles models to specialized models, e.g. for the charging programs or the gas flow [30]. The authors point out that in the end, it is decisive that the models are integrated into the plant control and data acquisition system (cf. [30], p. 214ff). Similarly, Simon Lekakh and colleagues identify the three levels of modeling [50]:

Level 1: process understanding,
Level 2: designing better processes and equipment,
Level 3: controlling the entire steelmaking process online "when several models need to be integrated in a fast-working algorithm, communicating with sensors and process control devices."

The majority of work is about levels 1 and 2. Most models are based on engineering knowledge and are not verified on real data. Some approaches support the model building by machine learning methods. For instance, an artificial neural network was trained to predict the yield of steel, i.e. the efficiency of the steelmaking process [48]. Another approach trains the energy cost prediction on the basis of data. 4 years of monthly reports from a steelmaking process allow to map the input raw material and the steel output value to energy costs [5]. Another approach integrates a neural network with qualitative reasoning for predicting the silicon content in pig iron as an indicator of the thermal state of a blast furnace [14]. Note, that although the authors train neural networks on real world data, the learned model serves as input into a human decision process and is not integrated into the production process.

Some use of machine learning has been integrated more closely into the process. Training an artificial neural network on real data was successful for predicting how much oxygen should be blown and what the coolant requirements are in the end-blow period of the steelmaking process [18]. An approach to end-point prediction of basic oxygen furnace (BOF) turns the visual perception of the flame into a set of features which are input to a support vector machine (SVM) that estimates the carbon in the molten iron [89]. Others have combined fuzzy network models with a variant of the SVM. The fuzzy classifier decides whether cooling material is to be added and the fuzzy regression model determines how much oxygen and coolant should be given [31]. These papers argue in favor of a certain learning method, be it SVM or robust relevance vector machine or neural networks. They show that their method delivers one model that shows a good performance on a certain data set. In this chapter, we move beyond learning *one* model. The feature extraction is more important than the particular learning method for the model's quality. Each learning algorithm favors different features. Several studies have shown that changing the data representation changes the ranking of learning methods, substantially, e.g. [62, 63]. Hence, there is not one best method, but one best pair of method and representation. Moreover, we believe that processes change over time. In the BOF process, for instance, the refractory linings of the converter are slowly reduced and the lance frays over a series of uses. This changes the process. Then, a different model is better suited for prediction. Also the sensors may wear off so that the model should be changed. Hence, we propose a method of learning, storing, and using several models. Thus, we provide users with *a system that learns and stores models*, not just with one learned model. Although general, the system is not a general learning toolbox as is RapidMiner [61], which it may use, but specialized to the application, including sensor readings, feature extractions, modeling, model storage, and model application in real time.

1.3 Outlook

This chapter is about making industrial processes sustainable through embedding data analysis directly into the production process. On the one hand, the production process must be ready for embedded data analysis. Section 2 describes the challenges for the production domain. We'll show steps of a preparation process that is often underestimated. On the other hand, data analysis methods from machine learning or data mining must be ready for the real-world data and control. Section 3 looks into the steps of data analysis. The data acquisition, cleansing, the choice of its representation and extracting the right features are at least as important as the model building (i.e. the learning) step. Particular challenges arise from distributed and streaming data. Our contributions in model building are:

- Distributed data mining: In contrast to the common horizontally partitioned data, where each partition has exactly the same set of features, the data from diverse sources are heterogeneous, each with its own set of features. They realize a stream of vertically partitioned data. We present a general method for the analysis of vertically partitioned data that is resource-aware in its own right.
- Label proportions: A special challenge in hot mill applications is that the quality prediction refers to groups and not to individuals, e.g. the quality evaluation states how many steel blocks of some customer order had defects, but not which of them. The labels *not deficient* and *deficient* are assigned by the quality control as proportions for a set of blocks, not as a label for an individual block. This means, that learning is to be based on label proportions or aggregations. We present an efficient algorithm that solves this new data analysis task.
- A framework for processing data streams offers the right level of abstraction for developing real-time processing. It eases the integration of diverse computations executed at different compute nodes, e.g. at the factory and at the engineer's office. Based on user-given configurations, it *glues* diverse contributions of data analysis and model application together to form an application.

We are happy to present real-world case studies in Sect. 4, starting with the production process and the data acquisition until the model building and model application. The case studies started as academic studies, but have moved far beyond over the years. The first case study is about the BOF end point prediction (Sect. 4.1). For the first time, not only one appropriate model is created, but a system that covers all the steps of data analysis, eases to learn models, and stores them. The system is embedded into the steel production where a selected model is applied in real time to the converter process.

The second case study is about the hot rolling mill process (Sect. 4.2). This study guides the development of decentralized methods for the real-time analysis of sensor data. It underpins again the importance of feature extraction and selection. The system is developed in close collaboration with experts in machining, production, and the steel mill.

The conclusion summarizes our findings and gives an outlook to further research.

2 Challenges for the Domain

Before any data about a production process can be analyzed at all, it must be made available and transformed into a format that is suitable for the analysis task. Based on experiences from the two case studies presented in Sects. 4.1 and 4.2, the following sections discuss typical problems that are usually encountered when trying to embed data analysis in a production environment.

2.1 Proprietary Systems and Heterogenous Data Sources

Quality assurance, certification according to well-defined standards and customer demands require the documentation of steps that were taken in manufacturing. Therefore, data about the processing of products is usually already recorded. However, the data sometimes is not directly accessible for the automated processing required by real-time analysis. In some cases, the data might not even be available electronically. In other cases, proprietary systems might allow for entering and displaying data on screen, but not for exporting the data. Then, the accompanying systems don't necessarily provide an programmer's interface for direct access. Therefore, an important first step is to make as much data as possible electronically available and accessible. This potentially requires a change in existing IT infrastructures.

Another challenge may be the heterogeneity of data stemming from different sources. Steps in a production process are often maintained by different departments, where each can have their own terminology and standards for recording data. The integration of data from different sources is a common problem in data mining projects. For an offline analysis, it is often only done once and manually. For the continuous analysis of data during a running production it is even more demanding, because the process requires the fully automatic fusion of data. Here, open standards for data and meta data description, like XML and according schemas, support the unification of heterogenous data formats, making them easier accessible for data analysis.

2.2 Installation and Maintenance of Sensors

Some production processes are already equipped with monitoring sensors, but with respect to data analysis, it may turn out that not all of them are relevant for the prediction task, while others may be missing. The first step therefore is to gain an understanding of the process chain and assess the available sensors. The next step is the identification of potential decision points, i.e. points in the process that could allow for taking quality-related actions. Based on that information, it can be decided which sensors need to be exchanged, what additional types of sensors are needed and where they should be installed.

Maintenance tasks include the monitoring of sensors for deviations and failures, regular recalibration and potential reinstallation. Such maintenance tasks become especially challenging and expensive when leading to a temporary halt of the whole production process. The continuous gathering of high-quality sensor data thus requires new types of machinery which allow for a non-invasive maintenance and measurement of process parameters.

2.3 Definition and Measurement of Quality Deviations

If quality deviations can be detected early enough, parameters of subsequent processing steps may be changed in order to reach a desired quality level. If the detection is to be learned from data, *labeled* examples are necessary (see also Sect. 3.5.1), i.e. historical records of parameters for the processing of individual steel blocks and associated quality measurements. However, it is not always clear how the quality measurements on final products relate to intermediate production units.

In case of the hot rolling mill process described in Sect. 4.2, the quality is assessed on bars that are cutted from larger steel blocks. For each of the bars, different measures of quality are available. Predictions during the running process, however, are to be made on the larger steel blocks. For methods predicting a single label, it thus becomes necessary to aggregate information about the bars to a single label for each steel block. Which aggregations are meaningful and how to calculate them can only be decided by domain experts.

Even more complicated is the case in which, for technical or organizational reasons, the quality has only been assessed on a small sample of the bars or the relationship between steel blocks and bars is lost. Then, quality information is only available as group statistics, e.g. for whole customer orders, but not for individual units. Learning from such aggregate outputs or label proportions is an exciting new field of machine learning research and we have developed a method to solve the problem (see also Sect. 3.5.4). For a successful application of data analysis it is nevertheless advantageous to establish a continuous end-to-end tracking of units through the whole process chain.

2.4 Encoding of Domain Knowledge

Processes in the chemical and steel industry usually can be characterized by an abundance of partially interdependent process parameters which may potentially influence the quality of the final product, resulting in high-dimensional learning and optimization tasks. Both types of tasks may benefit from restricting the search space of relevant parameters according to already existing background knowledge, from experts or physical and chemical laws. Similarly, known models may guide unsupervised data mining methods to interesting patterns.

Currently, the incorporation of background knowledge, e.g. in the form of search constraints, is usually done manually and based on interviews with domain experts. However, a truly adaptive analysis of data would require the electronic storage and continuous maintenance of background knowledge according to well-defined formalisms. Similar to the problem of data representation facing heterogenous sources, standards should be established for the description of knowledge about production processes which then could automatically be accessed and used by data analysis methods.

3 Challenges and Methods for Embedded Data Analysis

Even if all of the aforementioned requirements were met, the nature of industrial processes or technical restrictions still poses challenges for data analysis [85]. Sensors and machines will fail at some time, processing steps may be optional and products may take different routes through a process chain. With such events, difficult questions arise how data should be preprocessed, represented and what features may be relevant for the prediction task. Moreover, the distributed nature of sensors and the real-time processing of all data may require the development of new data analysis methods. The following sections will discuss such problems and available methods in more detail, for each step of the data analysis process.

3.1 Data Acquisition

We illustrate process chains and associated sensor measurements by an example. Figure 1 shows how two different steel blocks, A and B, might have moved through

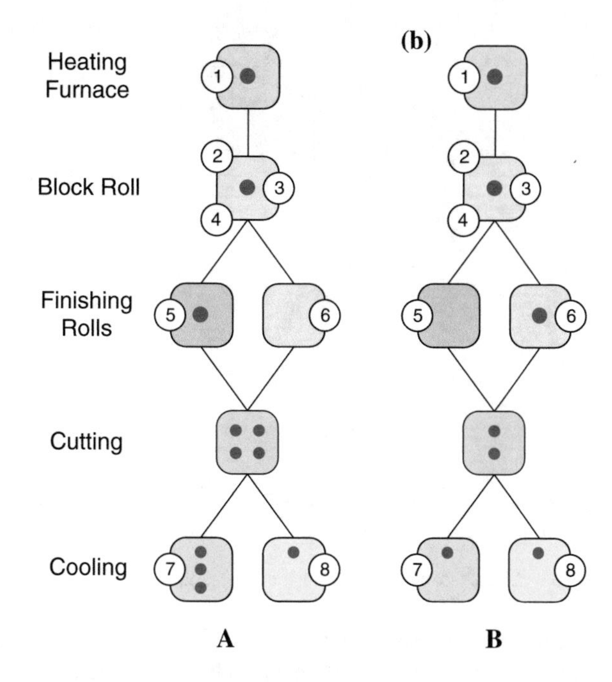

Fig. 1 Routes that *blocks A* and *B* could take through some process chain

several processing stations: a heating furnace, a block roll, two finishing rolls, a station for cutting the blocks, and two cooling stations. Each station, except the cutting station, has sensors attached. Steel block *A* is first heated up, rolled and then moves through the first finishing roll. It is then cutted into four parts. Three of such parts move through the first cooling station while one moves through the second one. Steel block *B* moves through the second finishing roll instead, is cutted into two parts afterwards, and such parts then move through the cooling stations in parallel. The measurements that might have been recorded by each sensor over time are shown in Fig. 2.

During time intervals $[t_1, t_2]$, $[t_5, t_6]$ and $[t_7, t_8]$, steel block *A* has moved from one processing station to the other, as well as steel block *B* during time intervals $[t_1, t_2]$, $[t_4, t_5]$ and $[t_6, t_7]$. For steel block *A*, the recording at the block roll starts with sensor 3, while sensor 2 has a time lag of $t_3 - t_2$ and sensor 4 a lag of $t_4 - t_2$. Similar lags occur during the processing of steel block *B*. Sensor 4 is an example of a sensor whose values are not continuous, but discrete.

The continuous measurements of sensors form an indexed stream of countably infinite data items x_i. Every data item contains an index, for example a timestamp, and can contain arbitrarily many different values and value types, like images, strings or numbers.

A segment of this stream with length n and only numerical values form a *value series* which can be defined [60] as a mapping $x : \mathbb{N} \to \mathbb{R} \times \mathbb{C}^m$, $m \in \mathbb{N}$ where each element x_i of a value series with length n is an ordered pair (d_i, w_i). $d_i \in \mathbb{N}$ is called

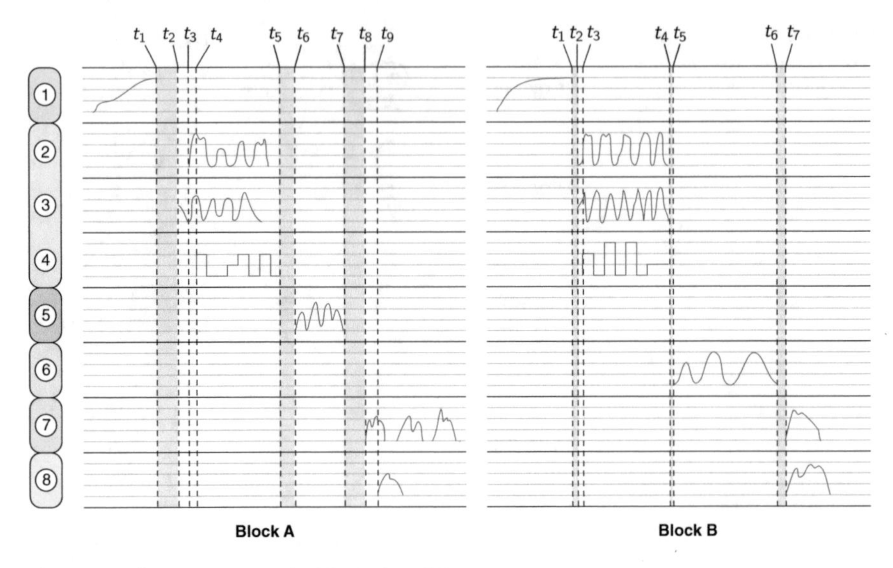

Fig. 2 Sensor measurements for blocks A and B

the index component and $w_i \in \mathbb{R} \times \mathbb{C}^m$ the value component. It is called a *time series* if the index dimension represents a temporal order.

The repeating patterns in many of the time series, for instance at the rolls, represent different rolling steps, i.e. a single steel block could move several times through a roll. For instance, according to the measurements of sensor 2, steel block A moved four times through the block roll and steel block B five times. Since A moved through the first finishing roll, there are no measurements for the second roll, while for B, there are values for the second roll, but none for the first.

3.2 *Cleansing of Sensor Measurements*

In real production environments, sensors might provide wrong readings or might fail entirely [68]. At least, their readings are noisy. Equipment could be exchanged and machines will fatigue. Time series may contain irrelevant readings, be wrongly aligned or have different resolutions. When analyzing the data offline, such cases can simply be excluded from the analysis. In contrast, the embedded real-time analysis of data must somehow detect such cases automatically and react accordingly. The first analysis step therefore usually consists of cleaning the sensor readings.

3.2.1 Detection and Handling of Faulty Sensor Readings

Faulty sensor readings can only be handled if they are detected. Such detection is easiest in cases in which sensor readings lie outside physically meaningful ranges, as defined by accompanying meta data. But there are also non-trivial cases, in which faulty readings overlap with the normal data, requiring the automatic detection of faulty patterns. If such patterns cannot be defined based on knowledge about the underlying hardware [35], or based on visual inspection, they might be derived automatically by supervised learning methods (see also Sect. 3.5.1). However, if the faults are highly irregular or not frequent, it is difficult to learn their detection on the basis of given training examples. Models for the detection of anomalies in production settings often describe only the normal data, marking patterns as anomalies that deviate from the learned description (see also Sect. 3.5.2). Nevertheless, the correct definition of parameters, like threshold values, remains difficult with only a few negative examples. Moreover, it can be difficult even for domain experts to identify such negative examples correctly.

Once faulty readings or missing values are detected, there are different possible ways to handle them. A simple strategy for the replacement of single or only a few faulty values is to replace them by their predecessor value or based on ARMA (autoregressive moving average) models [9]. In other cases, faulty values can be imputed based on prediction models that were trained on other existing values. However, if many relevant values are missing or whole sensors fail, the quality of the predictions may either decrease or it might become impossible to provide a prediction at all. The challenge here is to estimate the confidence of predictions correctly, since it is not always clear how missing or faulty values in the raw sensor data will influence later preprocessing and model building steps. Another challenge is that different types of sensors may require different strategies for the handling of faults and that knowledge about the best strategy is often scarce.

Finally, even correctly working sensors usually have some level of *noise*, which may also be introduced by the production process itself, like sensors moving due to vibrations. If the underlying noise model is known, it should be used. Otherwise, measurements can be filtered and smoothed. Finding the correct parameters for filtering is not necessarily trivial, since it also interacts with subsequent preprocessing and model building steps.

3.2.2 Detection and Handling of Changes

Sensor readings that deviate from what is already known may also be caused by *intended changes* in the underlying hardware, like new production equipment and new or differently calibrated sensors. It is vital that the prediction engine is informed about such changes and, if possible, also provided with new meta data. Otherwise, it may happen that correct new patterns are wrongly identified as faulty ones. Moreover, based on the type of changes, it must be decided which of the previously trained models need to be updated and how.

Unintended changes are usually caused by machine fatigue. Without any models describing such changes, methods for concept drift detection are needed and it has to be decided if already trained prediction models must be updated, or alternative models be applied. In contrast, incremental training methods, like streaming methods (see Sect. 3.5.6), can incorporate such changes into their models, automatically.

3.2.3 Irrelevant Readings and Different Alignments

Sensor readings may be entirely *irrelevant* for the prediction. Such parts can then be stripped from recorded time series or be skipped in the online processing. For example, production processes may include phases in which materials are just transported or posititioned for the next processing step, but sensors nevertheless provide readings. Ideally, the data management would clearly mark such events. Whenever this is not possible, however, such phases must be detected automatically, facing similar challenges as the detection of faulty sensor readings.

Similarly, the time series or data streams of different sensors may have different *resolutions*, *lengths* and *offsets*. Depending on the prediction task and methods, it can be necessary to scale and align time series correctly before they are further processed. This task is not to be underestimated.

3.3 *Representation of Value Series*

As shown in Sect. 3.1, a single run through the process chain can be represented by a set of time series with different lengths and offsets, which may overlap in time, may contain different numbers of segments at different levels of granularity, may stem from different machines and sensors and may also be entirely missing for processing steps that are optional.

Many common data analysis methods cannot work directly on such sets of time series, but expect all observations to be represented by fixed-length feature vectors. Instead of inventing specialized methods for the handling of multivariate sensor readings, the question arises if the raw data can be transformed into a representation that can be used by standard data analysis methods. While Sect. 3.4 deals with the extraction of features from value series, not necessarily resulting in value series again, here we discuss how the raw series values can be rearranged and transformed for the use with different distance measures.

3.3.1 Mapping of Series Values to a Fixed-Length Vector

We want to represent the sensor data in fixed-length vectors in order to make standard learning techniques applicable. The basic idea is to reserve enough space for the readings of each sensor in a single fixed-length numerical vector. Original series values are then projected to appropriate (predefined) positions in this vector, as shown in Fig. 3. The mapping works, if the maximum number of time series recorded is fixed and known beforehand. This is usually the case, as long as the process

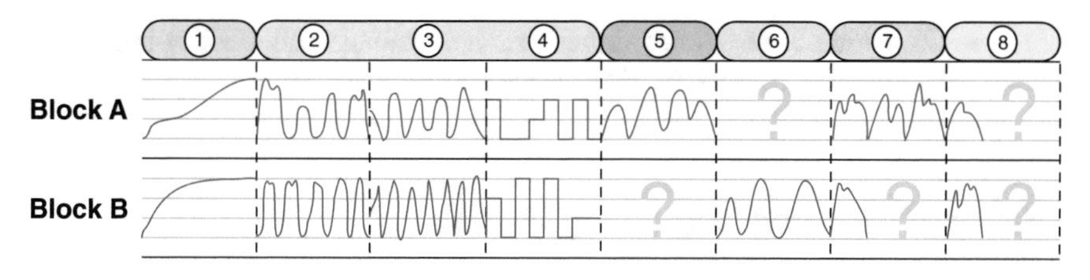

Fig. 3 The value series from different sensors as a single fixed-length vector

and sensor setups do not change. For the projection, the original time series might need to be rescaled, e.g. by interpolation, and possibly shifted. Once mapped, it becomes possible to use the resulting vectors with most distance-based data analysis algorithms.

The most difficult question is what values to assign to portions where no processing happened (the question marks in Fig. 3), e.g. for optional processing steps. A simple approach might fill missing portions with zeros or the last recorded value. However, filling with zero values can easily lead to problems with several distance measures. For example, how similar are two value series, where steel block A moved through a different finishing roll than did steel block B? When filling with zeros, both series would be marked as highly dissimilar by Euclidean distance, although both blocks could well lead to a similar final quality of the steel blocks. In such a case, the desired correspondence between similar feature vectors and similar labels would be lost. Reserving the same portion for both finishing rolls (sensors 5 and 6) in the fixed-length vector seems to solve the problem, but it doesn't take into account that both finishing rolls might have somewhat different properties, e.g. value scales, which usually require a careful normalization. Moreover, the solution would not be transferable to a situation where parts of a steel block are processed in parallel, like at the cooling stations (sensors 7 and 8).

3.3.2 Concatenation of Value Series

Instead of mapping all series values to a fixed-length vector by rescaling, another option is to use distance measures that can handle value series with different lengths, like the Dynamic Time Warping (DTW) [66] or Longest Common Subsequence (LCSS) [19] distance.

In principle, there are two approaches for transforming the original time series appropriately. The first approach simply concatenates all time series belonging to the processing of a single steel block. The resulting series might then be compared with one of the aforementioned distance measures. Given data about the processing of two steel blocks, A and B, the second approach calculates distance values for each time series of each sensor independently and then sums them up to a total distance. However, also these approaches do not indicate how to handle optional or parallel processing steps.

A transformation of raw series values to coarse grained patterns like processing modes, offers another path to further analysis (see Sect. 4.2.3).

3.4 Extraction and Selection of Features

Instead of working on the raw data stream directly, we can describe production processes by a restricted set of features which are calculated from the raw data, a step that is called *feature extraction*.

In general, the problem of feature extraction consists of finding a transformation of the raw data into a feature vector such that the prediction performance on the feature vectors of unknown observations is maximized. The problem is very challenging, since the search space of possible transformations in general is infinite. Even restricting the types of allowed transformations to a finite set easily leads to search spaces of exponential size. Also the simpler problem of selecting the best subset of features for a given classification task has an exponential search space, namely that of all possible subsets of features. A fast ensemble method has been developed [78] and successfully applied in high volume data sets.

The following sections present a non-exhaustive list of transformation and feature extraction methods looking especially promising in the context of production processes.

3.4.1 Aggregation and Summarization

Aggregation and summarization methods for time series or data streams reduce the amount of raw data as much as possible, while at the same time trying to keep its most important characteristics. The amount of feasible reduction depends on the prediction task.

The simplest type of aggregation is the calculation of summary statistics, like minimum and maximum values, the mean, median, standard deviation, percentiles or histograms. Such simple global features can already be sufficient and oftentimes performed better than more sophisticated methods in the first case study (see Sect. 4.1.3).

More sophisticated methods try to represent a given time series as the combination of a (usually fixed number) of basis series, like the Discrete Fourier Transform (DFT) [25] or the Discrete Wavelet Transformation (DWT) [59]. Kriegel et al. [45] approximate time series Y by a model $Y = f(X, \boldsymbol{\alpha}) + \epsilon$, where f is an arbitrary mathematical function and X a fixed set of basis functions. The basis functions can be derived, for example, by clustering (see Sect. 3.5.3), while the coefficients $\boldsymbol{\alpha}$ can be determined, for example, by least squares, such that the random error ϵ is minimized. Instead of representing time series by their raw values, they can then be represented by a fixed-length coefficient vector. A disadvantage is that such coefficients are usually much harder to interpret than the aforementioned simpler summary statistics.

3.4.2 Segmentation

In the area of image classification, the images are often first divided into parts, a process that is called *segmentation*. There, the borders of segments usually indicate significant changes in basic properties of the pixels, e.g. their color. Once the segments are determined, features can be extracted from them, like their average, minimum and maximum color, gradients or textural features. SIFT features [53], which are translation, rotation and scale invariant, have almost become a standard for the meaningful description of images.

The salient features approach by Candan et al. [11] transfers ideas from the segmentation of two-dimensional images and the extraction of SIFT features to the space of one-dimensional value series. Salient points in the series, which are points that deviate much from their surrounding values, are used for segmenting the series. Then, from each segment, characterizing features are extracted. The method determines salient points at different resolutions, allowing for a description of value series at different levels of granularity. Another promising approach by Rakthanmanon et al. [72] not only automatically divides time series into their segments, but also clusters them.

A segmentation approach developed in the context of the second case study (see Sect. 4.2) determines segments based on expert knowledge and signals from machines in the process chain [51]. As shown in Fig. 4, the value series are divided into different meaningful segments, e.g. rolling steps.

In the first case study (see Sect. 4.1) the data stream is segmented in real-time. Multiple events are detected online, for example the start of the process itself or the start of the so called combustion phase. Every pair of event can define a segment of the data stream.

Once segmented, different statistics are calculated on the segments' values, like the mean, the standard deviation, minimum and maximum values. Other features are differences between values and histograms. Of course, it would also be possible to use the coefficient-based summerization methods introduced in Sect. 3.4.1. All values calculated are then assigned to predefined portions of a fixed length feature vector, similar to the mapping of raw values described in Sect. 3.3.1. In addition, statistics across segments and on the whole value series are calculated, for representing the series at different levels of granularity.

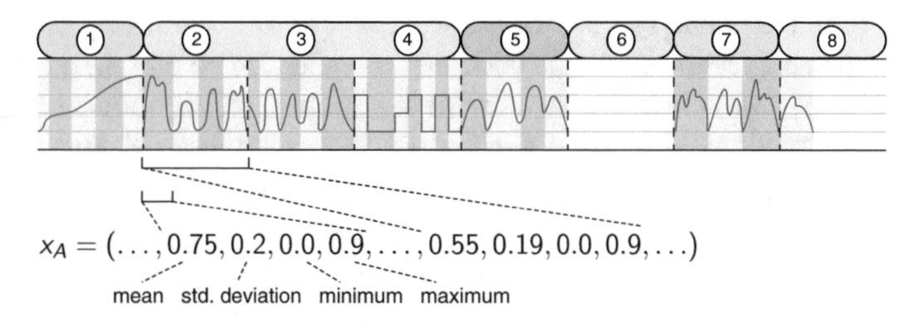

Fig. 4 Segmentation of value series and encoding of descriptive statistics about these segments in a fixed-length feature vector. Alternating *gray* values indicate the segments

The biggest advantage of such an approach is that it is multivariate in the sense that features from different value series and their parts, at different levels, may be combined in a highly interpretable manner. For example, a classification rule that is formed based on such features may read like "Predict a steel block as defect if it was heated less than one hour at 900 degree Celsius and the maximum rolling force in the first rolling step exceeds the value of 10,000". The approach already has been used successfully for the identification of coarse grained patterns, like processing modes (see Sect. 4.2.3).

3.4.3 Symbolic Representation

SAX (Symbolic Aggregate Approximation) [52] first determines the elements of a sequence $C = (c_1, \dots, c_n)$ by piece-wise aggregate approximation and maps them to a new sequence C' with $w < n$:

$$c'_i = \frac{n}{w} \sum_{j=\frac{n}{w}(i-1)+1}^{\frac{n}{w}i} c_j$$

The elements c'_i are then discretized by mapping them to a fixed number of symbols, preserving the upper bounded Euclidean distance between all series. A gradient-based approach for the symbolization of streaming sensor data was introduced by Morik and Wessel [64].

The symbolization of time series data bridges the gap between numerical methods and those that work on symbols, like frequent item set mining [1] or text processing. For example, frequencies of symbols, sequences or words are length-invariant features that already have been used successfully in areas such as text classification or intrusion detection.

3.4.4 Sequential Patterns and Motifs

Once time series are symbolized (see Sect. 3.4.3), several algorithms working on individual symbols or sequences of symbols can be applied. The KRIMP method by Siebes et al. [79] compresses a database of binary transactions by code tables. An open research question is if such code tables could also be used as a condensed representation of time series. The AprioriAll algorithm [2] finds *frequent sequential patterns* in transactions of items, e.g. subsequences in symbolized time series. Its extension GSP (General Sequential Patterns) [82] can also respect constraints on the patterns. Similarly, the WINEPI algorithm by Mannila et al. [57] can find *frequent episodes* in event sequences. While in a production setting, the most frequent patterns are not necessarily those that are relevant for the quality (see Sect. 4.2.3), they might help to define what is *not* interesting.

Algorithms for *motif discovery*, like a probabilistic one by Keogh et al. [16], not necessarily require a symbolic representation of time series, but try to find frequent reoccurring subsequences directly. An interesting new direction is the supervised discovery of motifs, like the shapelet approach [90], which can also take given class labels into account.

3.4.5 Method Trees

Potentially, all of the aforementioned methods, with different parameterizations and combinations, may be useful for the extraction of relevant features from value series. Instead of trying and combining all such methods and their parameters manually, Mierswa and Morik [60] propose to learn promising combinations of preprocessing methods and their parameters. The method is based on a structure of all possible features which offers a unifying framework. Basis transformations, filters, mark-ups and a generalized windowing cover elementary methods that can be combined in the form of a method tree. The tree applies the operators (nodes) in a breadth-first manner thus transforming a time series, or even more general, a value series. The root of each tree represents a windowing function, while the children of each parent node form operator chains consisting of basis transformations, filters and a finishing functional. Learning the feature extraction tree is done by a genetic programming algorithm. In each iteration, their algorithm generates a new population of method trees, by mutation and crossover operators that change and combine respective subtrees. The fitness of each method tree can be determined by an arbitrary inner classifier. The approach has been used successfully for the classification of music by genre or the personal music taste. In principle, it might also be used for the analysis of time series from production processes. However, its demand for stratified data sets with respect to the labels is not met by the production data. Also, the amount of production data is prohibitive.

3.5 Model Building

In the past, myriads of methods have been developed for data analysis and the automatic building of prediction models. The following sections give a short overview of commonly used methods and research areas that we consider to be especially relevant for the embedded data analysis in production processes.

3.5.1 Supervised Learning

The probably most common task of supervised learning is the derivation of *functions* from labeled *examples*, also known as *function learning*.

Definition 1 (*Function Learning from Examples*) Let X be a set of possible instance descriptions, Y be a set of possible target values and D be an unknown probability distribution on X, Y. Let further H be a set of possible functions. Given examples $(x, y) \in X \times Y$, drawn from D, where $y = f(x)$ for an *unknown* function f, find a function $h \in H : X \to Y$, such that the error $\mathrm{err}_D(h, f)$ is minimized.

In the case of production processes, an instance $x \in X$ could be a description of the sensor measurements recorded for a single steel block and $y \in Y$ would be a related quality label to be predicted. As already discussed in Sects. 3.3 and 3.4, instances are often expected to be represented by fixed-length vectors, consisting of discrete or numerical features. The prediction of numerical target values is also called *regression*, that of discrete values *classification*.

Rule-based learners, among them Decision Trees, decide on the target value by testing for the values of different features and connecting such tests by logical operators. Given an unknown instance x, *instance-based* methods, like k-NN, search for the k most similar instances they have previously seen, according to a given distance function, and predict the average or majority label of such k nearest neighbors. While k-NN needs to store the whole training set, the Support Vector Machine (SVM) only stores the support vectors of a hyperplane that separates the examples of both classes with a *large margin*. Naïve Bayes, a *probabilistic method*, makes assumptions about the probability distribution D and tries to fit parameters based on the given training examples and labels. For a detailed in-depth overview over such methods, see Hastie et al. [32].

In comparison to the incremental and streaming methods described in Sect. 3.5.6, the aforementioned methods (at least in their original form) expect all training examples to be available on batch. As such, they cannot be trained during the running process, and their models need to be retrained if concepts change. However, in cases were all sensor measurements are recorded and stored at a central location and concepts do not change too often, models can be trained offline, but used online.

3.5.2 One Class Learning

The previously mentioned supervised classifiers are trained on two or more classes. However, their accuracy may suffer if the distribution of observations over the classes is highly imbalanced. For example, production processes with high quality standards usually can be expected to output more high quality goods than ones with defects. Similarly, certain events, like machine or sensor failures, may only occur very seldom. In such cases, many positive examples are available, but only few or even no examples of the negative class.

The task of one class learning [65] is to find a model that well describes a set of observations. Tax and Duin [86] propose a Support Vector Data Description (SVDD) which computes a spherical boundary around the given data points. The diameter of the enclosing ball and thereby the volume of the training data falling within the ball can be chosen by the user. Observations inside the ball are then classified as normal whereas those outside the ball are treated as outliers or anomalies.

More formally, given a vector space X and a set $T = \{\mathbf{x}_1, \ldots, \mathbf{x}_n\} \subseteq X$ of training instances, the primal problem is to find a minimum enclosing ball (MEB) with radius R and center $\mathbf{c}$ around all data points $\mathbf{x}_i \in T$:

$$\min_{R,\mathbf{c}} R^2 : ||\mathbf{c} - \mathbf{x}_i||^2 \leq R^2, \ i = 1, \ldots, n$$

For non-spherical decision boundaries, all observations may be implicitly mapped to another feature space by the use of kernel functions [77]. The center in this feature space is then described by a linear combination of support vectors lying on the boundary or outside of the ball. The MEB problem is solved as a quadratic optimization problem.

Schölkopf et al. [76] instead have proposed the 1-class ν-SVM, which separates all training examples with a maximum margin from the origin. For certain kernel functions, however, 1-class ν-SVM and SVDD can be shown to yield equivalent results.

Tsang et al. [88] proposed a fast $(1 + \epsilon)$ approximation algorithm, the Core Vector Machine (CVM). A set of points $C \subseteq T$ is called a core set if its MEB is only $R\epsilon$ smaller than the MEB around T. Such a core set can be constructed by first choosing two points far away in feature space and then iteratively adding the point that is furthest away from the current center. If no points in C lie outside the current MEB, the algorithm stops. It can be shown that the algorithm reaches a $(1 + \epsilon)$-approximation of the MEB around T with constant time and space requirements. The algorithm can also work incrementally. Moreover, a distributed version of this algorithm has been developed in the context of the second case study (see Sect. 3.5.5).

3.5.3 Unsupervised Learning

If no labels are given at all, unsupervised learning methods may be used to find most striking patterns in the given data. *Cluster analysis* [34] tries to group observations, usually by their distance or density, such that similar observations lie in the same group and dissimilar observations lie in different groups. Often, the number of clusters k to be found must be specified by the user. Well-known clustering algorithms are, for example, k-Means [55] and DBSCAN [24]. *Frequent item set mining* [1] assumes a binary database of transactions and finds frequent patterns and relationships in the data. For use with numerical data, like sensor measurements, the data usually must be discretized (see Sect. 3.4.4). *Dimensionality Reduction* techniques, like principal component analysis (PCA) [37], aim at simplifying high-dimensional data sets. Some of them may be used for visualization, like Self-Organizing Maps (SOMs) [43], which map a set of high-dimensional input vectors to a low-dimensional grid. If successful, vectors that are similar to each other in the input space are lying close to each other on the grid. SOMs have also been used for analysing the data in the second case study (see Sect. 4.2).

The biggest disadvantage of unsupervised methods is that, without *any* labeled data, their results often can only be validated by domain experts. As such, they are not well-suited for the automatic real-time analysis of data, especially when concepts change. If at least *some* labeled examples and many unlabeled observations are available, *semi-supervised* learning methods [13], like the transductive SVM [36], can however perform very well.

3.5.4 Learning from Aggregate Outputs and Label Proportions

The problem of learning from *aggregate outputs* (which is the regression task) or from *label proportions* (which is the classification task) is different from all of the aforementioned settings in so far as labels are not given for individual observations, but summarized label information is given for *groups* of observations. For example, it might be known how many steel blocks belonging to some customer order had defects, but not which of them. Given several of such customer orders and features of the related blocks, the goal is to train a prediction model that assigns the correct label to each individual steel block.

Definition 2 (*Learning from Label Proportions*) Let X be an instance space composed of a set of features $X_1 \times \cdots \times X_m$ and $Y = \{y_1, \ldots, y_l\}$ be a set of categorical class labels. Let $P(X, Y)$ be an unknown joint distribution of observations and their class label. Given is a sample of unlabeled observations $U = \{x_1, \ldots, x_n\} \subset X$, drawn i.i.d. from P, partitioned into h disjunct groups $G_1, \ldots, G_h$. Further given are the proportions $\pi_{ij} \in [0, 1]$ of label y_j in group G_i, for each group and label. Based on this information, we seek a function (model) $g : X \to Y$ that predicts $y \in Y$ for observations $x \in X$ drawn i.i.d. from P, such that the expected error

$$\mathrm{Err}_P = E[L(Y, g(X))] \tag{1}$$

for a loss function $L(Y, g)$ is minimized. The loss penalizes the deviation between the known and predicted label value for an individual observation x.

Since its introduction by Musicant et al. [67], several methods for the problem have been developed. The authors of [67] present first modified versions of well-known regression algorithms, like k-NN, neural networks and the linear SVM.

For the classification from label proportions, Quadrianto et al. [71] proposed the *Mean Map* method, which models the conditional class probability $P(Y|X, \theta)$ by an exponential model. The parameter vector θ is estimated by maximizing the likelihood $\log P(Y|X, \theta)$, which depends on the unknown labels by the empirical mean μ_{XY}. The mean is estimated from the groups and given label proportions. It is shown that *Mean Map* reaches a higher prediction accuracy than kernel density estimation, discriminative sorting and MCMC [46]. Rüping [73] proposes an *Inverse Calibration* method which scales the outputs of a regression SVM (SVR) such that they can be interpreted as probabilities. The constraints of the SVR optimization problem are modified to include information about the given label proportions. The empirical results demonstrate that *Inverse Calibration* yields significantly higher accuracy than *MeanMap* on different standard datasets from the UCI repository. Under the assumption that clusters correspond to classes, the *AOC* Kernel *k*-Means algorithm by Chen et al. [15] instead extends the optimization problem of kernel k-Means, trying to find an assignment of labels to clusters such that the difference to the given label proportions is minimized in each iteration. Newer algorithms are the ∝-SVM [91] and a Bayesian network approach [33].

The *LLP* (Learning From Label Proportions) algorithm [84] developed in the context of the second case study (see Sect. 4.2) follows a similar idea as *AOC* and is described in more detail here. Figure 5 shows the label proportions π_{ij}, conveniently written as a $h \times l$ matrix $\Pi = (\pi_{ij})$. The proportion of label y_j over sample U can be calculated from Π:

$$\eta(\Pi, y_j) = \frac{1}{n} \sum_{i=1}^{h} |G_i| \cdot \pi_{ij} \tag{2}$$

By multiplication of π_{ij} with its respective group size $|G_i|$, one gets the frequency counts μ_{ij} of observations with label $y_j \in Y$ in group G_i.

For any model $g(X)$ applied to all $x_i \in U$, the label proportions induced by the model can be calculated by counting the number of observations x_i with $g(x_i) = y_j$ for each label $y_j \in Y$ in each group and dividing the counts by the particular group size. This results in a matrix Γ_g:

$$\Gamma_g = (\gamma_{ij}^g), \quad \gamma_{ij}^g = \frac{1}{|G_i|} \sum_{x \in G_i} I(g(x), y_j), \quad I = \begin{cases} 1 : g(x) = y_j \\ 0 : g(x) \neq y_j \end{cases} \tag{3}$$

The deviance (i.e. loss) between the given label proportions Π and Γ_g can then be defined, for instance, as the average squared error over all matrix entries:

$$\text{Err}_{MSE}(\Pi, \Gamma_g) = \frac{1}{hl} \sum_{i=1}^{h} \sum_{j=1}^{l} (\pi_{ij} - \gamma_{ij}^g)^2 \tag{4}$$

Labeled examples (unknown)

$G_1 = \{(x_1, 1), (x_3, 1), (x_7, 0)\}$
$G_2 = \{(x_2, 0), (x_4, 0), (x_5, 1), (x_6, 1)\}$
$G_3 = \{(x_8, 0), (x_9, 0)\}$

Sample U (known)

$G_1 = \{x_1, x_3, x_7\}$ $\quad n = 9$
$G_2 = \{x_2, x_4, x_5, x_6\}$ $\quad h = 3$
$G_3 = \{x_8, x_9\}$ $\quad l = 2$

Label proportions (known)

$Y = \{0, 1\}$

$$\Pi = \begin{pmatrix} 0.33 & 0.67 \\ 0.50 & 0.50 \\ 1.00 & 0.00 \end{pmatrix} \quad \begin{matrix} |G_1| = 3 \\ |G_2| = 4 \\ |G_3| = 2 \end{matrix}$$

(columns y_1, y_2)

η: 0.56 0.44

Fig. 5 Example for a given label proportion matrix Π

Algorithm 1: The LLP algorithm

Data: Label proportion matrix Π, sample U, groups $G_1, \ldots, G_h$, parameters (e.g. k)
Result: Model of labeled centroids and feature weight vector
create new population of random weight vectors ;
repeat
    copy population ;
    apply mutation and crossover operators on copy ;
    foreach *weight vector in population* **do**
        cluster instances in U by k-Means with weighted Euclidean distance ;
        foreach *possible labeling of centroids* **do**
            calculate Err_{MSE} ;
            update model if current one is the best seen so far ;
        end
        fitness of current weight vector $\leftarrow$ min Err_{MSE} ;
    end
    prune population according to fitness values ;
until *fitness can't be improved*;

Instead of incorporating the loss as additional term into the k-Means optimization problem, as *AOC* does, *LLP* first clusters the given observations as usual and then minimizes the Err_{MSE} by trying all possible combinations of labels. Such an exhaustive search is feasible, because often it can be assumed that $k \propto |Y|$ and the number of classes $|Y|$ is small. The advantage in comparison to *AOC* is that k may also differ from the number of classes, allowing for an explicit control of the trade-off between bias and variance: The higher the number of clusters k, the better irregular class borders may be captured.

It is well-known that clustering results can much differ based on different weightings of features. The aforementioned labeling step is therefore combined with an evolutionary weighting of attributes, significantly improving the accuracy of *LLP*, whose basic steps are shown in Algorithm 1. Despite the exhaustive search for the best labeling of centroids, empirically the algorithm shows a much lower run-time than *MeanMap*, *Inverse Calibration* and *AOC*, while providing a similar or even better accuracy [84].

3.5.5 Distributed Data Mining

Several methods for distributed data mining follow the paradigm of speeding up computations by solving subproblems in parallel and merging their results. Their focus is mostly on solving *big data* problems with a large cluster of machines or in the cloud, were the size of the data is so big that it cannot be handled anymore by a single machine. In such a scenario, usually a large bandwidth and an unlimited amount of energy are assumed, the main focus being on speeding up computations. In contrast, distributed data mining algorithms in wireless sensor networks (WSNs) focus mostly on a reduction of energy [6], being the most valuable resource of battery powered devices.

Though in most production settings, the available bandwidth is high and the energy consumption of wired sensors is neglectable, centralized IT infrastructures can be hard and costly to maintain when facing changes such as new sensors being attached or reconfigured. Therefore, at least over the long run, we would expect an increase in wireless and more flexible sensors, which ideally could configure and find themselves and communicate with each other automatically. With battery powered sensors, however, distributed methods are needed that are energy-aware and resemble those used in WSNs. Since sending or receiving data is usually among the most energy draining operations [12], the reduction of communication between sensor nodes may lead to a longer lifetime of the whole sensor network.

Regarding the previously presented process chains, which and how much data needs to be sent will depend on how far the final quality is determined by patterns occuring at different processing stations. In the best case, it might be possible to decide about the quality only based on locally available features, requiring no communication at all. In the worst case, the value series from all processing stations might be required for correct predictions. If the exact dependencies between the label and features were already known, only the relevant features would need to be communicated. However, since such dependencies are usually found as a *result* of the data analysis step, methods are needed that can work in a distributed fashion, without communicating too much data with other nodes.

The distribution of sensors across different processing stations is mostly *vertical* (see Fig. 6), meaning that the processing of a single steel block is characterized by measurements from different network nodes. Considering potential dependencies between features from different nodes, reducing communication in this scenario can be particularly difficult.

Definition 3 (*Vertically Partitioned Data Scenario*) Let $P_1, \ldots, P_k$ be a set of local nodes and P_0 be a central node, potentially with more computational power. The data at node P_i ($\forall i > 0$) is denoted by $T_i = [\mathbf{x}_1^{(i)} \ldots \mathbf{x}_n^{(i)}]^T$ consisting of n rows where $\mathbf{x}_j^{(i)} \in \mathbb{R}^{m_i}$ and m_i is the number of features recorded by the ith node. The global set of features A is the vertical concatenation of all $m = \sum_{i=1}^{k} m_i$ features over all nodes and is defined as $A = [A_1\ A_2 \ldots A_k]$. Hence, the global data T is the $n \times m$ matrix defined as the union of all data over all nodes, i.e. $T = [\mathbf{x}_1 \ldots \mathbf{x}_n]^T$ with $\mathbf{x}_j \in \mathbb{R}^m$.

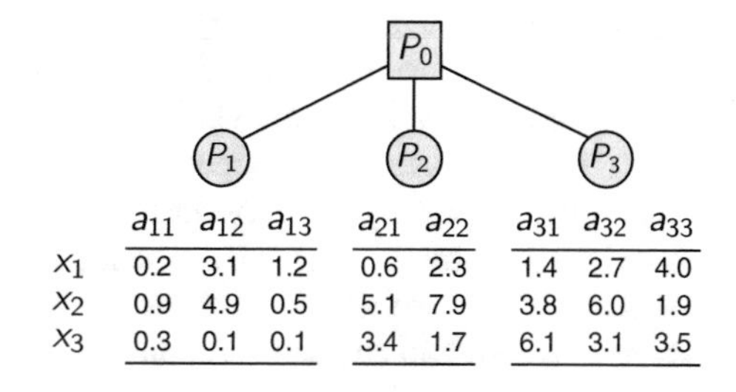

	a_{11}	a_{12}	a_{13}	a_{21}	a_{22}	a_{31}	a_{32}	a_{33}
x_1	0.2	3.1	1.2	0.6	2.3	1.4	2.7	4.0
x_2	0.9	4.9	0.5	5.1	7.9	3.8	6.0	1.9
x_3	0.3	0.1	0.1	3.4	1.7	6.1	3.1	3.5

Fig. 6 Vertically partitioned data scenario

The challenge here is to learn a global model of the data (e.g. for classification, regression or anomaly detection) without transferring all available (or even more) data to other nodes.

The vertically partitioned data scenario has been especially examined in the context of privacy preserving data mining. For example, Kumbhar and Kharat [47] review distributed association rule mining for the scenario, while Kianmehr and Koochakzadeh [41] present a ranking method. Mangasarian et al. [56], Yu et al. [92] as well as Yunhong et al. [93] present a privacy preserving support vector machine. However, since the focus of such methods is on the preservation of privacy, they care less or not at all about the amount of communication between network nodes.

Teffer et al. [87] learn a temporal model for Gaussian mixture models. Forero et al. [26] propose a distributed SVM, based on the alternating direction method of multipliers (ADMM), but for the horizontally partitioned scenario. ADMM in principle can also work on vertical data partitions, at least with separable kernel functions. However, as Kim et al. [42] remark, due to its large communication overhead, ADMM is not well-suited for wireless sensor networks. In general, distributed optimization methods based on gradient descent usually need several iterations over the whole dataset until convergence, potentially leading to large communication among the nodes.

Lee et al. [49] approach the scenario by solving the primal SVM problem locally at each node with stochastic gradient descent. The global prediction is then a weighted sum of the local predictions. During training, no data needs to be communicated, and during application, only the local predictions need to be transferred. However, optimizing the weights of the local classifiers would require such local predictions to be sent in several iterations.

Das et al. [20] have proposed a synchronized distributed anomaly detection algorithm based on the 1-class ν-SVM. A local 1-class model is trained at each node and points identified as local outliers are sent to the central node P_0, together with a small sample of all observations. A global model trained on the sample at node P_0 is used to decide if the outlier candidates sent from the data nodes are true global outliers or not. The method cannot detect outliers which are global due to a combination of attributes. However, the algorithm shows good performance if global outliers are also local outliers. Moreover, in the application phase, the algorithm is highly communication efficient, since the number of outlier candidates is often only a small fraction of the data. A drawback is that the fixed-size sampling approach gives no guarantees or bounds on the correctness of the global model. Moreover, during training, no other strategies than sampling are used for a reduction of communication costs.

The Vertically Distributed Core Vector Machine (VDCVM) [83], developed in the context of the second case study, addresses such issues. Here, the global 1-class SVM is replaced by the CVM, which incrementally samples from the network only as many data points as needed to reach a $(1 + \epsilon)$ approximation of the MEB with high probability. Replacing the non-linear RBF kernel by a sum of local RBF kernels allows for a distribution of the furthest point calculation, effectively reducing the communication overhead in each iteration to a single value per node, instead of having to transfer all attributes. Empirical results on several synthetic and real-world data sets show that the VDCVM may reach a similar accuracy as the distributed 1-class ν-SVM, but with lower communication during training.

3.5.6 Stream Mining

To build and apply models on data streams every step of the analysis should be executed in an online fashion, for example all features should be extracted and all events should be detected online. Usually, learning algorithms will either update the model with every new data item that arrives or learn a model on a batch of the most actual data items [29]. In addition to online algorithms, offline learning algorithms, like the Support Vector Machine [80], could be used.

The biggest difference between classical stream mining scenario and production processes is, that the labels are usually only measured at the end of the process. For example, in the context of the first case study, quality measures at the end of the process, like the temperature, have to be predicted.

The obvious approach to learn models in this setting is to redefine the data stream as a stream of pairs $(x, y) \in X \times Y$ of stream segments or time series $x \in X$ and labels $y \in Y$. The model will then describe a function $f(\{\phi_1(x), \ldots, \phi_p(x)\}) = y$ of p extracted features of the stream segment or time series $x \in X$. As a consequence, predictions and model updates are only possible when all relevant features of the actual process p_i are extracted.

There are several approaches to soften the restriction on the possibility of in process prediction. Let t_{i0} be the start, t_{ie} the end and t_{i*} the actual index of the process p_i.

One possible approach is to segment the time series and learn an individual model for every segment. If it is possible to identify events that exists in every process, for example the start of the combustion phase of the BOF-process t_{ic}, every feature that can be extracted from the segment $[t_{i0}, t_{ic}]$ can be used to learn a model. Therefore it would be possible to do a first prediction of the label if $t_{i*} > t_{ic}$.

A second approach is to only use a combination of static and statistical features, like the minimum, maximum or average, of the time series. The static data won't change over the complete process and the probability of change of the statistical features will decrease to the end of the process. That means, that there exists an index t_{is}, where the prediction error is bounded by $(\hat{f}(\{\phi_1([t_{i0}, t_{is}]), \ldots, \phi_k([t_{i0}, t_{is}])\} - y)^2 < \epsilon$.

Another approach would be to use one or a set of algorithms, that predict the set of features for the yet unseen segment of the process $[t_{i*}, t_{i.e.}]$. The model on the full feature set could be used to predict the labels at the end of the actual process. The overall prediction quality will therefore be strongly influenced by the quality of the feature prediction.

4 Case Studies

The content of the previous sections is based on our experiences with improving the sustainability of two real-world steel production processes, being the focus of two case studies presented in the following.

4.1 Real-Time Quality Prediction in a Basic Oxygen Furnace Process

The first case study is taken from a project with SMS Siemag,[2] in which a novel data-driven prediction model for BOF Endpoint [75] was developed.

There are several routes to produce steel products. For details on the most common steel production processes and the metallurgical backgrounds see [27]. The primary route, with almost 70 % of the world steel production, includes four major processes and is usually conducted in an so called integrated steel mill (Fig. 7). The given primary energy consumptions are from an energy analysis based on a hypothetical reference plant [22]. This should only give a brief insight on the relations of energy consumption of the different steps in the production processes.

In the first process step, iron ore is smelted to pig iron by the addition of preheated blast air, coke and slagging agents. The coke is used as source of energy and as an reduction agent. The main chemical reaction in the blast furnace is: $Fe_2O_3 + 3CO \rightarrow 2Fe + 3CO_2$. In this redox reaction the oxygen of the oxides in the ore are bounded to the carbon from the added coke. Additionally a slagging agent, e.g. dolomite is used to bind the unwanted contents of the ore. At the end of the process liquid raw iron is produced, but with a too high amount of carbon, phosphor and sulfur (amongst other things). In addition to a high primary energy consumption (Fig. 7) the process produces a lot of CO_2, because carbon is used as the reduction agent. In the second step, the amount of the unwanted contents in the raw iron is reduced in a Basic Oxygen Furnace (Fig. 8). The BOF is charged with around 150 tons of raw iron, around 30 tons of different types of scrap metal and a slagging agent. The amount of unwanted contents will be reduced by blowing pure oxygen (99 %) on the raw iron and scrap metal with an oxygen lance for 15–20 min. In this process the raw iron will be heated from around 1,200 °C up to 1,600 °C. The needed energy is created through the combustion of the contained carbon in the raw iron. With an intelligent off gas treatment, the process could even produce energy (Fig. 7). Further ways of controlling the temperature is the insertion of heating agents, like FeSi or cooling agents, like scrap metal. The reduction process is enhanced by bottom gas injection or purging with nitrogen or argon. In the third process the crude steel is refined further and casted into large blocks in a continuous caster. These blocks are rolled and shaped in the last process to flat or long products.

[2] http://www.sms-siemag.com.

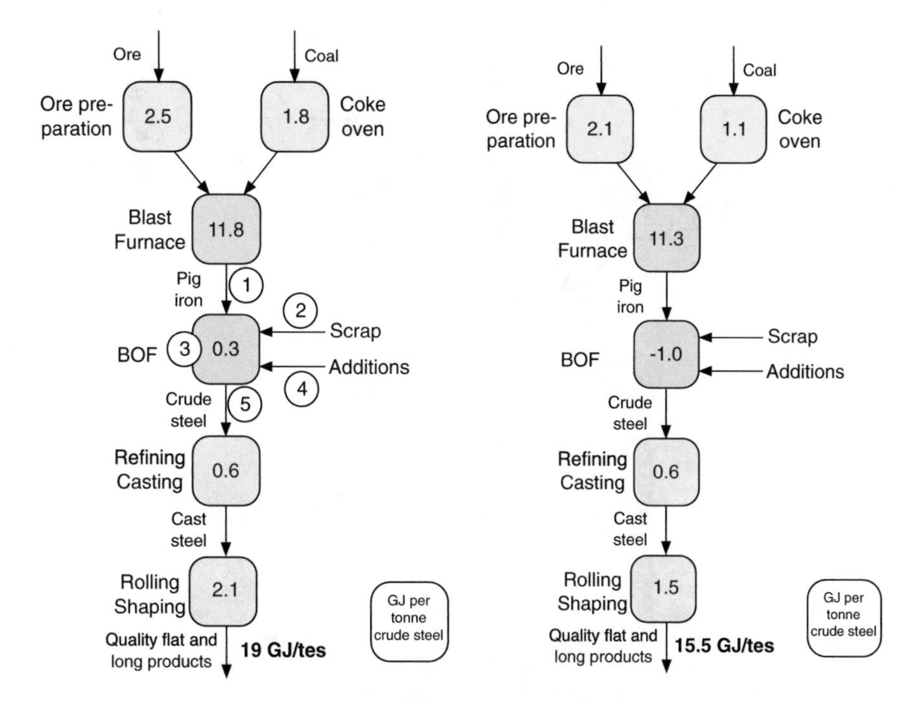

Fig. 7 Primary energy consumption for a integrated primary steel mill 3–5 million tonnes per year before and after improvement [22]

4.1.1 The BOF-process and Data Analysis

The BOF-process has a high impact on the efficiency and the overall quality of the steel products. There are possibilities to measure the quality during the process, like so called quick bombs or sub-lances [17], but in the given example the quality is determined at the end of the process, only. The most important quality measures are the temperature, the carbon content, the phosphorus content of the steel, and the iron content of the slag. The following steps in the production process usually need a pre-specified quality of the crude steel, and therefore the overall quality of the BOF-process is defined by the difference of the target values and the measured values.

If the process is stopped too early, the steel could have too low a temperature or the wrong chemistry and the process has to be restarted. In a similar BOF shop from ArcelorMittal (shop 7 in [17]), 22.6 % of the processes are restarted. This so called *reblow* reduces the productivity of the BOF and increases the energy consumption. If the process is stopped too late, the combustion of iron will start. This will reduce the amount of produced steel and the productivity of the BOF. It is also possible that the temperature of the crude steel is too high. This will delay the next process, because the raw steel needs to cool down.

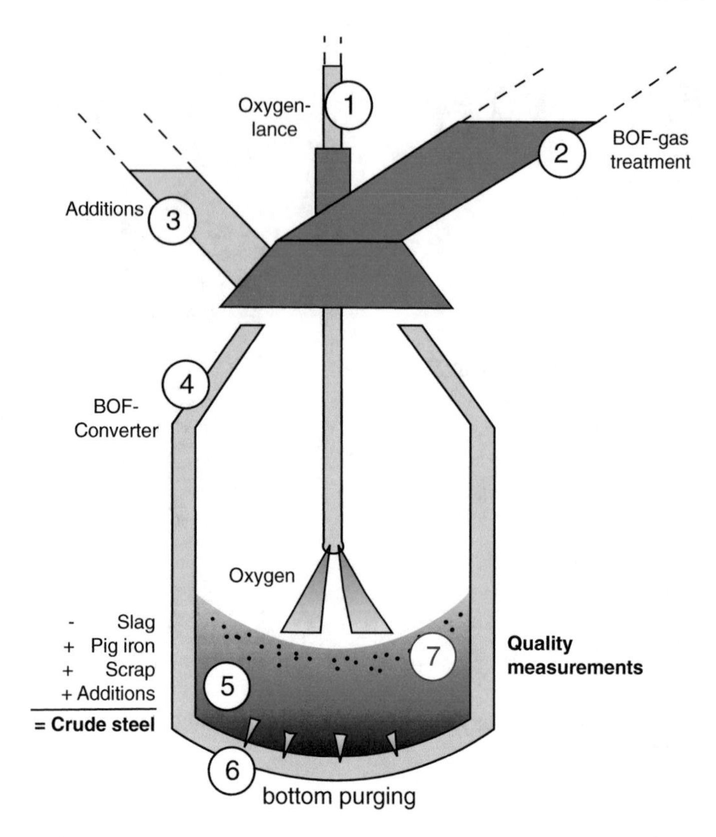

Fig. 8 Basis Oxygen Furnace (BOF)-converter

The optimal control of the BOF-endpoint plays therefore a crucial role for the quality of the BOF-process. Additional possibilities to control the quality of the process are the amount of cooling or heating agents, the height of the oxygen lance, or the bottom purging strategy. To reduce the duration and the addition of heating or cooling agents it is necessary to intervene as early as possible in the process, based on a well funded prediction/expectation of the outcome at the end of the process. It is standard to combine a thermodynamic model for determining the needed amount of oxygen, that results in the desired chemical composition, and a set of rules that monitor whether or not the BOF-endpoint is reached. Usually, the thermodynamic models can only use information of the composition of the used materials, like the contents of the pig iron and the heating and cooling agents. They incorporate only the physics of the BOF-process itself, but no information about the context of the process. The context of the process, e.g. the wear of the used plant components, time differences between process steps or other sensor data, can not be used. Contrary to thermodynamic models, data mining models can cope with almost arbitrary features and data.

We now go through the steps described above. It starts with the sensor readings. In particular, data from different sensors and subprocesses with different sampling rates need be aligned.

The sensor readings can be divided into three groups. The first group are event based data. These data will only be measured when an event will occur, like the analysis of the temperature and the contents of the pig iron after the transport from the blast furnace or when cooling, heating or slagging agents are added. The second group are sensor readings with a fixed sample rate data, like the oxygen rate or the off gas temperature. The third group of data contains all the data concerning the used plant components, like the id and the age of the used oxygen-lance and the age of the refractory lining of the BOF. To complicate it further, the huge dimensions of the factory and the sensor distribution over the factory makes it impossible to co-relate the different sensor readings directly. For example, the analysis of the BOF-gas will take longer than the measurement of the temperature of the oxygen-lance cooling water and will be conducted dozens of meters above the BOF. Therefore, the available sensor readings will represent the process at different points in time (Fig. 9).

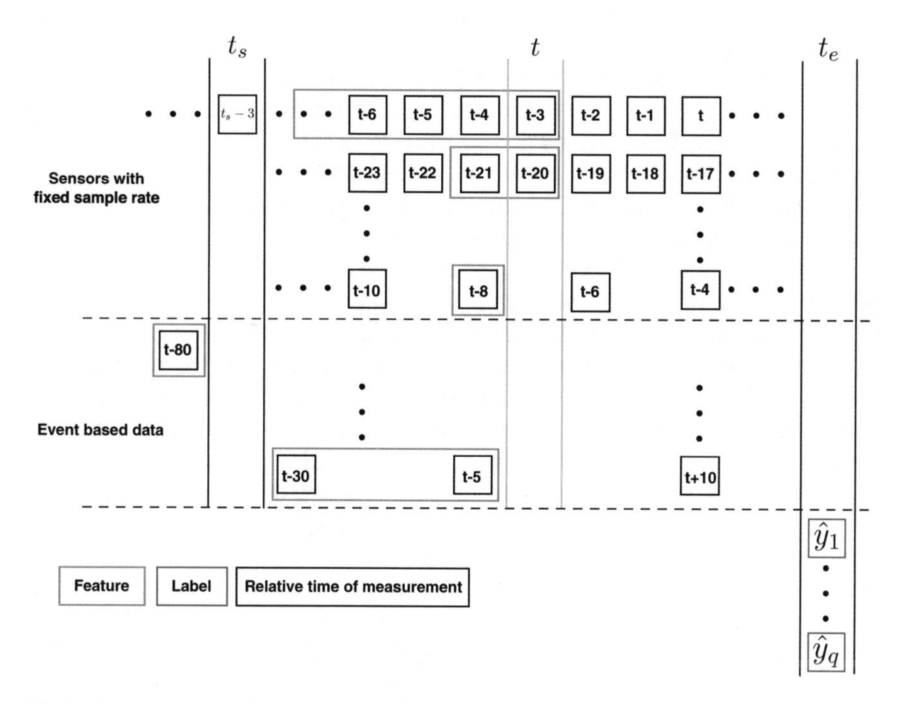

Fig. 9 Sensor streams with different sample rates and frequencies

4.1.2 Data Processing and Storage

In the first phase of the project a prototype of online application was implemented. It had to cope with the common problems of the integration of legacy systems. The application and the learning of data mining models had two completely different sets of interfaces to the data. In a first step, data were gathered to find good features for. The event based data and the plant component data where exported manually via a legacy system as a CSV-file. Due to security compliances, the high frequency data could not be accessed directly. A read-only connection was established via a ibaPDA[3] and the data were stored in a proprietary file per BOF-process. After manually preprocessing the proprietary files with a processing chain in Java and Matlab the data were stored in a CSV file per process. The time information was lost in the transformation process and the correct co-relation of events and the high frequency data was nearly impossible. After that, the data from the two different files could be loaded into popular data mining tool RapidMiner[4] [61] and the data could be preprocessed, features could be extracted and a model could be learned.

In a second step a prototype of the online application of the learned model was implemented. The high frequency data are accessed via the OPC-interface.[5] This interface is only stable for sample rates under 5 Hz. The event based data are accessed via key-value files. That are two completely different data access methods than for the offline access of the data. The two data sources are combined and integrated in a data stream framework. The *streams*-Framwork [8] can easily extend with arbitrary data sources and can handle sample rates over 100 kHz [28]. To apply the learned model all the used feature extractions and preprocessing steps had to be reimplemented for the online application and prediction.

The implementation of the second phase of the project will remove the manual process steps, the problems with the co-relation of events and high frequency data in the offline processing and the redundant feature extractions. As an additional requirement, sensors with a higher sample rate should be integrated. To achieve all of this, the offline processing was reduced to a minimum and all raw data are stored in plain files. As a new interface to the sensor data, memory mapped files were used and the maximal sample rate was increased to 1 kHz. This limit is imposed by Windows and could be extended with additional hardware to 20 kHz. Each line of the file represents at least the sensor readings with the highest sample rate and all the other sensor readings and event based data, that where available at that point in time (see Fig. 9). In addition to this sparse file format, the files are compressed to reduce the file size. The initial file size of 4 GB per hour of data was reduced to 40 MB.

For an easy access of the data, the files are indexed with the ids of the contained BOF-processes. This index was implemented within a PostgreSQL[6] database system.

---

[3]http://www.iba-ag.com/.

[4]http://rapidminer.com.

[5]http://openscada.org/projects/utgard/.

[6]http://www.postgresql.org.

The data are stored on the local file system, but the system could be easily extended with a distributed file system or object storage and a distributed data base system.

In this new implementation, preprocessing steps and feature extractions have to be implemented just once. If a new feature should be extracted from historical data, the files are read line by line and the whole recorded data stream of the historical BOF-process can be accessed in the original order. The newly implemented feature extraction and preprocessing steps can be used directly in the online application of the learned model.

4.1.3 Feature Extraction and Model Learning

In difference to the usual data mining problems, the BOF-process has two heterogenous but not independent data sources. The possible influence of events on the sensor data, like the addition of heating agents will influence the off gas temperature, is not negligible. Sensor readings with a fixed sample rate and event based data have to be combined into one data mining model. The features have to have minimal latency, minimal time difference between them and be extracted online and as early as possible in the BOF-process. Especially, the time dependency of the addition of heating, cooling or slagging agents impose additional complexity on the modeling process. In general, the task is to extract features online to build a batch/offline learning model, which is applied online again.

The basic features for all the models are the amounts of pig iron, the chemical composition of the pig iron, the amounts of different types of scrap and the amount of used oxygen. In addition to preexisting events, like the addition of heating and slagging agents, new ones were created. Especially durations of periods, for example tab-to-tab times or the duration from the measurement of the temperature of the pig iron and the start of BOF-process, helped to characterize the process further.

Even though multiple common time series feature extractions have been tried, the best prediction results have been achieved with the extraction of “global” features over the whole BOF-process. These global features, like the average deviation of the median of a time series or the total amounts of used heating and cooling agents will lead to good prediction results, but only if the BOF-process is almost finished. Features like the position and value of a maximum of a time series, like the off gas temperature, can be influenced by the addition of heating agents. This contradicts the efforts to be able to have a relatively good prediction of the quality as early as possible in the BOF-process. Further work will include the development of a time-invariant phase model or time series segmentation [40]. Each BOF-process has a different duration and therefore a time-based segmentation is not possible. If it is possible to develop an online segmentation that is reproducible over multiple BOF-processes, it could be possible to learn a prediction model for each segment and start to predict as early as possible.

If the features are extracted once, they could be stored directly for a later use in new data mining models. In the second phase of the project the value and the extraction time of each feature in relation to the id of the BOF-process is stored automatically in the aforementioned database. In combination with the new raw data storage arbitrary new features can be added afterwards.

The first phase of the project has shown, that the selection of features had an higher impact on the prediction quality than the used data mining method. Due to highly non-linear relations of the used features and the quality measurements, the used method should be able to model non-linear relationships. For the sake of simplicity a Support Vector regression [80] was used for each of the quality measurements.

4.1.4 Feature and Data Selection and Model Evaluation

In the first phase of the projects, the features were only selected in the learning step via a cross validation and the resulting prediction quality. One model was selected and evaluated online. Due to the two different sets of interfaces and the slightly different sets of features in the offline and the online processing of the data, it is not certain if the expected quality of the prediction will be reached in the online application of the model. Since the online application was only a prototype, the prediction quality had to be analyzed manually.

An important feature of the new implementation of the online application is, that without any manual steps new models can be learned, applied and evaluated. If a query of a set of features of a set of BOF-processes is stated, the corresponding data will be exported as a CSV-file. Automatically a RapidMiner process will be started and as a result a new data mining model will be exported and indexed in the aforementioned database. If the learning process is finished, the new model will be applied online and evaluated automatically.

In combination with the interpretation of the production process as a stream of differences between target quality measurements and actual quality measurements, this approach presents a wide variety of new opportunities. In the first phase of the project it was shown, that the validity of the one learned prediction model was limited. If the creation of new data mining models is not scheduled manually, but, for example, by Concept Drift algorithms [69], the prediction models should have a higher prediction quality and no manual intervention is necessary. Additionally, it is possible to evaluated multiple data mining models in parallel and to select the one model with the best prediction quality over the last BOF-processes. Even online ensemble methods [7] are applicable. Furthermore, it is possible to select arbitrary subsets of BOF-processes. If, for example, a clustering of the BOF-processes based on the scrap types or the position in the maintenance cycle is created, the models can be adapted much better to the actual conditions of the BOF-process. In general, the in-stream evaluation of data mining models will reflect the characteristics of the BOF-process much better than the offline cross validation and will lead to better average prediction quality.

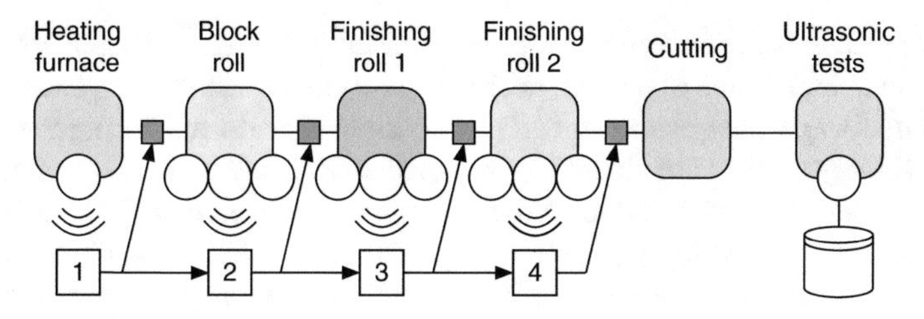

Fig. 10 Hot rolling mill process with prediction and decision/control modules

4.2 Real-Time Quality Prediction in a Hot Rolling Mill Process

In project B3 of the Collaborative Research Center SFB 876,[7] the Artificial Intelligence Group (LS 8) and the chair of Production Systems and Industrial Engineering (APS) of the institute for production systems (IPS) at TU Dortmund University research new data mining and machine learning techniques for interlinked production processes. The current main focus is on a hot rolling mill case study, i.e. the last processing step in Fig. 7. Here, steel blocks move through a process chain as the one shown in Fig. 10. Already casted blocks are first heated for up to 15 h in five different heating zones of a furnace. They are then rolled at the block roll and the first finishing roll. The rolling in the second roll is optional. Each block usually moves back and forth through a single roll for several times, where each of the rolling steps takes only about a few seconds. The blocks are finally cutted into smaller bars whose quality is assessed by ultrasonic tests several days later.

Different sensors attached along the process chain provide online measurements about how a steel block is currently processed. For example, in the furnace, every 5 min sensors measure the air temperature in each of the five zones. From such measurements, the core temperature of the blocks can be estimated by an already existing mathematical model. At each roll, sensors provide measurements such as rolling force, rolling speed and the height of the roll, with 10 values per second. Additional signals provide meta information about the process itself, like the current number of rolling steps. The ultrasonic test results indicate the number of bars tested and, for each bar, the amount of material containing defects, though not their exact position. Moreover, due to technical reasons, most often it is not possible to reconstruct which of the final bars belonged to which of the cutted steel blocks.

According to the current technical state of the art, it is impossible to assess the physical quality of hot steel blocks or smaller bars at intermediate steps of the process chain. The blocks first must cool down before their final quality can be tested. In cases where some of the blocks are, for example, already wrongly heated, energy, material and human work force are wasted if blocks below a desired quality threshold nevertheless move through the whole process chain. Considering potential material

[7] http://sfb876.tu-dortmund.de/SPP/sfb876-b3.html.

and energy savings for the rolling step as estimated from Fig. 7, the goal of the case study is the identification of quality-related patterns in the sensor data, and to predict the final quality of steel blocks as early as possible in real-time during the running process. Energy savings are already to be expected if, depending on the predictions, blocks with defects could be sorted out of the process early enough. For one thing, all of the following processing steps could be spared. For another thing, blocks might be reinserted into the heating furnace while still being hot, sparing the energy needed for a complete reheating. A reinsertion into the heating furnace might even be entirely spared if, depending on the predictions, parameters of subsequent processing stations could be adjusted such that the aimed-at final quality level would still be reached. Concepts for the integration of prediction models with control have already been developed by our project partners [44].

The main research focus of project B3 is the development of decentralized methods for the real-time analysis of sensor data. Another research area are methods for learning from label proportions. As such methods (LLP and VDCVM) already have been presented in Sect. 3.5, in the following it is described in more detail instead which sensor measurements are recorded and how they are stored, preprocessed and analyzed in the context of the given case study.

4.2.1 Assessment and Storage of Sensor Data

During the period of 1 year, over one billion measurements from 30 different sensor types have been recorded during the processing of about 10,000 steel blocks, together with according quality information. Among the readings are the air temperature for each furnace zone, the rolling speed, force, position and temperature, which domain experts consider to be the most relevant quality-related parameters. For validation purposes and guaranteeing the reproducibility of results, all data has been stored in a single SQL database.

Figure 11 shows an excerpt of the database schema, representing the most important tables and relationships. The steel blocks resulting from a single cast can be divided according to different customer orders. Steel blocks from a single order are usually inserted into the furnace together. For each order, quality information about the bars that were cutted from each block is available. Each row consists of the test results for several bars. In only a few cases it is possible to relate the bars back to the steel block they originally were cutted from, based on the last two digits of their ID. For such cases, our project partners have introduced a mapping that derives a single label from multiple types of quality information available for all bars [44].

A tool developed in the Java programming language allows for reading in the raw data delivered in different files and formats and transforming them into the shown database schema. Once imported, sensor measurements can be exported based on filters written in SQL. Exported are several CSV files, where each contains all measurements recorded by a particular sensor during the processing of a single steel block.

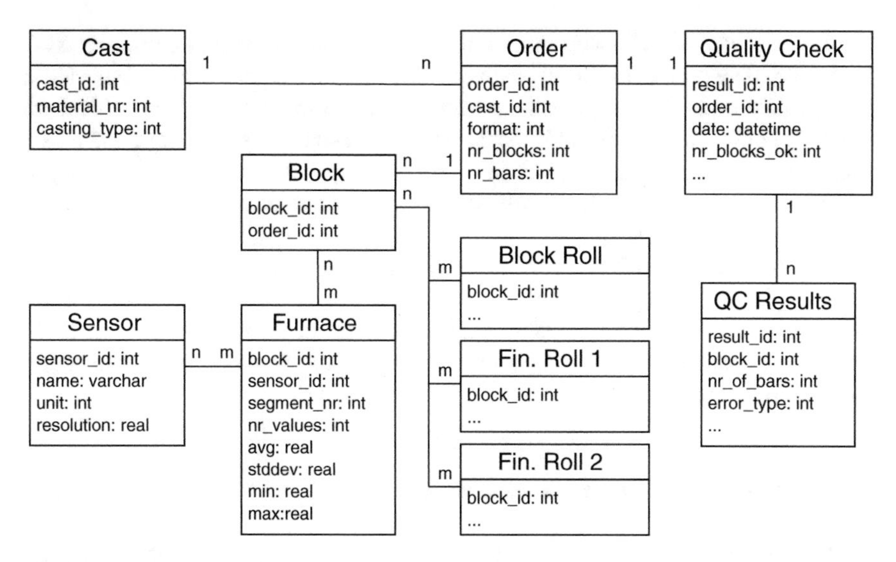

Fig. 11 Database schema for hot rolling mill case study

For the preprocessing of time series in production environments, a highly modular process has been developed in collaboration with our project partners [51] and implemented with RapidMiner. The process handles most time series stored in the CSV files independently from each other and is thus well-suited for a decentralized preprocessing of time series. The following sections provide a summary of the procedure and results already presented in [51].

4.2.2 Preprocessing and Feature Extraction

At first, all value series are cleansed. Cleansing consists of cutting away irrelevant parts where no processing happened, as discussed in Sect. 3.2. Then, measurements lying outside meaningful value ranges are marked as outliers and replaced by their predecessor value. Optionally, values are rescaled and normalized to the [0, 1] interval.

Afterwards, the value series are segmented based on background knowlegde, as already described in Sect. 3.4.2. In case of the heating furnace, for instance, the five different heating zones make up natural borders for the segments. Similarly, individual rolling steps seem to be natural divisions for all series stemming from the three different rolls. At the block roll, a change in the signal counting the number of rolling steps directly indicates the beginning of a new division. At the finishing rolls, due to the aforementioned signal not being available, the rolling force can be used accordingly, as longer segments with zero force indicate the period of no processing between rolling steps. It should be noted here that in practice, even seemingly simple tests like the ones described are not always easy to implement. For example, the rolling force sensor will catch vibrations of the roll, even without any processing happening. Therefore, it will not deliver values exactly equaling zero, but values that oscillate around zero instead. In such cases, it sometimes can be difficult to manually devise global thresholds that separate valid signals from background noise. Therefore, even despite having background knowledge, it can become necessary to use learning methods already at early stages of the whole data mining process, like the data preprocessing step.

Once segmented, each segment is described by several statistics, as described in Sect. 3.4.2, and mapped to portions of a fixed-length vector. Thereby, the up to 60,000 raw series values recorded for each steel block are aggregated to about 2,000 features. The resulting data table can then be input into common feature selection and learning algorithms.

4.2.3 Data Analysis and Prediction Results

The feature vectors of 470 processes for which the relation between steel blocks and the bars cutted from them could be established were first analyzed with different learning methods, like Naïve Bayes, Decision Trees, k-NN and the SVM. It soon turned out that including features about the individual segments decreases accuracy in comparison to only including global information about the value series and segments. Features of individual segments were therefore excluded for the following analysis, resulting in 218 remaining features. However, even with the reduced feature set, none of the classifiers mentioned before could reach a significantly better prediction accuracy than the baseline, which predicts the majority label.

For getting a better impression of the data, the feature vectors were mapped to a two dimensional SOM and colored according to different types of meta information (see Fig. 12). As mentioned in Sect. 3.5.3, points lying close to each other on the map have similar feature vectors. The shading indicates a weighted distance between the points, where lighter shades represent a larger distance.

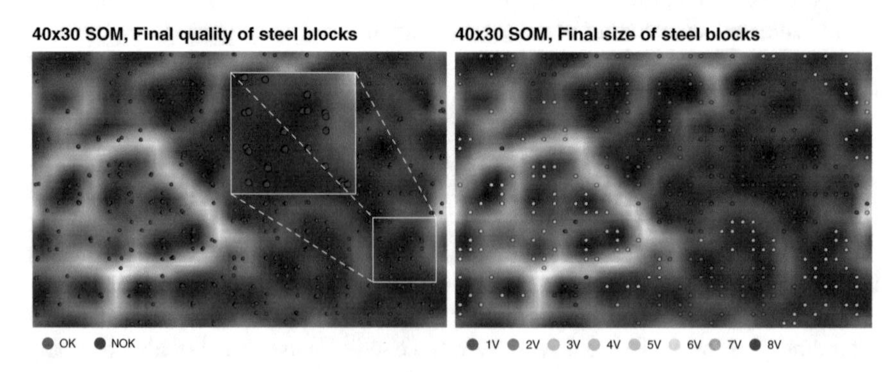

Fig. 12 Similarity relationships between feature vectors

In the SOM on the left hand side, the points represent the feature vectors of production processes and their color the final quality of the resulting steel bars as discretized values, "okay" (OK) and "not okay" (NOK). In many cases, processes leading to a low final quality of the bars are lying very close to processes resulting in a high quality (see also the zoomed area in Fig. 12), meaning they have highly similar feature vectors. As it seems, the features extracted so far do not suffice to distinguish well between low and high quality processes, explaining the previously mentioned prediction results.

In comparison, the SOM on the right hand side of Fig. 12 shows the final size of the resulting steel bars. Here, processes resulting in the same size form large continuous areas on the SOM, i.e. their feature vectors are similar. As it seems, the features extracted are thus highly correlated with distinct operational modes for the different bar sizes produced. The hypothesis could be verified by training a decision tree on features of the first finishing roll (see Fig. 13). The accuracy as estimated by a tenfold cross validation is 90 %, while k-NN ($k = 11$) even achieves 97 %. Most important for the decision is the position of the roll (sensor 501). Domain experts have verified that the results reflect the real modes of operation in the rolling mill. Similar results have also been achieved later, by concatenating raw series values and comparing them with Dynamic Time Warping, as described in Sect. 3.3.2. While the analysis of raw series values has the advantage of not requiring any background knowledge about the segmentation, the results were a lot harder to interpret. Moreover, the distance-based approach took much longer than summarizing the values of segments.

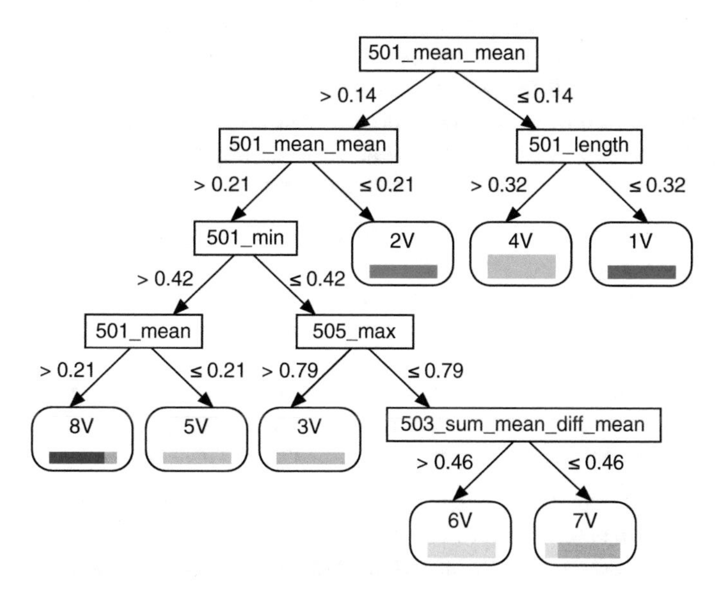

Fig. 13 Decision tree for predicting the final size of steel bars

As the results demonstrate, data analysis methods are able to detect meaningful patterns in production processes. As the results also show, finding exactly those features that are relevant for the prediction task sometimes is not straightforward, as already discussed in Sect. 3.4. Especially, results from the first case study, where global features were successfully used, are not transferable to the rolling process, where deviations from the global patterns might play a bigger role. Another approach that has been tried without success so far is the genetic programming of method trees, extended by a SAX operator and a frequency representation of the resulting symbols.

However, each insight into the data might provide new ideas for the extraction of better features. Future experiments will show if the newly gained ability to quantify the most striking patterns might help with normalization and the extraction of features mainly representing differences to the normal patterns that were detected. Hopefully, with improved prediction results, it will also become possible to quantify potential energy savings in the hot rolling mill process more exactly.

5 Conclusion

In this chapter, we gave an overview of data analysis for sustainability in steel processing and described our contributions. The first sections offer guidelines for applications similar to ours. In Sect. 1, we presented the needs and requirements for embedding data analysis into production processes in order to reduce material and energy consumption. In particular, real-time quality prediction allows to save resources. The clear and comprehensive summary of how to prepare a factory for the good use of data analysis gives readers a guideline for new applications (Sect. 2). Carefully going through all the steps of a data analysis process in Sect. 3 presents a large variety of methods for each step. We stress the importance of feature extraction and selection and show some transformation methods. The modeling step includes not only classification learning but also one class learning and unsupervised learning. Three algorithmic contributions are described here:

- Learning from aggregate outputs, especially our learning from label proportions, is an important learning method for factory processes where the object identity is not given over all steps of the process as is the case in, e.g. a hot mill process.
- The streams framework for integrating diverse subroutines and sensor readings allows to apply learning results in real-time.
- Distributed mining of several sensor data, each with different features, is the vertically partitioned data scenario. Our method of Vertically Distributed Core Vector Machines approximates the minimum enclosing balls using little communication.

These methods are particularly relevant for the production scenario, but are general and applicable to other domains, as well.

The chapter presents two real world case studies in Sect. 4. Along the steps described, the real-time quality prediction is characterized for the two real world applications: the endpoint prediction of the Basic Oxygen Furnace process (Sect. 4.1) and the quality prediction in a hot mill process (Sect. 4.2).

The case studies make possible investigations at the macro level: since all data and models are stored, the fit of prediction and actual values, the benefit for the resource saving can be investigated in more detail. We shall inspect regularities in changes of fit. Are they due to machine fatigue or sensor faults or special situations which were not yet included in the training data? Further work will use the experience from the case studies to the management of several models. Since the machine fatigue will occur in cycles, the models from different parts of the cycle can be used when appropriate. The concept drift will be recognized and the appropriate model be selected. Also typical failure situation, e.g. missing sensor reading, can be coped with by particular models. The management of several models is new and promising. It presupposes the completion of all the steps which we characterized in this chapter.

Acknowledgments This work has partially been supported by the DFG, Collaborative Research Center 876, projects B3 and A1, http://sfb876.tu-dortmund.de/. The case study on BOF end time prediction has partially been supported by SMS Siemag and developed in collaboration with Hans-Jürgen Odenthal, Jochen Schlüter and Norbert Uebber. The application is pushed forward by Markus Reifferscheidt and Burkhard Dahmen from the SMS group. We thank *AG der Dillinger Hüttenwerke*, particularly Helmut Lachmund and Dominik Schöne, for testing our methods and insightful discussions. The hot rolling mill application has been developed in cooperation with Jochen Deuse, Daniel Lieber, Benedikt Konrad and Fabian Bohnen from the TU Dortmund University. We thank Ulrich Reichel and Alfred Weiß from the Deutsche Edelstahl Werke.

References

1. Agrawal, R., Imielinski, T., Swami, A.: Mining association rules between sets of items in large databases. In: Proceedings of the ACM SIGMOD Conference on Management of Data, pp. 207–216. Washington, D.C. (1993)
2. Agrawal, R., Srikant, R.: Mining sequential patterns. In: Proceedings of the 11th International Conference on Data Engineering (ICDE), pp. 3–14. IEEE, Washington, DC, USA (1995)
3. AlGhazzawi, A., Lennox, B.: Model predictive control monitoring using multivariate statistics. J. Process Control **19**(2), 314–327 (2009)
4. OECD/IEA: International energy outlook 2011. Technical Report DOE/EIA-0484(2011), U.S. Energy Information Administration (2011)
5. Bai, Z., Wei, G., Liu, X., Zhoa, W.: Predictive model of energy cost in steelmaking process based on BP neural network. In: Proceedings of 2nd International Conference on Software Engineering, Knowledge Engineering and Information Engineering, pp. 77–80 (2014)
6. Bhaduri, K., Stolpe, M.: Distributed data mining in sensor networks. In: Aggarwal, C. (ed.) Managing and Mining Sensor Data, chap. 8. Springer, Berlin, Heidelberg (2013)
7. Bifet, A., Holmes, G., Pfahringer, B., Kirkby, R., Gavaldà, R.: New ensemble methods for evolving data streams. In: Proceedings of the 15th ACM SIGKDD International Conference on Knowledge Discovery and Data Mining, pp. 139–148. ACM (2009)
8. Bockermann, C., Blom, H.: The streams framework. Technical report, Technical Report 5, TU Dortmund University, 12, 2012 (2012)
9. Box, G., Jenkins, G., Reinsel, G.: Time Series Analysis. Forecasting and Control, 3rd edn. Prentice Hall, Englewood Cliffs (1994)

10. Brock, W., Mäler, K., Perrings, C.: chap. Resilience and sustainability: the economic analysis of non-linear dynamic systems. In: Panarchy: Understanding Transformations in Human and Natural Systems. Island Press (2001)
11. Candan, K., Rossini, R., Wang, X., Sapino, M.: sDTW: Computing DTW distances using locally relevant constraints based on salient feature alignments. Proc. VLDB Endow. **5**(11), 1519–1530 (2012)
12. Carroll, A., Heiser, G.: An analysis of power consumption in a smartphone. In: Proceedings of the 2010 USENIX Conference on USENIX Annual Technical Conference, USENIXATC'10. USENIX Association, Berkeley, CA, USA (2010)
13. Chapelle, O., Schölkopf, B., Zien, A. (eds.): Semi-Supervised Learning. MIT Press, Cambridge, MA (2006)
14. Chen, J.: A predictive system for blast furnaces by integrating a neural network with qualitative analysis. Eng. Appl. Artif. Intell. **14**(1), 77–85 (2001)
15. Chen, S., Liu, B., Qian, M., Zhang, C.: Kernel k-Means based framework for aggregate outputs classification. In: Proceedings of the International Conference on Data Mining Workshops (ICDMW), pp. 356–361 (2009)
16. Chiu, B., Keogh, E., Lonardi, S.: Probabilistic discovery of time series motifs. In: Proceedings of the 9th ACM SIGKDD International Conference on Knowledge Discovery and Data Mining, pp. 493–498. ACM, New York, NY, USA (2003)
17. Chukwulebe, B., Robertson, K., Grattan, J.: The methods, aims and practices (map) for BOF endpoint control. Iron Steel Technol. **4**(11), 60–70 (2007)
18. Cox, I., Lewis, R., Ransing, R., Laszczewski, H., Berni, G.: Application of neural computing in basic oxygen steelmaking. J. Mater. Process. Technol. **120**(1), 310–315 (2002)
19. Das, G., Gunopulos, D., Mannila, H.: Finding similar time series. In: Principles of Data Mining and Knowledge Discovery. LNCS, vol. 1263, pp. 88–100. Springer, Berlin, Heidelberg (1997)
20. Das, K., Bhaduri, K., Votava, P.: Distributed anomaly detection using 1-class SVM for vertically partitioned data. Stat. Anal. Data Min. **4**(4), 393–406 (2011)
21. De Beer, J.: Future technologies for energy-efficient iron and steel making. In: Potential for Industrial Energy-Efficiency Improvement in the Long Term, pp. 93–166. Springer (2000)
22. De Beer, J., Worrell, E., Blok, K.: Future technologies for energy-efficient iron and steel making. Annu. Rev. Energy Environ. **23**(1), 123–205 (1998)
23. Deb, K., Agrawal, S., Pratap, A., Meyarivan, T.: A fast elitist non-dominated sorting genetic algorithm for multi-objective optimization: Nsga-ii. Lect. Notes Comput. Sci. **1917**, 849–858 (2000)
24. Ester, M., Kriegel, H.P., Sander, J., Xu, X.: A density-based algorithm for discovering clusters in large spatial databases with noise. In: Proceedings of the 2nd International Conference on Knowledge Discovery and Data Mining (KDD), pp. 226–231. AAAI Press (1996)
25. Faloutsos, C., Ranganathan, M., Manolopoulos, Y.: Fast subsequence matching in time-series databases. In: Proceedings of the 1994 ACM SIGMOD International Conference on Management of Data, vol. 23, pp. 419–429. ACM Press, New York, NY, USA (1994)
26. Forero, P., Cano, A., Giannakis, G.: Consensus-based distributed support vector machines. J. Mach. Learn. Res. **11**, 1663–1707 (2010)
27. Fruehan, R., et al.: The Making, Shaping, and Treating of Steel. AISE Steel Foundation Pittsburgh, PA, USA (1998)
28. Gal, A., Keren, S., Sondak, M., Weidlich, M., Blom, H., Bockermann, C.: Grand challenge: the techniball system. In: Proceedings of the 7th ACM International Conference on Distributed Event-Based Systems, pp. 319–324. ACM (2013)
29. Gama, J., Gaber, M.: Learning from Data Streams: Processing Techniques in Sensor Networks. Springer (2007)
30. Ghosh, A., Chatterjee, A.: Iron Making and Steelmaking: Theory and Practice. PHI Learning Pvt Ltd. (2008)
31. Han, M., Zhao, Y.: Dynamic control model of BOF steelmaking process based on ANFIS and robust relevance vector machine. Exp. Syst. Appl. **38**(12), 14786–14798 (2011)

32. Hastie, T., Tibshirani, R., Friedman, J.: The Elements of Statistical Learning: Data Mining, Inference, and Prediction, Statistics, 2nd edn. Springer (2009)
33. Hernández-González, J., Iñza, I., Lozano, J.: Learning Bayesian network classifiers from label proportions. Pattern Recogn. **46**(12), 3425–3440 (2013)
34. Jain, A.K., Murty, M.N., Flynn, P.J.: Data clustering: a review. ACM Comput. Surv. **31**(3), 264–323 (1999)
35. Jeffery, S., Alonso, G., Franklin, M., Hong, W., Widom, J.: Declarative support for sensor data cleaning. In: Pervasive Computing. LNCS, pp. 83–100. Springer, Berlin (2006)
36. Joachims, T.: Transductive inference for text classification using support vector machines. In: Proceedings of the 16th International Conference on Machine Learning (ICML), pp. 200–209. Morgan Kaufmann, San Francisco, CA (1999)
37. Jolliffe, I.: Principal Component Analysis, 2nd edn. Springer (2002)
38. Kano, M., Nakagawa, Y.: Data-based process monitoring, process control, and quality improvement: recent developments and applications in steel industry. Comput. Chem. Eng. **32**(1), 12–24 (2008)
39. Kaplan, R., Norton, D.: Balanced Scorecard. Springer (2007)
40. Keogh, E., Chu, S., Hart, D., Pazzani, M.: Segmenting time series: a survey and novel approach. Data Mining Time Ser. Datab. **57**, 1–22 (2004)
41. Kianmehr, K., Koochakzadeh, N.: Privacy-preserving ranking over vertically partitioned data. In: Proceedings of the 2012 Joint EDBT/ICDT Workshops, pp. 216–220. ACM (2012)
42. Kim, W., Yoo, J., Kim, H.: Multi-target tracking using distributed SVM training over wireless sensor networks. In: IEEE International Conference on Robotics and Automation (ICRA), pp. 2439–2444 (2012)
43. Kohonen, T.: Self-Organization and Associative Memory. Springer, Berlin (1989)
44. Konrad, B., Lieber, D., Deuse, J.: Striving for zero defect production: Intelligent manufacturing control through data mining in continuous rolling mill processes. In: Robust Manufacturing Control (RoMaC). LNCS, pp. 215–229. Springer, Berlin, Heidelberg (2013)
45. Kriegel, H.P., Kröger, P., Pryakhin, A., Renz, M., Zherdin, A.: Approximate Clustering of Time Series Using Compact Model-based Description, LNCS, vol. 4947, pp. 364–379. Springer, Berlin, Heidelberg (2009)
46. Kueck, H., de Freitas, N.: Learning about individuals from group statistics. In: Uncertainty in Artificial Intelligence (UAI), pp. 332–339. AUAI Press, Arlington, Virginia (2005)
47. Kumbhar, M., Kharat, R.: Privacy preserving mining of association rules on horizontally and vertically partitioned data: A review paper. In: 12th International Conference on Hybrid Intelligent Systems (HIS), pp. 231–235. IEEE (2012)
48. Laha, D.: Ann modeling of a steelmaking process. In: Panigrahi, B., Suganthan, P., Das, S., Dash, S. (eds.) Swarm, Evolutionary, and Memetic Computing. Lecture Notes in Computer Science, vol. 8298, pp. 308–318. Springer International Publishing (2013). doi:10.1007/978-3-319-03756-1_28. http://dx.doi.org/10.1007/978-3-319-03756-1_28
49. Lee, S., Stolpe, M., Morik, K.: Separable approximate optimization of support vector machines for distributed sensing. In: Machine Learning and Knowledge Discovery in Databases. LNCS, vol. 7524, pp. 387–402. Springer, Berlin, Heidelberg (2012)
50. Lekakh, S.N., Robertson, D.: Application of the combined reactors method for analysis of steelmaking process. In: Celebrating the Megascale: Proceedings of the Extraction and Processing Division Symposium on Pyrometallurgy in Honor of David GC Robertson, pp. 393–400. Wiley (2014)
51. Lieber, D., Stolpe, M., Konrad, B., Deuse, J., Morik, K.: Quality prediction in interlinked manufacturing processes based on supervised and unsupervised machine learning. In: Proceedings of the 46th CIRP Conference on Manufacturing Systems (CMS), vol. 7, pp. 193–198. Elsevier (2013)
52. Lin, J., Keogh, E., Wei, L., Lonardi, S.: Experiencing sax: a novel symbolic representation of time series. Data Mining Knowl. Discov. **15**(2), 107–144 (2007)
53. Lowe, D.: Distinctive image features from scale-invariant keypoints. Int. J. Comput. Vis. **60**(2), 91–110 (2004)

54. Lytvynyuk, Y., Schenk, J., Hiebler, M., Sormann, A.: Thermodynamic and kinetic model of the converter steelmaking process. Part 1: The description of the BOF model. Steel Res. Int. **85**(4), 537–543 (2014)
55. MacQueen, J.: Some methods for classification and analysis of multivariate observations. In: Proceedings of the 5th Berkeley Symposium on Mathematical Statistics and Probability, vol. 1, pp. 281–297. University of California Press (1967)
56. Mangasarian, O., Wild, E., Fung, G.: Privacy-preserving classification of vertically partitioned data via random kernels. TKDD **2**(3) (2008)
57. Mannila, H., Toivonen, H., Verkamo, A.: Discovery of frequent episodes in event sequences. Data Mining Knowl. Discov. **1**(3), 259–290 (1997)
58. Martins, A., Mata, T., Costa, C., Sikdar, S.: Framework for sustainability metrics. Ind. Eng. Chem. Res. **46**(10), 2962–2973 (2007)
59. Matias, Y., Vitter, J., Wang, M.: Dynamic maintenance of wavelet-based histograms. In: Proceedings of the 26th International Conference on Very Large Data Bases (VLDB), pp. 101–110. Morgan Kaufmann, San Francisco, CA, USA (2000)
60. Mierswa, I., Morik, K.: Automatic feature extraction for classifying audio data. Mach. Learn. J. **58**, 127–149 (2005)
61. Mierswa, I., Wurst, M., Klinkenberg, R., Scholz, M., Euler, T.: YALE: rapid prototyping for complex data mining tasks. In: Eliassi-Rad, T., Ungar, L.H., Craven, M., Gunopulos, D. (eds.) Proceedings of the 12th ACM SIGKDD International Conference on Knowledge Discovery and Data Mining (KDD 2006), pp. 935–940. ACM Press, New York, USA (2006)
62. Morik, K.: Tailoring representations to different requirements. In: Watanabe, O., Yokomori, T. (eds.) Algorithmic Learning Theory—Proceedings of 10th International Conference on ALT99, Lecture Notes in Artificial Intelligence, pp. 1–12. Springer (1999)
63. Morik, K., Köpcke, H.: Features for learning local patterns in time-stamped data. In: Morik, K., Boulicaut, J.F., Siebes, A. (eds.) Local Pattern Detection: International Seminar, Dagstuhl Castle, Germany, 12–16 Apr 2004, Revised Selected Papers, chap. 7, pp. 98–114. Springer (2005)
64. Morik, K., Wessel, S.: Incremental signal to symbol processing. In: Making Robots Smarter, pp. 185–198. Springer (1999)
65. Moya, M., Koch, M., Hostetler, L.: One-class classifier networks for target recognition applications. In: Proceeding of World Congress on Neural Networks, pp. 797–801. Int. Neural Network Society (1993)
66. Müller, M.: Dynamic time warping. In: Information Retrieval for Music and Motion, pp. 69–84. Springer, Berlin, Heidelberg (2007)
67. Musicant, D., Christensen, J., Olson, J.: Supervised learning by training on aggregate outputs. In: Proceedings of the 7th IEEE International Conference on Data Mining (ICDM), pp. 252–261. IEEE, Washington, DC, USA (2007)
68. Ni, K., Ramanathan, N., Chehade, M.N.H., Balzano, L., Nair, S., Zahedi, S., Kohler, E., Pottie, G., Hansen, M., Srivastava, M.: Sensor network data fault types. ACM Trans. Sensor Netw. (TOSN) **5**(3), 1–29 (2009)
69. Nishida, K., Yamauchi, K.: Detecting concept drift using statistical testing. In: Discovery Science, pp. 264–269. Springer (2007)
70. Pendelberry, S., Ying Chen Su, S., Thurston, M.: A Taguchi-based method for assessing data center sustainability. In: Proceeding of the iEMSs 4th Biennial Meeting: International Congress on Environmental Modelling and Software. Int. Environ. Modelling and Software Society (2010)
71. Quadrianto, N., Smola, A., Caetano, T., Le, Q.: Estimating labels from label proportions. J. Mach. Learn. Res. **10**, 2349–2374 (2009)
72. Rakthanmanon, T., Keogh, E., Lonardi, S., Evans, S.: Mdl-based time series clustering. Knowl. Inf. Syst. **33**(2), 371–399 (2012)
73. Rüping, S.: SVM classifier estimation from group probabilities. In: Proceedings of the 27th International Conference on Machine Learning (ICML) (2010)

74. Ryman, C., Larsson, M.: Reduction of CO_2 emissions from integrated steelmaking by optimised scrap strategies: application of process integration models on the BF-BOF system. ISIJ Int. **46**(12), 1752–1758 (2006)
75. Schlueter, J., Odenthal, H.J., Uebber, N., H., B., K., M.: A novel data-driven prediction model for bof endpoint. In: AISTech Conference Proceedings. Association for Iron & Steel Technology, Warrendale, PA, USA (2013)
76. Schölkopf, B., Platt, J., Shawe-Taylor, J., Smola, A., Williamson, R.: Estimating the support of a high-dimensional distribution. Neural Comput. **13**(7), 1443–1471 (2001)
77. Schölkopf, B., Smola, A.J.: Learning with Kernels—Support Vector Machines. Optimization, and Beyond, Regularization. MIT Press (2002)
78. Schowe, B., Morik, K.: Fast-ensembles of minimum redundancy feature selection. In: Okun, O., Valentini, G., Re, M. (eds.) Ensembles in Machine Learning Applications, pp. 75–95. Springer (2011)
79. Siebes, A., Vreeken, J., van Leeuwen, M.: Item sets that compress. In: Proceeding of the 6th SIAM International Conference on Data Mining, pp. 395–418 (2006)
80. Smola, A., Schölkopf, B.: A tutorial on support vector regression. Stat. Comput. **14**(3), 199–222 (2004)
81. Spengler, T., Geldermann, J., Hähre, S., Sieverdingbeck, A., Rentz, O.: Development of a multiple criteria based decision support system for environmental assessment of recycling measures in the iron and steel making industry. J. Clean. Prod. **6**(1), 37–52 (1998)
82. Srikant, R., Agrawal, R.: Mining sequential patterns: Generalizations and performance improvements. In: Proceedings of the 5th International Conference on Extending Database Technology. LNCS, vol. 1057, pp. 3–17. Springer, London, UK (1996)
83. Stolpe, M., Bhaduri, K., Das, K., Morik, K.: Anomaly detection in vertically partitioned data by distributed core vector machines. In: Machine Learning and Knowledge Discovery in Databases. Springer (2013)
84. Stolpe, M., Morik, K.: Learning from label proportions by optimizing cluster model selection. In: Machine Learning and Knowledge Discovery in Databases. LNCS, vol. 6913, pp. 349–364. Springer, Berlin, Heidelberg (2011)
85. Stolpe, M., Morik, K., Konrad, B., Lieber, D., Deuse, J.: Challenges for data mining on sensor data of interlinked processes. In: Proceeding of the Next Generation Data Mining Summit (NGDM) (2011)
86. Tax, D.M., Duin, R.P.: Support vector data description. Mach. Learn. **54**(1), 45–66 (2004)
87. Teffer, D., Hutton, A., Ghosh, J.: Temporal distributed learning with heterogeneous data using Gaussian mixtures. In: IEEE 11th International Conference on Data Mining Workshops (ICDMW), pp. 196–203 (2011)
88. Tsang, I., Kwok, J., Cheung, P.M.: Core vector machines: fast SVM training on very large data sets. J. Mach. Learn. Res. **6**(1), 363–392 (2005)
89. Xu, L.F., Li, W., Zhang, M., Xu, S.X., Li, J.: A model of basic oxygen furnaceBOF end-point prediction based on spectrum information of the furnace flame with support vector machine (SVM). Optik—Int. J. Light Electron Opt. 594–598 (2011)
90. Ye, L., Keogh, E.: Time series shaplets: A new primitive for data mining. In: Proceeding of the 15th ACM SIGKDD International Conference on Knowledge Discovery and Data Mining, pp. 947–956. ACM, New York, NY, USA (2009)
91. Yu, F., Liu, D., Kumar, S., Jebara, T., Chang, S.F.: $\propto$-SVM for learning with label proportions. arXiv:1306.0886 (2013)
92. Yu, H., Vaidya, J., Jiang, X.: Privacy-preserving SVM classification on vertically partitioned data. In: Proceedings of the 10th Pacific-Asia Conference on Advances in Knowledge Discovery and Data Mining (PAKDD), pp. 647–656. Springer, Berlin, Heidelberg (2006)
93. Yunhong, H., Liang, F., Guoping, H.: Privacy-preserving SVM classification on vertically partitioned data without secure multi-party computation. In: 5th International Conference on Natural Computation (ICNC), vol. 1, pp. 543–546 (2009)

Relational Learning for Sustainable Health

Sriraam Natarajan, Peggy L. Peissig and David Page

Abstract Sustainable healthcare is a global need and requires better value–better health–for patients at lower cost. Predictive models have the opportunity to greatly increase value without increasing cost. Concrete examples include reducing heart attacks and reducing adverse drug events by accurately predicting them before they occur. In this paper we examine how accurately such events can be predicted presently and discuss a machine learning approach that produces accurate such predictive models.

1 Introduction

Details of sustainable processes likely look very different for varying aspects of the interrelated issues of energy, transportation, health, food production, housing, commerce, and government. Therefore, disparate creative processes are likely needed in order to develop different aspects of a sustainable planet in what many scholars are now calling the "anthropocene" era—the era in which human decision-making is the single largest influence on the state of the planet. Are there any guiding principles for these disparate processes? We claim there are at least two: prediction and evaluation. To develop a sustainable process, we must be able to predict the effects of current and proposed processes on scales varying from microscopic to global, and beyond.

S. Natarajan (✉)
School of Informatics and Computing,
Indiana University, Bloomington, USA
e-mail: natarasr@indiana.edu

P.L. Peissig
Center for Human Genetics, Marshfield Clinic Research Foundation,
Marshfield, USA
e-mail: Peissig.Peggy@mcrf.mfldclin.edu

D. Page
Department of Biostatistics and Medical Informatics,
University of Wisconsin-Madison, Madison, USA
e-mail: page@biostat.wisc.edu

J. Lässig et al. (eds.), *Computational Sustainability*,
Studies in Computational Intelligence 645,
DOI 10.1007/978-3-319-31858-5_11

Once a process is implemented, we need to evaluate how accurate our predictive models are, how well the process is working, and the extent to which we need to revise our predictive models and the process itself. This paper studies these issues of prediction and evaluation specifically in the context of sustainable health.

For an individual, the most important health-related predictions are: "What diseases or events am I likely to suffer if I continue living as I do," "What changes (lifestyle or treatment decisions) can I make to possibly improve this future," and "What other events might these changes cause, both desirable and undesirable?" For a healthcare system, a nation, or the world, the questions are similar but involve more focus on cost. Notice, though, that cost for a nation or the planet necessarily takes into account providing good care to individuals, because good health should result in higher satisfaction and productivity. Important questions are: "What is healthcare likely to cost us over the next 10 years operating as we do," "What changes can we make to increase profit—reduce costs while increasing health and income?" and "What unanticipated effects could these changes have?" For questions of the first type, we show that we can use existing machine learning algorithms to analyze existing electronic health record (EHR) and clinical data to answer reasonably accurately in some cases. Nevertheless, questions of the second and third type require reasoning about causality. We do not focus on causality in this chapter.

From a clinical perspective, accurate predictive models of major health events have many more potential applications. First, such models can be incorporated into the EHR to provide prompts to clinicians such as, "your patient is at high risk for an heart attack and is not currently on an aspirin regimen." Second, the models themselves can be inspected to identify surprising connections, such as a correlation between the outcome and the use of certain drugs, which might in turn provide important clinical insights. Third, these models can be used in research to identify potential subjects for research studies. For example, if we want to test a new therapy for its ability to prevent an event such as heart attack, it would be most instructive to test it in a population of high-risk subjects, which a predictive model can accurately identify.

We focus on three different clinical prediction problems that can have a long-term impact on developing sustainable health care. First is the problem of predicting cardiovascular events in older adults by using the data from their youth. It allows the subject to take control of their cardiovascular health early in adulthood thus preventing serious risks later. Developing plans early to mitigate or minimize the risk will significantly reduce the costs associated with the treatment for the individual and society. Second is the problem of predicting if a subject is at risk for Alzheimer's disease. This prediction is performed using MRI images of the subjects' brain. As with the previous case, detecting early if a patient will potentially have Alzheimer's disease can have a significant impact in reducing treatment costs and potentially improve the cognitive ability of the subject. The final problem is that of predicting adverse events of drugs from Electronic Health Records (EHRs). If made with reasonable accuracy, this prediction has the potential of developing patient specific plans (i.e., personalized medicine) by considering their medical history thus reducing costs associated with specialized tests. Mining adverse events from EHRs can also potentially shape clinical trials to focus on specific diseases making it possible

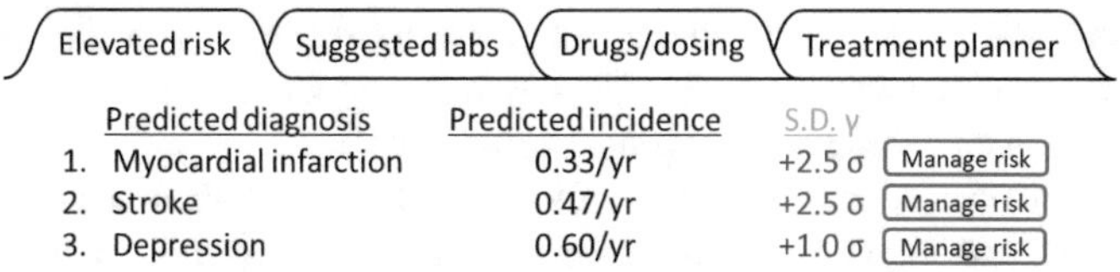

Elevated risk | Suggested labs | Drugs/dosing | Treatment planner

	Predicted diagnosis	Predicted incidence	S.D. γ	
1.	Myocardial infarction	0.33/yr	+2.5 σ	Manage risk
2.	Stroke	0.47/yr	+2.5 σ	Manage risk
3.	Depression	0.60/yr	+1.0 σ	Manage risk

Fig. 1 This figure shows a possible future EHR interface for the physician that includes AI predictions and recommendations. The system suggests that the patient is at elevated risk for specific diagnoses. In other tabs the system recommends the collection of additional health information such as laboratory assays, provides optimal drug regimens, and reports details on temporally-extended treatment plans

to reduce major events that might otherwise prove costly to the subject in particular and society in general.

We envision immense potential for such prediction problems and present a potential interface for EHRs of the future in Fig. 1. When a patient meets with physicians and has a diagnosis, the system can potentially make predictions based on the patient's medical history and current lab measurements, then present these predictions unobtrusively to the physician. In addition, it can also potentially suggest further lab tests, drugs and doses and even assist in developing patient-centric treatment plans. This potential scenario arguably reduces the costs associated with developing treatments.

In this chapter, we focus on employing advanced machine learning methods that operate on relational data. Collectively known as *Statistical Relational Learning (SRL)* [15], these methods go beyond the standard learning methods by relaxing the need for a single feature vector and allow for modeling complex relationships between objects of interest. The advantage of these models is that they can succinctly represent probabilistic dependencies among the attributes of different related objects leading to a compact representation of learned models. The presence of rich, relational noisy data in health care problems motivates the use of SRL models. Along with exhibiting superior empirical performance, these methods also result in easily interpretable models and perform automatic feature selection.

The rest of the chapter is organized as follows: after explaining the need for relational models, we consider the three case studies—predicting early onset of cardiovascular events from a clinical study data, predicting occurence of Alzheimer's from MRI data and predicting adverse drug events from EHRs. We conclude by summarizing the lessons learned and motivating future research directions.

2 Need for Richer Analysis

Why statistical relational analysis? Consider a classic Artificial Intelligence dream of developing a clinical decision-support system that can aid a clinician in developing personalized treatment plans for patients. Such a system must represent and reason using EHRs that include data about drugs, diagnoses, lab tests, imaging data such as CT scans, MRIs, and even genetic data. Conventionally, one could design such a system using either statistical AI [28] (such as support vector machines [5]) or logical AI (such as inductive logic programming [24]).

Purely statistical approaches [1, 5, 28, 29, 40] assume data is in the form of feature vectors, forcing homogeneity on the data and ignoring their natural representation [14, 16, 23]. Real-world domains such as health care problems, in fact, contain inherent structure—they are multimodal (e.g., lab tests, prescriptions, patient history) and highly relational. A flat feature representation for learning is highly limiting as it ignores structure and does not faithfully model this task. Advances in logical AI address these issues by explicitly representing structure through trees, graphs, etc., and reasoning with propositional and first-order logic [24, 38] in a principled manner. However, these approaches assume a deterministic, noise-free domain and cannot faithfully model noise or uncertainty. Again, in the decision-support system example, patient data is noisy and can even be incomplete or missing. The world is relational and noisy, and this is inherent in the tasks we model everyday: they consist of objects with diverse and non-uniform properties, which interact with each other in complex and noisy ways. Such domains are ubiquitous: information mega-networks, linked open data and triple stores, streaming data, heterogeneous bibliographic, organizational and social networks, drug-disease-gene interactions, complex molecules, human behavior and so on.

In recent years, Statistical Relational Learning (SRL) methods [15] (present at the top corner of Fig. 2) have been proposed that combine the expressiveness of first-order logic and the ability of probability theory to handle uncertainty. SRL approaches, unlike what is traditionally practiced in statistical learning, seek to avoid explicit state enumeration, through a symbolic representation of states. The advantage of these models is that they can succinctly represent probabilistic dependencies among

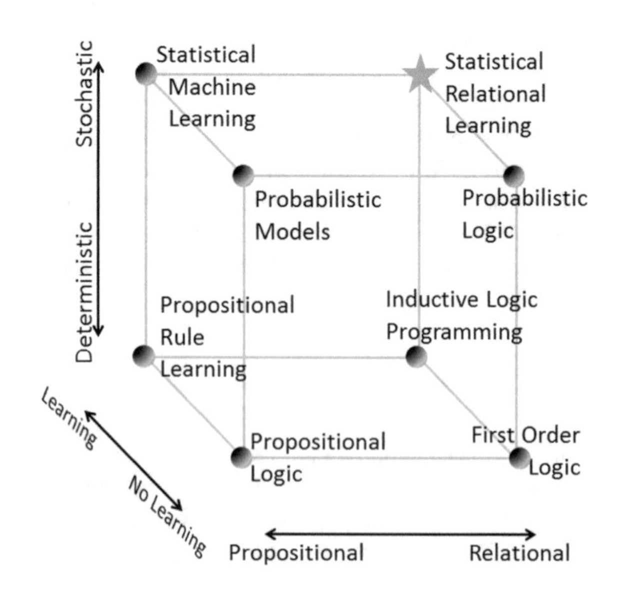

Fig. 2 Landscape of Artificial Intelligence research w.r.t three dimensions—learning, representation and uncertainty

the attributes of different related objects leading to a compact representation of learned models that allow sharing of parameters between similar objects. Given that the health care data is highly relational (multiple measurements and multiple data types corresponding to every patient) and temporal, these methods are very powerful in building predictive models for health care data.

As an example, consider a statement such as *Friends have similar smoking habits*. In any population, every person has a different number of friends. To model this using classical methods, we need to consider the specific identities of the individuals. On the other hand, it can be simply captured using a rule in predicate logic:

$$\forall x, y Friends(x, y) \Rightarrow (Somkes(x) \iff Smokes(y)) \tag{1}$$

The above equation specifies that if x and y are friends, then they have exactly the same smoking habit. Of course, since this is not always true, we can "soften" this rule by adding a numeric weight. Higher the weight, higher is the probability of the rule being true in the domain. An example of such a formulation is Markov Logic Networks [10]. There are efficient learning algorithms for learning these weights (parameters) and the rules themselves.

There are several justifications for adopting statistical relational analyses for the purposes of analyzing health care data. First, the data consists of several diverse features (e.g., demographics, psychosocial, family history, dietary habits) that interact with each other in many complex ways making it *relational*. Without extensive feature engineering, it is difficult—if not impossible—to apply propositional approaches to such structured domains. Second, the data could have been collected as part of a longitudinal study, i.e., over many different time periods such as 0, 5, 10, years etc., making it *temporal*. Third, most data sets from biomedical applications, contain missing values i.e., all data are not collected for all individuals. Fourth, the nature of SRL algorithms allows more complex interactions between features. Clearly, a single, flat feature representation will not suffice here. Fifth, the learned models must be generalized across different sub-groups of populations and across different populations themselves. Finally, the relational models are very easily interpretable and hence enable the physician and policy-maker to identify treatable risk factors and plan preventative treatments.

Next, we present the three case studies where relational methods have led to highly predictive models and provide reason and hope for sustainable health care.

3 Case Study I—Predicting Early Onset of Cardiovascular Conditions

The broad long-term objective of this case study is to reduce deaths and negative health consequences of cardiovascular diseases (CVD) which comprehensively include diseases related to the heart and blood vessels. Coronary heart disease (CHD) is a major cause of death and illness worldwide. Successful and established lifestyle intervention can prevent the development of risk factors especially when applied early in life. The National Heart, Lung and Blood Institute's Coronary Artery Risk Development in Young Adults (CARDIA) Study is a longitudinal cohort study with 25 years of data examining the development and determinants of clinical and sub-clinical cardiovascular disease and its risk factors in black and white Americans. CARDIA study is a longitudinal study of cardiovascular risk factors that began in 1985–86. There were several risk factors measured in different years (2, 5, 7, 10, 15, 20) respectively. The purpose of this project is to understand the relationship between the measured risk factors and the development of CVD and overall plaque burden. As the cohort ages and sufficient clinical events occur, this work will allow us to apply state-of-the-art machine learning techniques to hard clinical events such as heart attack, heart failure and premature death. In particular, our prior work [33] uses the longitudinal data collected from the CARDIA participants in early adult life (ages 20–50 years), to develop machine learning models that can be used to predict the Coronary Artery Calcification (CAC) amounts, a measure of subclinical CAD, at years 25 given the measurements from the previous years. CAC is a measure of advanced atheroma and has previously been demonstrated to add to risk factor in the prediction of heart attack in men, women and four major ethnic groups [8]. We present this work briefly in this section.

We used known risk factors such as *age, sex, cholesterol, bmi, glucose, hdl level and ldl level of cholesterol, exercise, trig level, systolic bp and diastolic bp* that are measured between years 0 and 20 across the patients. Our goal is to predict if the CAC-levels of the patients are above 0 for year 20 given the above mentioned factors over all the years. Predicting the CAC-levels for year 20 using the measurements from previous years allows us to identify sub-groups of populations that are to be monitored early and identified for treatment and counseling. Any CAC-level over 0 indicates the presence of advanced coronary atheroma and elevated risk for future heart disease. So, we are in a binary classification setting of predicting 0 versus non-0 CAC levels. In our data set, most of the population had CAC-level of 0 (less than 20 % of subjects had significant CAC-levels) in year 20. Hence, there is a huge skew in the data set where there is a very small number of positive examples.

We used data from 3600 subjects and performed 5-fold cross validation. We compare the results (area under the curve of the ROC curves) of the SRL methods (presented below) against traditional regression methods such as linear and logistic regression. We also compare against standard machine learning methods such as Naive Bayes [20], Support Vector Machines [6], Decision trees [36] (J48) and a propositional boosting method (AdaBoost) [11]. We employed two versions of SRL models that have been proven to be very successful.

In the first version, we learned relational regression trees (RRTs) [2]. These trees upgrade the attribute-value representation used within classical regression trees. Each RRT can be viewed as defining several new feature combinations, one corresponding to each path from the root to a leaf. These regression trees are learned directly from relational data. In the second version, we boosted a set of RRTs using functional-gradient boosting (RFGB) [9, 12, 18, 32, 34]. The benefits of a boosting approach

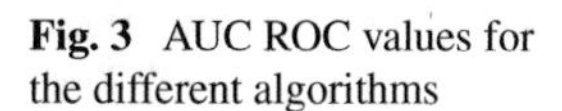

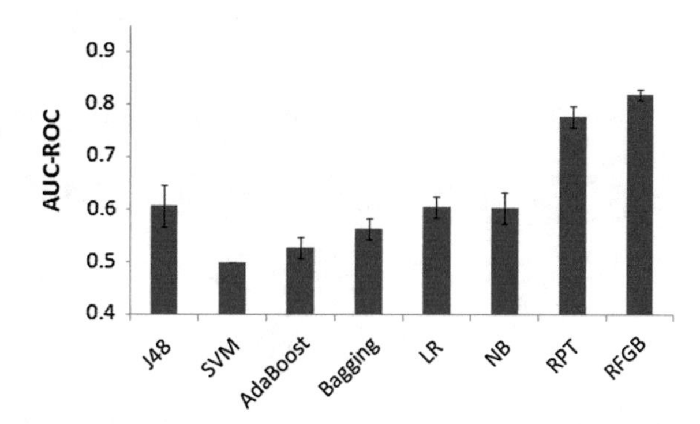

Fig. 3 AUC ROC values for the different algorithms

are: First, being a nonparametric approach the number of parameters grows with the number of training episodes. In turn, interactions among random variables are introduced only as needed, so that the potentially large search space is not explicitly considered. Second, such an algorithm is fast and straightforward to implement. Existing regression learners can be used to deal with propositional, continuous, and relational domains in a unified way. Third, the use of boosting for learning SRL models makes it possible to learn the structure and parameters simultaneously, which is an attractive feature as structure learning in SRL models is computationally quite expensive. Finally, given the success of ensemble methods in machine learning, it can be expected that our method is superior in predictive performance across several different tasks compared to the other relational probabilistic learning methods. As we had demonstrated in our earlier work [21, 32, 34] such a learning method has been successfully employed in social network prediction, citation analysis, movie ratings predictions, discovering relationships, learning from demonstrations, and natural language processing tasks. We employed this highly successful technique for learning to predict CAC levels in adulthood.

This is essentially a "rediscovery experiment" in that it uses known risk factors for predicting CAC levels. Preliminary results on CAC prediction task are presented in Fig. 3. Measuring accuracy over the entire data set can be misleading [19], hence, we also compute the area under the curve for the Receiver Operating Characteristics curv (AUC-ROC). The AUC-ROC has long been viewed as an alternative single-number measure for evaluating the predictive ability of learning algorithms. This is because the AUC-ROC is independent to the decision threshold and invariant to the priors on the class distribution. As can be easily observed, the relational methods (RPT and RFGB) have a superior performance over the rest of the methods. In particular, the gradient boosting method (RFGB) exhibits more than 20 % increase over the best standard method. Our results are consistent with the published results [4, 11, 37] in that with approximately 25 trees, we can achieve the best empirical performance. We refer to our work [33] for more details.

Figure 4 presents a part of one tree learned. We are not presenting the entire tree and indicate the missing branches by dots. The first argument *a* of every predicate is the subject's ID and the last argument of every predicate (except *sex*) indicates the

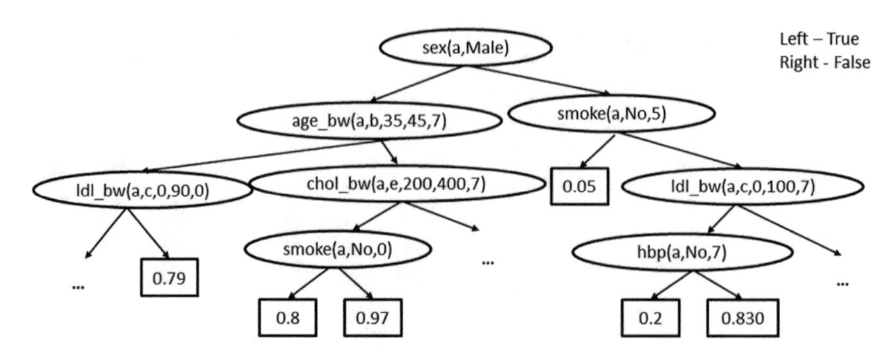

Fig. 4 Learned Tree for predicting CAC-level greater than 0. The leaves indicates $P(cac(a)) > 0$. The left branch at any test corresponds to test returning *true* while the right branch corresponds to *false*

year of measurement. The left branch out of every node is the *true* branch, the right branch the *false* branch. The leaves indicate the probability of CAC-level (say p) being greater than 0. We use *_bw* in predicates to indicate that the value of a certain variable is between two values. For instance, $ldl\_bw(a, b, 0, 100, 10)$ indicates that the LDL level of the person a is b and is between 0 and 100 in year 10. The leaves indicate the probability (p) of that subject having a non-zero CAC level in year 20. For example, the left branch states that if a person is male, he is of middle age in year 7 (i.e., between 35 and 45 years) and has a high ldl level, $p = 0.79$. Similarly the right branch indicates that if the subject is a female and has not smoked in year 5, $p = 0.05$.

We performed an additional experiment—we first used only year 0 data and learned a single tree. Now using this tree, we learned the second tree using year 5 data and so on. So the goal is to see how AUC-PR changes with adding more observations in future yeas and can be seen as the progress of the risk factor over time. The results are presented in Fig. 5 (solid). As expected from the previous experiment, year 0 has a substantial leap and then adding individual years increases performance until year 7 and then plateaus beyond that. This is again a significant result. Our initial results show that beyond ages 25–37 of a person, there is not much significant information from the risk factors.

Only family history and drug use data: Since the above experiments were essentially "rediscovery experiments", we were interested in finding how non-standard risk factors such as family history and drug use can affect the CAC-levels i.e., can we unearth a new discovery? Thus, we used only these two sets of features in our next experiment. These diverse features included the age of the children, whether the participant owns or rents a home, their employment status, salary range, their smoking and alcohol history, etc. There were approximately 200 such questions that were considered. Again, as with the previous case, we used the 3600 subjects and performed 5-fold cross validation. Initial experiments showed that we were able to predict the CAC-levels reasonably with an AUC-ROC value of around **0.75**. An example rule

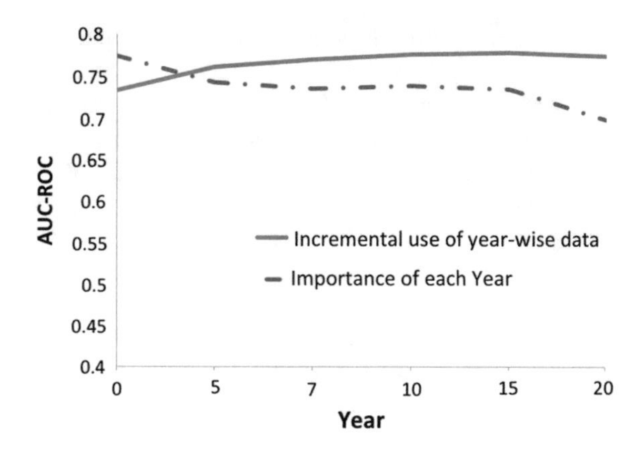

Fig. 5 The impact of the measurements in different years in the CAC-level at year 20

says that if the person was in school 5 years earlier, had smoked regularly in their 20s, was not paying for their residence then and had children at home, then their chance of having a significant CAC-level is higher than someone who never smoked and was married during that time. While the results are preliminary, they reveal **striking socioeconomic impacts on the health state of the population**, factors that have long been speculated on, but which can now be conclusively quantified. Future investigations are necessary to uncover more informative and correct relationships between social and economic factors on the cardiovascular health state of populations.

Application to a Real EHR: We have made an initial attempt of evaluating our algorithm on a real EHR [44, 45]. EHRs are an emerging data source of great potential use in disease prevention. An EHR effectively tracks the health trajectories of its patients through time for cohorts with stable populations. We analyzed de-identified EHR data on 18,386 subjects enrolled in the Personalized Medicine Research Project (PMRP) at Marshfield Clinic [26, 27]. The PMRP cohort is one of the largest population-based bio-banks in the United States and consists of individuals who are 18 years of age or older, who have consented to the study and provided DNA, plasma and serum samples along with access to their health information in the EHR. Most of the subjects in this cohort received most, if not all, of their medical care through the Marshfield Clinic integrated health care system. We included major risk factors such as cholesterol levels (LDL in particular), gender, smoking status, and systolic blood pressure, as well as less common risk factors such as history of alcoholism and procedures for echocardiograms and valve replacements. The best cross-validated predictor of primary MI according to AUC-ROC was the RFGB model as with the earlier case. It is of note that the RFGB and RPT models significantly outperformed their direct propositional analogs (Boosted Tree and Tree models, respectively) emphasizing the need for richer relational models for such challenging tasks.

Summary: In the U.S., heart disease is responsible for approximately one in every 6 deaths with a coronary event occurring every 25 s and about 1 death every minute based on data current to 2007 [39]. Heart diseases are the number one cause of death and disability. This results in an estimated annual U.S. expenditure of $425 billion. This case study provides an opportunity to investigate the possibility of predicting later cardiac events by analyzing the young adulthood data. Our experiments reveal that the relational learning methods are more suited for this task due to their ability to handle multi-relational data. It also appears that the risk factors from the early adulthood of the subjects seem to be the most important ones in predicting risks at later years. This allows the populations to take control of their cardiovascular health and develop preventive treatment plans that avoids potential cardiovascular events later. Initial experiments with socioeconomic risk factors appear to demonstrate that these risk factors are as predictive as clinical risk factors. This opens up another potential avenue for developing sustainable health care for cardiovascular risks—improving the living standards of the society as a whole.

4 Case Study II—Predicting Mild Cognitive Impairment

Alzheimer's disease (AD) is a progressive neurodegenerative condition that results in the loss of cognitive abilities and memory, with associated high morbidity and cost to society [42]. Accurate diagnosis of AD, as well as identification of the prodromal stage, mild cognitive impairment (MCI) is an important first step towards a cure and has been a focus of many neuroimaging studies. Structural MRI has been widely used to identify changes in volume and size of specific brain regions, as well as regional alterations in gray matter, white matter and cerebrospinal fluid (CSF) on a voxel-by-voxel basis [46]. Recently, there are several approaches that either employ network analysis [42, 43] or machine learning [3, 46] on the voxel data. These approaches, however, only consider the binary classification problem, that of AD versus CN (cognitively normal), in which a clear decision boundary between these categories can be easily obtained. In reality, the progression to Alzheimer's disease is a continuum, with subjects spanning different stages from being normal to MCI to AD, making classification much more difficult. In fact, this distinction is most important, as identifying the subjects who are MCI but at a higher risk of moving to AD can potentially have a high impact in developing sustainable health care plan for these subjects. This is more effective compared to waiting until the onset of AD to begin treatment for the individual.

We have recently proposed a pipeline approach that performs three-way classification—AD versus MC versus CN [30, 31]. The pipeline is presented in Fig. 6 and consists of three stages—first is the *MRI segmentation stage* that takes volumetric brain MRI data as an input and segments the brain into clinically relevant regions. Second is a *relational learning stage* that considers the segmented brain to be a graph to build a series of binary classifiers. The final stage is the *combination stage* that combines the different classifiers. The idea underlying this pipeline is simple and is based on the idea of classical mixture of experts: rather than choose a single segmentation technique, we combine multiple segmentation techniques and different imaging data. In our previous work, we used two different

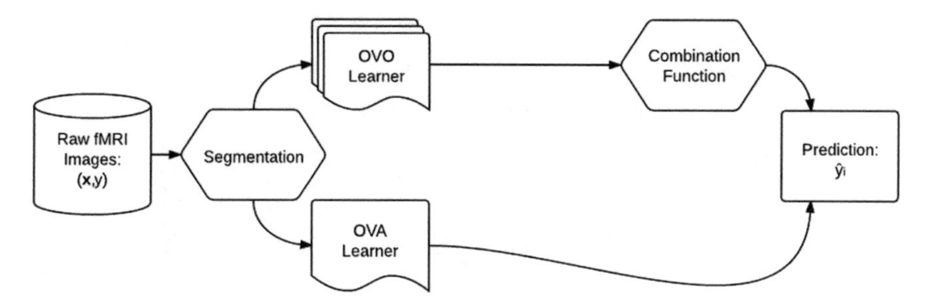

Fig. 6 Graphical representation of the pipeline

types of segmentation algorithms—atlas based segmentation (AAL) which divides the brain into 116 regions and EM [7] based unsupervised segmentation that could result in different number of segments for different subjects depending on their brain characteristics.

In order to handle the relational (graph) data, we employed the previously outlined RFGB algorithm [34]. Given the importance of the brain network connectivity in identifying AD, this particular SRL algorithm becomes a natural choice due to its ability to model relations such as neighborhood information and the fact that we learn the parameters and structure of the graphical model simultaneously. Note that if we employ a propositional classifier, we have to assume that all the subjects have equal number of segments, which is not the case in EM segmentations. As illustrated, our methods outperform propositional classifiers. Also, the ability to use domain knowledge is one of the attractive features of SRL algorithms and is an essential attribute from a medical imaging perspective since the knowledge gained from decades of medical research can be incredibly useful in guiding learning/mining algorithms.

SRL approaches based on first-order logic mostly employ predicate logic that essentially performs binary classification. Our goal, on the other hand, is the more challenging three-way classification. We took the popular approach of converting this classification as a series of binary classification tasks (i.e., AD vs. CN, AD vs. MCI and MCI vs. CN) also called One-versus-One (OvO) classification approach [13, 22]. The results are compared against a One-versus-all strategy (OvA) where a classifier is learned for each class separately and each class is discriminated from the others. The key idea in OvO is to divide the multi-class classification problem into a series of binary classification problems between pairs of classes, then combine the outputs of these classifiers in order to predict the target class. We use SRL-based classifiers for each binary classification and later combine them using a few different techniques (weighted combination, a meta-classifier, etc.). The results are compared against an OVA where a classifier is learned for each class separately and each class is discriminated from the other classes.

We evaluate the pipeline on a real-world dataset, namely the Alzheimer's Disease Neuroimaging Initiative (ADNI) database of 397 subjects. It should be mentioned that in the experiments we report no subject selection took place (i.e., we did not carefully choose the subjects for the study) and instead we used the complete set of

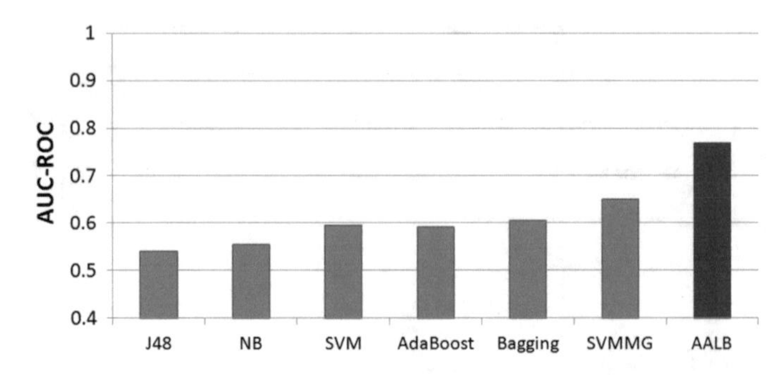

Fig. 7 Classification performances in terms of "Area under the ROC curve" of the different algorithms: **a** propositional classifiers (*blue*) compared against the relational AALB (*red*) algorithm

subjects. This particular group was selected based upon having both structural MRI and functional metabolic positron emission tomography data as part of a separate study. Similarly, we do not employ a careful feature selection but rather simply use resulting average tissue-type volume measurements obtained from the segmentation algorithms as features for our classification. Our results demonstrate that we have comparable or better performance than the current methods based upon individual binary and collective classification tasks with minimal feature engineering.

To illustrate the need for relational models, we compared the propositional classifiers with AAL segmentation. We now present the results in Fig. 7. All these classifiers used AAL segmentation to construct a feature vector and then performed classification using one of the standard machine learning methods (J48—decision tree, NB—Naive Bayes, SVM—Support Vector Machine, AdaBoost and Bagging—ensemble classifiers). We used Weka's multi-class classification setting. For each of the classifiers, we used 5-fold CV to select the best parameters. As can be seen, the propositional algorithms do not show a good performance when compared to AALB which is the SRL method (RFGB) used on top of the AAL segmentation method. We also present the results of running LibSVM on the voxel data (i.e., without any segmentation—SVMMG). The results demonstrate that the performance is slightly better but still is not comparable to the performance of AALB. As with the previous case study, we present AUC-ROC instead of accuracy.

Recall that we converted a 3-class classification problem into 3 binary classification problems. To understand how the successful methods performed on individual classification tasks (AD vs. CN, AD vs. MCI, MCI vs. CN), we also present the confusion matrices in Table 1. We include a single confusion matrix for each of the three binary classifiers that used RFGB as their classifier while employing AAL segmentation. These matrices clearly show that, while we can achieve a relatively high true positive rate (TPR) and true negative rate (TNR) when classifying AD v CN and AD v MCI, classification of MCI v CN is a more difficult task. Hence, we see a proportionally larger number of false negatives in the third confusion matrix. It appears that while we are tackling the difficult challenge of 3-class classification, it

Table 1 Confusion matrices for the three classifiers

Confusion matrices						
	AD v CN		AD v MCI		MCI v CN	
	Pos	Neg	Pos	Neg	Pos	Neg
Pos	64	18	27	60	149	44
Neg	16	86	30	168	76	26

also helps in the two-class classification case. More precisely, learning in the more difficult task helps the classifiers to improve on the less difficult task.

Summary: The total costs of care of Americans with Alzheimers above the age of 65 is projected to increase to $1.08 trillion per year in the next 40 years.[1] Hence, it is crucial to identify subjects that are at risk for Alzheimer's early and isolating the subjects who appear to have MCI currently is a first step in this process. To this effect, we have shown that relational models are quite effective in isolating the MCI subjects. The next logical step is in estimating the number of these MCI subjects who would go on to develop Alzheimer's. This is a challenging problem as the human expertise in predicting this is still not fully realized. Nonetheless, this problem has potentially a very important effect in developing a sustainable society whose costs are balanced between the treatment to the individual patients and keeping the costs to the society as low as possible. Identifying and developing patient centric Alzheimer's treatment plans can largely contribute towards realizing this goal.

5 Case Study III—Predicting Adverse Drug Events

Adverse drug events (ADEs) are estimated to account for 10–30 % of hospital admissions, with costs in the United States alone between 30 and 150 billion dollars annually [25], and with more than 180,000 life threatening or fatal ADEs annually, of which 50 % could have been prevented [17]. Although the U.S. Food and Drug Administration (FDA) and its counterparts elsewhere have preapproval processes for drugs that are rigorous and involve controlled clinical trials, such processes cannot possibly uncover every aspect of a drug. While a clinical trial might use only a thousand patients, once a drug is released on the market it may be taken by millions of patients. As a result, additional information about possible risks of use is often gained after a drug is released on the market to a larger, more diverse population.

This particular task poses several key challenges to machine learning techniques that require development of more advanced relational learning techniques—(1) The adverse events associated with the drugs are not known in advance (unanticipated adverse events) and hence it is not possible to learn from labelled data. (2) The

[1] http://www.alz.org/alzheimers_disease_trajectory.asp.

data is multi-relational since they are learned from EHRs. (3) The data is non-i.i.d., as the data cannot be represented using a fixed feature vector. (4) The number of features associated with a data point are arbitrary—the number of diagnosis, visits and prescriptions. (5) Finally, there is a necessity to explicitly model time. Some drugs taken at the same time can lead to events while in some cases, drugs taken after one another cause adverse events.

The task of identifying previously unanticipated ADEs is similar to an unsupervised learning task: without a hypothesized ADE, how can we run a supervised learner? As an example, without knowing in advance that Myocardial Infraction (MI) is an ADE for Cox2 inhibitors, how can we provide supervision such that the algorithm will predict that MI risk is raised by these drugs? In our prior work [35], we show that the problem can be addressed by running supervised learning in reverse, to learn a model to predict who is on a Cox2 inhibitor (Cox2ib). This seems counterintuitive, but if we can predict some subgroup of Cox2ib patients based on the events occurring after they start Cox2ib, this can provide evidence that the subgroup might be sharing some common effects of Cox2ib. When addressed by covering-based Inductive Logic Programming systems such as Aleph [41], it is equivalent to relational subgroup discovery. We anticipate this same reverse ML approach can also be applied to other situations where the real class variable of interest is not measured. We referred to this as *reverse machine learning*.

Suppose we did not know that Cox2ib doubled the risk of MI, but we wondered if these drugs had any associated ADE. Our reverse ML approach is to treat patients on Cox2 inhibitors as our cases, or positive examples, and to treat age- and gender-matched controls as the negative examples. Specifically, for each positive example, a control is a patient of the same age and gender who is not on a Cox2ib. (Controls could be selected to be similar to the cases in other ways age and gender are just the most common such features in clinical studies). For example, if we have on the order of 200 positive (P) patients who suffer an MI, we expect on the order of 100 negative (N) patients who suffer an MI. The following rule would have a strong score of $P - N = 100$ and hence would be returned by Aleph unless a different rule scores even better.

cox2ib(Patient) IF mi(Patient)

This rule says that a patient was likely on a Cox2ib if they suffered an MI.

Another advantage of relational subgroup discovery, compared with ordinary subgroup discovery, is that the body (precondition) of the rule does not have to be a single condition, but it could be combination of conditions and lab results, possibly in a particular temporal order. Hence, in principle ADEs can be discovered that do not neatly correspond to an exact pre-existing diagnosis code. Furthermore, the body of the rule can involve other drugs, so that in principle ADEs caused by drug interactions can be captured. For example, it has recently been observed that patients on Plavix have an increased risk of stroke (ordinarily prevented by Plavix) if they are also on Prilosic. This can be represented by the following rule plavix(Patient) IF prilosic(Patient) $\wedge$ stroke(Patient)

We evaluated the approach on Marshfield Clinics Personalized Medicine Research Project [27] (PMRP) cohort consisting of approximately 19,700+ subjects. The PMRP cohort included adults aged 18 years and older who reside in the Marshfield Epidemiology Study Area. Marshfield has one of the oldest internally developed EHRs (Cattails MD) in the US, with coded diagnoses dating back to the early 1960s. Prior to running Aleph, the de-identified EHR database was converted into Prolog format. For analysis, medicated subjects were identified (referred to as positive cases **P**) by searching through the medications database and selecting those subjects having an indicated use of Cox2ib. An equal number of subjects (referred to as negative controls **N**) were randomly selected from a pool of controls that were matched to the cases based on gender and age. The controls could not have any indicated use of Cox2ib drug in the medications database.

We ran on two drugs, Warfarin and Vioxx. For Warfarin the approach easily rediscovered the known ADE of bleeding, together with the common treatment for Warfarin-induced bleeding (Phytonadione, or Vitamin K1).

$$warfarin(X) \leftarrow bleeding(X, D1) \wedge phytonadione(X, D2) \wedge \ after(D1, D2)$$

Vioxx is a drug that was recalled from the market because it was found to double the risk of heart attack, or myocardial infarction (MI). Our aim was to test whether Aleph would uncover this link with MI if the link were unknown. Vioxx belongs to a larger class of drugs called Cox2 inhibitors. The overall goal was to identify possible ADEs caused by Cox2ib. In our reverse ML approach, the specific goal of the Aleph run was to learn rules to accurately predict which patients had an indicated use of Cox2ib. These rules would then be vetted by a human expert to distinguish which were merely associated with indications of the drug (diseases or conditions for which the drug is prescribed) and which constituted possible ADEs (or other interesting associations, such as off-label uses for the drug). We validated our methodology with a run in which only diagnoses are used and rules are kept as short as possible—one body literal (precondition) per rule.

Myocardial infarction (MI) is a known adverse event of Cox2ib we wanted to test if the method would uncover MI automatically. In Table 2, we show the ten most significant rules identified by Aleph for a single run. Note that the penultimate rule (highlighted) identifies the diagnosis of 410 (MI) as a possible ADE of Cox2. The fact that this ADE can be learned from data demonstrates that our method is capable of identifying important drug interactions and side-effects.

In some cases, a drug may cause an ADE that does not neatly correspond to an existing diagnosis code (e.g., ICD9 code), or that only occurs in the presence of another drug or other preconditions. In such a case, simple 1-literal rules will not suffice to capture the ADE. We now report a run in which all of the background knowledge was used, including labs, vitals, demographics and other drugs. Table 3 shows the top ten most significant rules. The use of ILP yields interpretable rules.

Table 2 Aleph rules generated for Cox2 inhibitor use (Single Diagnosis)

Rule	Pos	Neg	Total	P-value
diagnoses(A,_,'790.29','Abnormal Glucose Test, Other Abn Glucose',_)	333	137	470	6.80E-20
diagnoses(A,_,'V54.89','Other Orthopedic Aftercare ',_)	403	189	592	8.59E-19
diagnoses(A,_,'V58.76','Aftercare Foll Surg Of The Genitourinary Sys',_)	287	129	416	6.58E-15
diagnoses(A,_,'V06.1','Diphtheria-Tetanus-Pertussis,Comb(Dtp)(Dtap)',_)	211	82	293	2.88E-14
diagnoses(A,_,'959.19','Other Injury Of Other Sites Of Trunk ',_)	212	89	301	9.86E-13
diagnoses(A,_,'959.11','Other Injury Of Chest Wall',_)	195	81	276	5.17E-12
diagnoses(A,_,'V58.75','Aftercare Foll Surg Of Teeth, Oral Cav, Dig Sys',_)	236	115	351	9.88E-11
diagnoses(A,_,'V58.72','Aftercare Following Surgery Nervous Syst, Nec',_)	222	106	328	1.40E-10
diagnoses(A,_,'410','Myocardial Infarction',_)	212	100	312	2.13E-10
diagnoses(A,_,'790.21','Impaired Fasting Glucose ',_)	182	80	262	2.62E-10

Rule	+	-	
+	838	333	1171
-	987	1492	2479
	1825	1825	3650

Fisher's exact test indicated that many rules demonstrated a significant difference in identifying positive cases over chance.

Summary This case study resulted in several important lessons to be learned. First, learning in reverse is a good alternative for unsupervised learning problems—when the labels are not known in advance as in the case of adverse events. Censoring of the data based on the learning task is important. It is important to censor (omit) data about patients before they started the drug. Left censoring is important but not guaranteed that our learned models describe only ADEs. Finally, given the multi-relational nature of the data, it is necessary for algorithms that directly operate on relational data without flattening the data. Hence, relational methods are a natural choice for modeling such tasks which are essential in developing a sustainable society. Identifying potential side-effects of drugs in the first few years after they are released to a wider population has the possibility of reducing treatment costs, focusing clinical studies, reducing hospital admissions and preventing catastrophic events on individuals thus realizing the goal of sustainable health care.

Table 3 Aleph rules generated for Cox2 inhibitor use

Rule	Pos	Neg	Total	P-value
gender(A,‘Female’), hasdrug(A,_,‘IBUPROFEN’),	509	177	686	4.25E-38
diagnoses(A,_,‘305.1’,‘Tobacco Use Disorder’,_)				
diagnoses(A,B,‘462’,‘Acute Pharyngitis’,_),	457	148	605	1.27E-37
hasdrug(A,B,‘IBUPROFEN’)				
hasdrug(A,_,‘NORGESTIMATE-ETHINYL ESTRADIOL’),	339	88	427	8.12E-36
gender(A,‘Female’)				
diagnoses(A,_,‘V70.0’,‘Routine Medical Exam’,_),	531	199	730	1E-35
hasdrug(A,B,‘IBUPROFEN)				
diagnoses(A,B,‘724.2’,‘Lumbago’,_)	433	144	577	1.44E-34
diagnoses(A,_,‘462’,‘Acute Pharyngitis’,_),	502	186	688	2.02E-34
gender(A,‘Male’)				
diagnoses(A,_,‘89.39’,‘Nonoperative Exams Nec’,_),	415	135	550	4.12E-34
diagnoses(A,_,‘305.1’,‘Tobacco Use Disorder’,_)				
hasdrug(A,_,‘CYCLOBENZAPRINE HCL’), gender(A,‘Male’)	493	189	682	3.6E-32
hasdrug(A,_,‘FLUOXETINE HCL’)				
gender(A,‘Female’). l_observations(A,B,‘Calcium’,9.8),	487	189	676	3.28E-31
diagnoses(A,B,‘724.5’,‘Backache Nos’,_)				
diagnoses(A,_,‘V71.89’,‘Other Specified Suspected Condi10/00’,_),	492	193	685	5.35E-31
gender(A,‘Male’)				

Rule	+	-	
+	1729	708	2345
-	96	1119	1215
	1825	1825	3650

6 Discussion and Conclusion

For a sustainable society, it is important to develop personalized health plans for the population. In order to do so, it is crucial to develop methods that can predict the onset of critical events before they happen. In this chapter, we addressed three important such problems—that of predicting cardiovascular events years in advance, predicting the subset of population that needs to be monitored for the onset of Alzheimer's and predicting adverse events of drugs based on the attributes of the patients and the drug regiment that they are on. The three prediction problems all focus on isolating the set of subjects that are at risk rather than waiting for the event to happen and then develop a treatment plan. Planning after an event has occurred can prove to be costly to both the individual and society as a whole.

In order to perform such predictions, it is crucial to develop learning methods that can naturally handle multi-relational, uncertain, noisy and missing data. To this effect, we proposed the use of advanced relational learning algorithms including statistical relational learning that can faithfully model the rich structured data while being able to handle noise and uncertainty. We also demonstrated how a rule learning algorithm can be used to handle imperfect observations in EHRs. The experimental results were consistent across all the case studies—relational learning algorithms were able to outperform the propositional (standard) machine learning algorithm with minimal feature engineering. We claim that these initial results along with other results from our groups and others demonstrate the potential of these rich algorithms on developing models tailored to subgroups of populations possibly leading to realizing the grand vision of personalized medicine.

In the future, it is important to develop learning algorithms that handle multiple modalities of data—EHRs, drug tests, X-rays, MRI, CT scans, genetic information and natural language text. Machine learning algorithms in general and relational learning algorithms in particular have been applied to individual modalities in the past. They must be extended to handle sub-sets of the data types in order to make effective predictions. The lessons learned from one study or one group of a population must be validated across multiple groups to draw useful inferences that can be generalized across multiple populations. With successes in other fields when employing machine learning, it appears that it will not be long before we realize the grand vision of personalized medicine leading to sustainable health care practices.

References

1. Bishop, C.: Pattern Recognition and Machine Learning. Information Science and Statistics. Springer, Secaucus (2006)
2. Blockeel, H.: Top-down induction of first order logical decision trees. AI Commun. **12**(1–2), 119–120 (1999)
3. Chen, K., Reiman, E.M., Alexander, G.E., Bandy, D., Renaut, R., Crum, W.R., Fox, N.C., Rossor, M.N.: An automated algorithm for the computation of brain volume change from sequential mris using an iterative principal component analysis and its evaluation for the assessment of whole-brain atrophy rates in patients with probable Alzheimer's disease. Neuroimage **22**(1), 134–143 (2004)
4. Craven, M., Shavlik, J.: Extracting tree-structured representations of trained networks. In: NIPS, pp. 24–30 (1996)
5. Cristianini, N., Shawe-Taylor, J.: An Introduction to Support Vector Machines: and Other Kernel-Based Learning Methods. Cambridge University Press, New York (2000)
6. Cristianini, N., Shawe-Taylor, J.: An Introduction to Support Vector Machines and Other Kernel-based Learning Methods. Cambridge University Press, Cambridge (2000)
7. Dempster, A.P., Laird, N.M., Rubin, D.B.: Maximum likelihood from incomplete data via the EM algorithm. J. Roy. Stat. Soc. B. **39**(1), 1–38 (1977)
8. Detrano, R., Guerci, A.D., Carr, J.J., et al.: Coronary calcium as a predictor of coronary events in four racial or ethnic groups. N. Engl. J. Med. **358**, 1338–1345 (2008)
9. Dietterich, T.G., Ashenfelter, A., Bulatov, Y.: Training conditional random fields via gradient tree boosting. In: ICML (2004)

10. Domingos, P., Lowd, D.: Markov Logic: An Interface Layer for AI. Morgan & Claypool, San Rafael (2009)
11. Freund, Y., Schapire, R.E.: Experiments with a new boosting algorithm. In: ICML, pp. 148–156 (1996)
12. Friedman, J.H.: Greedy function approximation: a gradient boosting machine. Ann. Stat. **29**, 189–1232 (2001)
13. Galar, M., Fernández, A., Barrenechea, E., Bustince, H., Herrera, F.: An overview of ensemble methods for binary classifiers in multi-class problems: experimental study on one-vs-one and one-vs-all schemes. Pattern Recogn. **44**, 1761–1776 (2011)
14. Getoor, L., Friedman, N., Koller, D., Pfeffer, A.: Learning probabilistic relational models. In: Dzeroski, S., Lavrac, N. (eds.) Relational Data Mining (2001)
15. Getoor, L., Taskar, B.: Introduction to Statistical Relational Learning. MIT Press, Cambridge (2007)
16. Glesner, S., Koller, D.: Constructing flexible dynamic belief networks from first-order probabilistic knowledge bases. In: Froidevaux, C., Kohlas, J. (eds.) Proceedings of the European Conference on Symbolic and Quantitative Approaches to Reasoning and Uncertainty (ECSQARU'95), pp. 217–226. Springer, Berlin (1995)
17. Gurwitz, J.H., Field, T.S., Harrold, L.R., Rothschild, J., Debellis, K., Seger, A.C., Cadoret, C., Fish, L.S., Garber, L., Kelleher, M., Bates, D.W.: Incidence and preventability of adverse drug events among older persons in the ambulatory setting. JAMA **289**, 1107–1116 (2003)
18. Gutmann, B., Kersting, K.: Tildecrf: conditional random fields for logical sequences. In: ECML (2006)
19. Huang, J., Ling, C.X.: Using auc and accuracy in evaluating learning algorithms. IEEE Trans. Knowl. Data Eng. **17**(3), 299–310 (2005)
20. John, G.H., Langley, P.: Estimating continuous distributions in bayesian classifiers. In: Eleventh Conference on Uncertainty in Artificial Intelligence, pp. 338–345. Morgan Kaufmann (1995)
21. Khot, T., Natarajan, S., Kersting, K., Shavlik, J.: Learning markov logic networks via functional gradient boosting. In: ICDM (2011)
22. Knerr, S., Personnaz, L., Dreyfus, G.: Single-layer learning revisited: a stepwise procedure for building and training a neural network. In: Soulié, F.F., Hérault, J. (eds) Neurocomputing: Algorithms, Architectures and Applications, vol. F68, pp. 41–50. Springer (1990)
23. Koller, D., Pfeffer, A.: Object-oriented Bayesian networks. In: Proceedings of the 13th Annual Conference on Uncertainty in AI (UAI), pp. 302–313, (1997). Winner of the Best Student Paper Award
24. Lavrac, N., Dzeroski, S.: Inductive Logic Programming—Techniques and Applications. Ellis Horwood Series in Artificial Intelligence. Ellis Horwood, New York (1994)
25. Lazarou, J., Pomeranz, B.H., Corey, P.N.: Incidence of adverse drug reactions in hospitalized patients: a meta-analysis of prospective studies. JAMA **279**, 1200–1205 (1998)
26. McCarty, C.A., Peissig, P., Caldwell, M.D., Wilke, R.A.: The marshfield clinic personalized medicine research project: 2008 scientific update and lessons learned in the first 6 years. Personalized Med. **5**(5), 529–542 (2008)
27. McCarty, C.A., Wilke, R.A., Giampietro, P.F., Wesbrook, S.D., Caldwell, M.D.: Marshfield clinic personalized medicine research project (pmrp): design, methods and recruitment for a large population-based biobank. Personalized Med. **2**(1), 49–79 (2005)
28. Mitchell, T.: Machine Learning, 1st edn. McGraw-Hill Inc., New York (1997)
29. Murphy, K.: Machine Learning: A Probabilistic Perspective. MIT Press (2012)
30. Natarajan, S., Joshi, S., Saha, B., Edwards, A., Khot, T., Moody, E., Kersting, K., Whitlow, C., Maldjian, J.: A machine learning pipeline for three-way classification of alzheimer patients from structural magnetic resonance images of the brain. In: IEEE Conference on Machine Learning and Applications (ICMLA) (2012)
31. Natarajan, S., Joshi, S., Saha, B., Edwards, A., Khot, T., Moody, E., Kersting, K., Whitlow, C., Maldjian, J.: Relational learning helps in three-way classification of alzheimer patients from structural magnetic resonance images of the brain. Int. J. Mach. Learn. Cybern. (2013)

32. Natarajan, S., Joshi, S., Tadepalli, P., Kersting, K., Shavlik, J.: Imitation learning in relational domains: a functional-gradient boosting approach. In: IJCAI, pp. 1414–1420 (2011)
33. Natarajan, S., Kersting, K., Ip, E., Jacobs, D., Carr, J.: Early prediction of coronary artery calcification levels using machine learning. In: Innovative Appl. AI (2013)
34. Natarajan, S., Khot, T., Kersting, K., Gutmann, B., Shavlik, J.: Gradient-based boosting for statistical relational learning. Relational Depend. Netw. Case MLJ (2012)
35. Page, D., Natarajan, S., Costa, V.S., Peissig, P., Barnard, A., Caldwell, M.: Identifying adverse drug events from multi-relational healthcare data. In: AAAI (2012)
36. Quinlan, J.: C4.5: Programs for Machine Learning (1993)
37. Quinlan, J.R.: Bagging, boosting, and c4.5. In: AAAI/IAAI, vol. 1, pp. 725–730 (1996)
38. De Raedt, L.: Logical and Relational Learning: From ILP to MRDM (Cognitive Technologies). Springer, New York (2008)
39. Roge, V.L., Go, A.S., et al., Lloyd-Jones, D.M.: Heart disease and stroke statistics-2011 update: a report from the american heart association. Circulation **123**, e18–e209 (2011)
40. Schapire, R., Freund, Y.: Boosting: Foundations and Algorithms. The MIT Press (2012)
41. Srinivasan, A.: The Aleph Manual (2004)
42. Sun, L., Patel, R., Liu, J., Chen, K., Wu, T., Li, J., Reiman, E., Ye, J.: Mining brain region connectivity for alzheimer's disease study via sparse inverse covariance estimation. In: KDD (2009)
43. Supekar, K., Menon, V., Rubin, D., Musen, M., Greicius, M.D.: Network analysis of intrinsic functional brain connectivity in Alzheimer's disease. PLoS Comput. Biol. **4**(6), e1000100 (2008)
44. Weiss, J., Natarajan, S., Peissig, P., McCarty, C., Page, D.: Statistical relational learning to predict primary myocardial infarction from electronic health records. In: Innovative Applications in AI (2012)
45. Weiss, J., Natarajan, S., Peissig, P., McCarty, C., Page, D.: Statistical relational learning to predict primary myocardial infarction from electronic health records. In: AI Magazine (2012)
46. Jieping, Y., Gene, A., Eric, R., Kewei, C., Wu, T., Jing, L., Zheng, Z., Rinkal, P., Min, B., Ravi, J., et al.: Heterogeneous data fusion for alzheimer's disease study. In: KDD, p. 1025 (2008)

ARM Cluster for Performant and Energy-Efficient Storage

Diana Gudu and Marcus Hardt

Abstract Low power hardware—such as ARM CPUs—combined with novel storage concepts—such as Ceph—promise scalable storage solutions at lower energy consumptions than today's standard solutions like network attached storage (NAS) systems. We have set up an ARM cluster, built of Cubieboards, which uses Ceph for storage management. We compare its performance as well as its energy consumption to typical NAS storages. The energy considerations include networking equipment, storage, as well as computing equipment. Ceph is a novel storage system which provides unified access to distributed storage units. It is a performant and fault tolerant solution for object and block storage, using commodity hardware. This allows a very flexible choice of hardware to support wide ranges of use-cases. The goal of the study is to outline paths for energy efficient storage systems.

1 Introduction

The goal of this chapter is to evaluate performance and energy consumption of storage. We will focus on the power of the storage controller. We will not include the power consumption of spinning disks, because in principle, any hard drive can be connected to all the different controllers under consideration. Furthermore, the power consumption of disks does not vary greatly across the models. We will therefore assume a fixed power consumption of "disk" and will relate power consumption on controllers and number of disks rather than storage capacity. That said, hard drive manufacturers are working on reducing the power consumption of their drives, like for example the HGST ultrastar He^6 drive [1].

According to a detailed analysis of power consumption [2] in data centres worldwide, the electrical power usage is growing over time. Due to power efficiency efforts, the overall data centre power consumption is growing slower than previously

D. Gudu (✉) · M. Hardt
Karlsruhe Institute of Technology, Zirkel 1, 76131 Karlsruhe, Germany
e-mail: Diana.Gudu@kit.edu

M. Hardt
e-mail: Marcus.Hardt@kit.edu

J. Lässig et al. (eds.), *Computational Sustainability*,
Studies in Computational Intelligence 645,
DOI 10.1007/978-3-319-31858-5_12

Table 1 Share and power consumption of storage devices in computer centres worldwide

Year	Share of storage (%)	Energy per year (TWh)
2000	7.9	5.6
2005	9.8	15.0
2010	17.1	38.2–68.0

Cooling is included in the figure given. The range given for 2010 is the boundary of the range given by four different prognoses [2]

anticipated. The share of the overall power required for storage is growing. Table 1 shows the values for 2000, 2005 and those projected for 2010 in 2007. Unfortunately, recent data is difficult to obtain, because especially large commercial data centres are reluctant to publish their power consumption numbers. However, the trend can clearly be seen. According to [2], storage equipment in data centres accounted for 38.2–68 TWh in 2010. Using the average price of electric power for industries within the European Union in the same year of 9.1 $\frac{\mathrm{ct}}{\mathrm{kWh}}$ [3], this corresponds to €3.4–6 billion for storage related electricity (including cooling). Every percent of power consumed less for storage corresponds to €34–60 million per year.

In an effort to make data centres more efficient by bringing application services closer to storage media, HGST has developed the Open Ethernet Drive Architecture [4]: the new disk drives also have integrated ARM-based CPU, RAM and Ethernet, being thus connected to the data centre fabric directly and appearing on the network as Linux servers. Software-defined storage solutions, such as Ceph, Gluster or OpenStack's Swift, can run unmodified on this new architecture. However, this is still in the development stage.

In this contribution we will therefore analyse if using low-energy storage servers based on ARM CPUs (such as found on the Raspberry Pi [5]) as storage servers can provide a viable alternative to existing network attached storage solutions. The distributed object store Ceph is the enabling technology we use for this, because it promises [6] and provides [7] a very good scalability over large amounts of resources. Based on this work, we want to find out if our initial test-ARM cluster will scale to the physical limits of the attached network, and if the power consumed for this is lower than for other storage services. We will compare the measured power and performance values to data from other NAS products obtained from literature.

Furthermore, big data applications and some data analysis cases, for example in high-energy physics, depend on the throughput per core rather than the CPU performance. In such cases, processing local data on an ARM CPU may be beneficial from the energy point of view.

The remainder of this chapter is organised into a coarse overview of existing storage technologies and their characteristic power consumption and throughput in Sect. 2, the description of our ARM cluster in Sect. 3, including energy and performance measurements in Sect. 4, and a discussion of the results in Sect. 5.

2 Network Attached Storage (NAS)

NAS [8] (Networked Attached Storage) is a network appliance that provides file-storage services to other devices over the network. It contains one or more hard drives that are combined internally via RAID controllers to provide an internal redundancy. Typical interfaces include FTP, NFS, SMB but also higher level interfaces based on HTTP.

2.1 QNAP

The NAS vendor [9] chosen to serve as an example is QNAP [10]. QNAP was chosen because they offer energy efficient devices together with well documented energy consumption and throughput numbers. The values shown in here are those published by QNAP. They offer a variety of NAS solutions ranging from 2 to 16 disks, combined to different RAID levels. Hardware specifications for four selected QNAP products are listed in Table 2.

2.2 SONAS

SONAS [13] is a highly scalable and performant NAS implementation developed by IBM. It is based on IBM's clustered filesystem GPFS [14]. The architecture consists of one controller and several disk enclosures, each enclosure containing multiple disks. Rough performance and power values for a SONAS set-up in use at the Karlsruhe Institute of Technology (KIT) are listed in Table 3. The controller overhead per disk amounts to 7.78 W per disk in the KIT installation chosen for comparison.

Table 2 Power consumption, performance and price (without disks) of selected QNAP NAS products [11, 12]

Product	Number of disks	Power (W)	Power with disks (W)	Read throughput (MB/s)	Write throughput (MB/s)	Price (€)
TS-EC1680U-RP	16	82.84	166.58	448	448	5499
TS-EC880U-RP	8	73.38	133.88	444	448	3464
TS-EC1679U-RP	16	89.1	229	445	448	4969.51
TS-421	4	13	26	105	79	399.9

Table 3 Power consumption of an IBM SONAS installation at KIT

Type	Number	Power (W)	Throughput (MB/s)
Enclosure	60 disks	400	
Controller	9 enclosures	600	5000

3 Cubieboard Cluster

We propose a scalable energy-efficient storage solution by combining low-power ARM-based hardware (Cubieboard [15]) with a novel storage technology (Ceph [16]).

Due to the power consumption as low as 4 W per Cubieboard, it is possible to build low power storage clusters. We therefore aim to obtain similar performance as current standard storage systems, but at lower energy consumption.

In our case, we have built a storage cluster consisting of 16 Cubieboards and 16 disks, as pictured in Fig. 1. Due to a broken board and the assignment of another board as monitor node (without disk storage), only 15 boards and 14 disks were in use throughout the measurements. The set-up also included a network switch and two power supplies. The price of the entire set-up (15 boards and a network switch, without disks) amounts to only €1705. We considered here the Cubieboard2 version of the ARM board and the Enterasys A4H124-24 network switch, which will be described in Sects. 3.1 and 3.3, respectively. We did not include the disks in the total price calculation, since we only focus on the storage controllers in our evaluation. However, unquantifiable costs are also involved but not included in the price calculation, such as certain skills required to assemble all the components into the cluster depicted in this chapter.

3.1 Cubieboards

Cubieboard is a low-cost ARM-based single-board computer that uses the AllWinner A20 SoC. The detailed hardware specification of the Cubieboard version that we used, Cubieboard2, is presented in Table 4. Although similar to Raspberry Pi [5] and other single-board computers, the key difference of the Cubieboard is the SATA interface, which is essential for a performant storage system. A 3.5-inch Seagate Baracuda ES hard disk drive (HDD) with a capacity of 750 GB is connected via SATA.

3.2 Ceph

Ceph is an emerging open-source distributed storage system that offers high scalability and reliability through its innovative design, while using commodity hard-

Fig. 1 The Cubieboard cluster, consisting of 16 Cubieboards (*left*), 16 disks (*right*), a network switch (*bottom*) and two power supplies (*bottom left*). Only 15 boards and 14 disks are in use

ware. Ceph is built on the premise that failures in large-scale storage systems are omnipresent; therefore, it uses different mechanisms to ensure fault tolerance, such as replication and erasure coding.

Ceph's core component is RADOS—the Reliable Autonomic Distributed Object store—which is scalable because it allows individual nodes to act autonomously in order to manage replication, failure detection and recovery.

There are different types of nodes in a Ceph cluster: Object Storage Daemons (OSD) and Monitors. OSDs are usually constructed from commodity hardware (consisting of CPU, network interface and disk). Objects are stored across a Ceph cluster on the OSDs' local filesystems using the pseudo-random distribution algorithm CRUSH (Controlled Replication Under Scalable Hashing). This allows OSDs and

Table 4 Full hardware specification of Cubieboard2 [18]

CPU	ARM® Cortex™-A7 Dual-Core
GPU	ARM® Mali400MP2, complies with OpenGL ES 2.0/1.1
Memory	1 GB DDR3 @960M
Storage	4 GB internal NAND flash, 8 GB microSD, 750 GB on 2.5 SATA disk
Power	5 V DC input 2 A or USB OTG input
Networking	10/100 Ethernet, optional wifi
USB	Two USB 2.0 HOST, one USB 2.0 OTG
Other	One IR
Price	€70 [17]

clients to compute object locations instead of looking them up in a centralised table, thus interacting directly with the OSDs for I/O. This promises, at least in theory, extreme scalability. Monitor nodes maintain a copy of the cluster map and work together to ensure high availability.

In addition to direct access to RADOS through different software libraries, Ceph offers a block device interface to the storage, built on top of RADOS—the RADOS Block Device (RBD). Block device images are stored as objects, automatically striped and replicated across the cluster. RBD is supported in several virtualisation platforms, such as OpenStack [19], OpenNebula [20] and CloudStack [21].

Ceph's architecture is illustrated in Fig. 2.

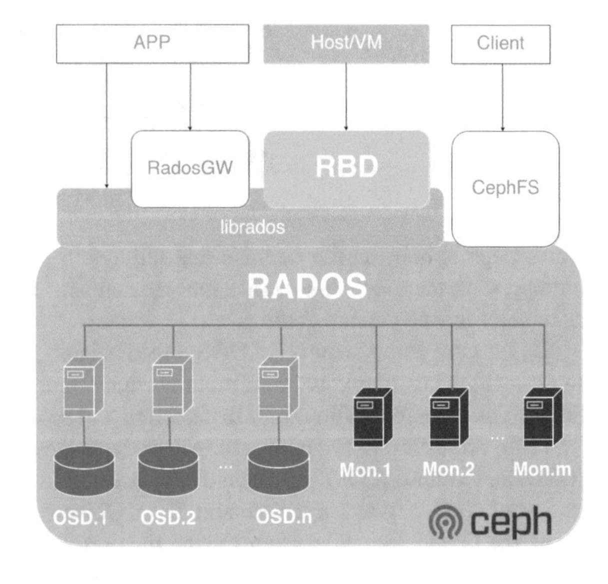

Fig. 2 A simplified diagram of the Ceph architecture. RADOS, the core component, consists of OSD and Monitor nodes. Clients can access the object storage directly through the librados library. Ceph offers three other access interfaces: the RBD block interface, the POSIX-compliant CephFS filesystem and the RESTful RadosGW object gateway

Table 5 Specification, power consumption and price for different network switches [24, 25]

Model	Number of ports	Power (W)	Price (€)
Cisco WS-C2960G-24	24	75	1487.5
Cisco WS-C2960G-48	48	140	2007.48
Enterasys A4H124-24	24	31	655.06
Enterasys A4H124-48	48	63	1218.21

From the point of view of maintenance costs, once installed, Ceph runs without attention, just like an enterprise system. However, it requires skills to be acquired at install time.

3.3 Network

The network gear used with the Cubieboard cluster was a Cisco [22] switch with 48 ports and 1 GE uplink (WS-C2960G-48), which has a power consumption of 140 W. However, several lower power alternatives are listed in Table 5.

Since the cluster consists of only 15 boards, a network switch with 24 ports would be sufficient. For production we would use an Enterasys [23] A4H124-24 switch, which consumes only 31 W. This is the power value we used for calculation in our evaluation.

4 Measurements

To evaluate the energy efficiency of the Cubieboard cluster, we measured the performance of the Ceph storage system, as well as the total power consumption (the power drawn by the boards and the disks) under different workloads.

We measured the I/O throughput using the same benchmarks described in our previous work evaluating Ceph [7]: multiple clients performing, in parallel, sequential read/write operations of 4 MB objects. Random I/O tests were also performed on the Ceph block device.

Figure 3 shows the results of these tests for both the underlying object store of Ceph (RADOS) and the block device interface (RBD), with up to 4 clients. The total read throughput that the Cubiecluster can sustain reached 133 MB/s in our tests, while the total write throughput reached 108 MB/s. These values should be read with respect to the maximum throughput that the underlying hardware can provide, i.e. the Cubieboards, the disk and the network. The limiting factor for the Cubiecluster's performance was the 100 Mbit Ethernet connection on Cubieboard2, which can maximally provide 12.5 MB/s throughput per OSD. Given our cluster configuration

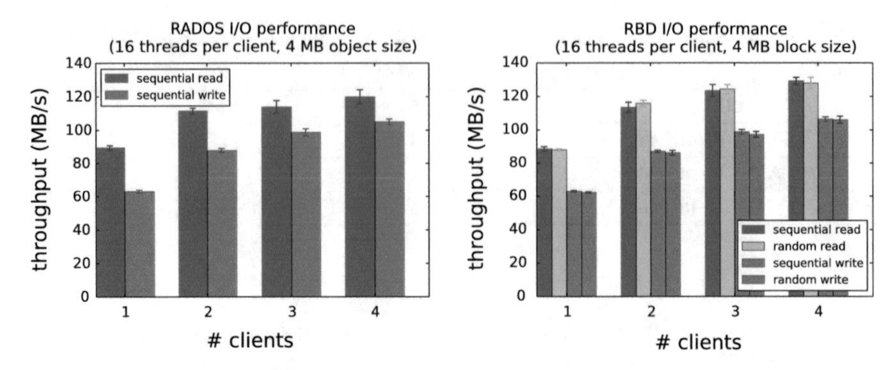

Fig. 3 I/O throughput of the Cubiecluster, measured at the Ceph RADOS and RBD levels, with different number of clients continuously writing/reading 4 MB objects

Table 6 Cluster throughput theoretical limits (with 14 OSDs) and the performance achieved with the Cubiecluster

Operation	Theoretical throughput limit (MB/s)	Maximum measured throughput (MB/s)	% of theoretical limit
Read	$14 \times 12.5 = 175$	133.5	76
Write	$14 \times 12.5 = 175$	108.9	62

of 14 OSDs, we would expect [7] a maximum theoretical throughput of 175 MB/s. It results that the Cubiecluster achieved 76 % and 62 % of the theoretical upper limit imposed by the hardware, for read and write, respectively. This is summarised in Table 6.

Moreover, we measured the total power consumption of the cluster under different workloads:

- **Idle:** only Ceph daemons were running on every node, but no other operations.
- **CPU intensive workload:** all the cores were under 100 % load, no I/O operations.
- **I/O intensive workload:** sustained read and write operations on every disk in the cluster using dd.
- **Ceph benchmark:** the performance benchmark described above, with 5 clients doing I/O operations in parallel to the cluster.

The power measurements are listed in Table 7. The switch power consumption of 31 W has to be added to these measurements in order to obtain the total power consumption of the storage cluster. For the Ceph benchmark, we also measured the power consumption of the boards only, separately from the disks. Therefore, in our next comparison of the Cubiecluster with NAS storages, we used the power consumption of the storage controller (only Cubieboards and network switch) from the Ceph benchmark workload ($41.7 + 31 = 72.7$ W).

Table 7 Power measurements for Cubieboard cluster under different workloads

Workload	Power with disks (W)	Power (W)
Idle	150.1	
CPU intensive workload	172.6	
I/O intensive workload	174.1	
Ceph benchmark	170.8	41.7

Table 8 Energy efficiency of the Cubiecluster and current storage solutions

Storage system	Capacity metric (GB/W)	Read throughput metric (MB/Ws)	Write throughput metric (MB/Ws)
Cubiecluster	144.43	1.84	1.50
QNAP TS-EC1680U-RP	144.86	5.41	5.41
QNAP TS-EC880U-RP	81.77	6.05	6.11
QNAP TS-EC1679U-RP	134.68	4.99	5.03
QNAP TS-421	230.77	8.08	6.08
SONAS	96.43	1.19	1.19

The literature defines several metrics to quantify the energy efficiency of a storage system from an operational point of view [26, 27]. We selected the following two metrics:

$$\text{Capacity metric} = \frac{\text{Storage space (GB)}}{\text{Power used (W)}} \tag{1}$$

$$\text{Data transfer throughput metric} = \frac{\text{Data transfer rate (MB/s)}}{\text{Power used (W)}} \tag{2}$$

These metrics allow for a fair comparison of the Cubiecluster's energy efficiency with existing storage solutions, as shown in Table 8. We used the above performance and power measurements of the Cubiecluster and compared with typical values for power consumption (of storage controller) and throughput for the other storage solutions. For the capacity metric, we assume all the storage systems have the same disks of 750 GB capacity; increasing the disk capacities would only increase the capacity metric by the same factor for all systems. Nevertheless, all the systems presented in this chapter (including the Cubieboard cluster) support any disk model of any capacity that can be connected via SATA. Moreover, the power consumption of disks does not vary greatly across models or disk capacities.

5 Discussion

Our measurements revealed that the Cubieboard cluster is a viable and performant storage solution, that can provide read and write throughput of 76 % and 62 % of the network bandwidth, respectively. The I/O throughput was primarily limited by the Cubieboard's 100 Mb network, therefore upgrading to 1 Gb Ethernet would bring a significant performance improvement. A newer version of the Cubieboard (Cubietruck or Cubieboard3 [28]) is now available with 1 Gb Ethernet.

Furthermore, the power draw measured during our benchmarks of the Cubiecluster without disks (72.7 W) was lower than the typical values—reported by vendors—of other storage solutions with a comparable number of disks (82.84 or 89.1 W for QNAP NAS servers with 16 disks).

Even though the low power consumption is a good indicator of energy efficiency, we used several other metrics for a more comprehensive comparison of different storage systems, which include both performance and power consumption: capacity metric (GB/W) and data transfer throughput metric (MB/Ws) [26]. Assuming disks of the same size in all six storage systems we compared, the Cubiecluster has a comparable capacity metric value, being surpassed only by two storage systems. The throughput metric values, although better than what SONAS offers and in the same order of magnitude as the other systems (at most 5 times lower), are relatively low due to the hardware limitations of the Cubieboard's network, as previously explained. A hardware upgrade to the newer, slightly more expensive Cubieboard3 [28] (€89 [29]), that would increase the board's network bandwidth 10 times, coupled with Ceph's inherent scalability [7], has the potential of bringing the Cubiecluster to the top of the list when using the throughput metric for energy efficiency.

6 Conclusions and Future Work

We proposed a novel storage solution that brings together low-cost, low-power ARM hardware and scalable distributed storage software, to achieve good performance and energy efficiency. We compared it to currently used NAS storages and found that it is similar or better in terms of power consumption and capacity metric for energy efficiency (GB/W). The limiting factor in achieving a good throughput metric value (MB/Ws) was the Cubieboard's network connection.

In the future, we will compare an upgraded version of the Cubiecluster (using newer Cubieboard hardware with better hardware specification, notably a 1 Gb network) to measured (not typical) performance and power values of different NAS solutions.

Nevertheless, our study shows that new technologies can be successfully combined to tackle the increasing energy consumption of storage in computer centres worldwide, in order to reduce operation costs.

References

1. Western Digital. Data sheet ultrastar he6 3.5-inch helium platform enterprise hard disk drives. http://www.hgst.com/tech/techlib.nsf/techdocs/83F507776F03BE7C88257C37000AED30/file/USHe6_ds.pdf
2. Koomey, J.: Growth in data center electricity use 2005 to 2010 (2011). http://www.analyticspress.com/datacenters.html
3. Eurostat. Electricity prices for industrial consumers, from 2007 onwards (2010). http://appsso.eurostat.ec.europa.eu/nui/show.do?dataset=nrg_pc_205&lang=en (last visited 2014-11-12)
4. Western Digital. Open ethernet drive architecture. https://www.hgst.com/science-of-storage/emerging-technologies/open-ethernet-drive-architecture (last visited 2015-06-18)
5. Eben Upton and Gareth Halfacree. Raspberry Pi user guide. Wiley (2013)
6. Weil, S.A., Brandt, S.A., Miller, E.L., Long, D.D.E., Maltzahn, C.: Ceph: a scalable, high-performance distributed file system. In: Proceedings of the 7th Symposium on Operating Systems Design and Implementation, pp. 307–320. USENIX Association (2006)
7. Gudu, D., Hardt, M., Streit, A.: Evaluating the performance and scalability of the ceph distributed storage system. In: 2014 IEEE International Conference on Big Data. IEEE (2014)
8. Gibson, G.A., Van, M.: Rodney: network attached storage architecture. Commun. ACM **43**(11), 37–45 (2000)
9. Top 2011 NAS vendors: Emc in sales, netgear in units. http://www.storagenewsletter.com/rubriques/market-reportsresearch/gartner-top-nas-vendors-in-2011/ (last visited 2014-11-27)
10. QNAP. http://qnap.com (last visited 2014-11-27)
11. Comparison of specs for various QNAP NAS products. https://www.qnap.com/i/en/product/contrast.php?cp%5B%5D=126&cp%5B%5D=124&cp%5B%5D=27&cp%5B%5D=193 (last visited 2015-06-17)
12. Cyberport: QNAP seller in Germany. http://www.cyberport.de/ (last visited 2015-06-24)
13. Lovelace, M., Boucher, V., Nayak, S., Neal, C., Razmuk, L., Sing, J., Tarella, J., et al.: IBM Scale Out Network Attached Storage: Architecture, Planning, and Implementation Basics. IBM Redbooks (2011)
14. Schmuck, F.B., Haskin, R.L.: GPFS: a shared-disk file system for large computing clusters. In: FAST, vol. 2, pp. 19 (2002)
15. Cubieboard. http://cubieboard.org (last visited 2014-11-26)
16. Ceph. http://ceph.com (last visited 2014-11-26)
17. Example seller and price of cubieboard2 in Germany. http://www.watterott.com/de/Cubieboard2-Open-ARM-Box?xcbcf3=4102494bb0acaedc5ab7d468993786b5 (last visited 2015-06-24)
18. Hardware specification of cubieboard2. http://cubieboard.org/2013/06/19/cubieboard2-is-here/ (last visited 2014-11-26)
19. Openstack. http://openstack.org (last visited 2014-11-26)
20. Opennebula. http://opennebula.org (last visited 2014-11-26)
21. Cloudstack. http://cloudstack.apache.org (last visited 2014-11-26)
22. Cisco systems. http://cisco.com (last visited 2014-11-27)
23. Extreme networks. http://www.extremenetworks.com/ (last visited 2014-11-28)
24. Metacomp: best price seller of cisco switches in Germany. http://xtreme.metacomp.de/ (last visited 2015-06-24)
25. Tim: best price seller of enterasys switches in Germany. http://www.tim.de/produkte/uebersicht.aspx?shop= (last visited 2015-06-24)

26. Chen, D., Henis, E., Kat, R.I., Sotnikov, D., Cappiello, C., Ferreira, A.M., Pernici, B., Vitali, M., Jiang, T., Liu, J., Kipp, A.: Usage centric green performance indicators. SIGMETRICS Perform. Eval. Rev. **39**(3), 92–96 (2011)
27. Belady, C., Rawson, A., Pfleuger, J., Cader, T.: Green grid data center power efficiency metrics: Pue and dcie. Technical report, Technical report, Green Grid (2008)
28. Hardware specification of cubieboard3 (cubietruck). https://linux-sunxi.org/Cubietruck (last visited 2014-11-26)
29. Example seller and price of cubieboard3 in Germany. http://www.watterott.com/de/Cubietruck?xcbcf3=4102494bb0acaedc5ab7d468993786b5 (last visited 2015-06-24)

Printed in the United States
By Bookmasters